Mathematical Logic

A Course with Exercises

Mathematical Logic

A Course with Exercises
Part I: Propositional calculus,
Boolean algebras, Predicate calculus

René Cori and Daniel Lascar

Equipe de Logique Mathématique
Université Paris VII

Translated by

Donald H. Pelletier

York University, Toronto

OXFORD
UNIVERSITY PRESS

OXFORD
UNIVERSITY PRESS

Great Clarendon Street, Oxford OX2 6DP

Oxford University Press is a department of the University of Oxford.
It furthers the University's objective of excellence in research, scholarship,
and education by publishing worldwide in

Oxford New York

Athens Auckland Bangkok Bogota Buenos Aires Calcutta
Cape Town Chennai Dar es Salaam Delhi Florence Hong Kong Istanbul
Karachi Kuala Lumpur Madrid Melbourne Mexico City Mumbai
Nairobi Paris São Paulo Singapore Taipei Tokyo Toronto Warsaw

with associated companies in Berlin Ibadan

Oxford is a trade mark of Oxford University Press
in the UK and in certain other countries

Published in the United States
by Oxford University Press, Inc., New York

This English Edition © Oxford University Press 2000
Published with the help of the Ministère de la Culture
First published in French as *Logique mathématique*
© Masson, Editeur, Paris, 1993

The moral rights of the author have been asserted

Database right (English edition) Oxford University Press (maker)

First published in English 2000

A catalogue record for this book is available from the British Library

Library of Congress Cataloging in Publication Data (Data available)

ISBN 0 19 850049 1 (Hbk)
ISBN 0 19 850048 3 (Pbk)

Typeset by Newgen Imaging Systems (P) Ltd., Chennai
using the translator's LaTeX files
Printed in Great Britain
on acid free paper by
Biddles Ltd.,
Guildford & King's Lynn

Foreword to the Original French edition

Jean-Louis Krivine

In France, the discipline of logic has traditionally been ignored in university-level scientific studies. This follows, undoubtedly, from the recent history of mathematics in our country which was dominated, for a long while, by the Bourbaki school for whom logic was not, as we know, a strong point. Indeed, logic originates from reflecting upon mathematical activity and the common gut-reaction of the mathematician is to ask: 'What is all that good for? We are not philosophers and it is surely not by cracking our skulls over modus ponens or the excluded middle that we will resolve the great conjectures, or even the tiny ones . . .' Not so fast!

A new ingredient, of some substance, has come to settle this somewhat byzantine debate over the importance of logic: the explosion of computing into all areas of economic and scientific life, whose shock wave finally reached the mathematicians themselves.

And, little by little, one fact dawns on us: the theoretical basis for this nascent science is nothing other than the subject of all this debate, mathematical logic.

It is true that certain areas of logic were put to use more quickly than others. Boolean algebra, of course, for the notions and study of circuits; recursiveness, which is the study of functions that are computable by machine; Herbrand's theorem, resolution and unification, which form the basis of 'logic programming' (the language PROLOG); proof theory, and the diverse incarnations of the Completeness theorem, which have proven themselves to be powerful analytical tools for mature programming languages.

But, at the rate at which things are going, we can imagine that even those areas that have remained completely 'pure', such as set theory, for example, will soon see their turn arrive.

As it ought to be, the interaction is not one-way, far from it; a flow of ideas and new, deep intuitions, arising from computer science, has come to animate all these sectors of logic. This discipline is now one of the liveliest there is in mathematics and it is evolving very rapidly.

So there is no doubt about the utility and timeliness of a work devoted to a general introduction to logic; this book meets its destiny. Derived from lectures for the Diplôme d'Etudes Approfondies (DEA) of Logic and the Foundations of Computing at the University of Paris VII, it covers a vast panorama: Boolean algebras, recursiveness, model theory, set theory, models of arithmetic and Gödel's theorems.

The concept of model is at the core of this book, and for a very good reason since it occupies a central place in logic: despite (or thanks to) its simple, and even elementary, character, it illuminates all areas, even those that seem farthest from it. How, for example, can one understand a consistency proof in set theory without first mastering the concept of being a model of this theory? How can one truly grasp Gödel's theorems without having some notion of non-standard models of Peano arithmetic? The acquisition of these semantic notions is, I believe, the mark of a proper training for a logician, at whatever level. R. Cori and D. Lascar know this well and their text proceeds from beginning to end in this direction. Moreover, they have overcome the risky challenge of blending all the necessary rigour with clarity, pedagogical concern and refreshing readability.

We have here at our disposal a remarkable tool for teaching mathematical logic and, in view of the growth in demand for this subject area, it should meet with a marked success. This is, naturally, everything I wish for it.

Foreword to the English edition

Wilfrid Hodges
School of Mathematical Sciences
Queen Mary and Westfield College
University of London

There are two kinds of introduction to a subject. The first kind assumes that you know nothing about the subject, and it tries to show you in broad brushstrokes why the subject is worth studying and what its achievements are. The second kind of introduction takes for granted that you know what the subject is about and why you want to study it, and sets out to give you a reliable understanding of the basics. René Cori and Daniel Lascar have written the second sort of introduction to mathematical logic. The mark of the book is thoroughness and precision with a light touch and no pedantry.

The volume in your hand, Part I, is a mathematical introduction to first-order logic. This has been the staple diet of elementary logic courses for the last fifty years, but the treatment here is deeper and more thorough than most elementary logic courses. For example the authors prove the compactness theorem in a general form that needs Zorn's Lemma. You certainly shouldn't delay reading it until you know about Zorn's Lemma – the applications here are an excellent way of learning how to use the lemma. In Part I there are not too many excitements – probably the most exciting topic in the book is the Gödel theory in Chapter 6 of Part II, unless you share my enthusiasm for the model theory in Chapter 8. But there are plenty of beautiful explanations, put together with the clarity and elegance that one expects from the best French authors.

For English students the book is probably best suited to Masters' or fourth-year undergraduate studies. The authors have included full solutions to the exercises; this is one of the best ways that an author can check the adequacy of the definitions and lemmas in the text, but it is also a great kindness to people who are studying on their own, as a beginning research student may be. Some thirty-five years ago I found I needed to teach myself logic, and this book would have fitted my needs exactly. Of course the subject has moved on since then, and the authors have included several things that were unknown when I was a student. For example their chapter on proof theory, Chapter 4 in this volume, includes a well-integrated section on the resolution calculus. They mention the connection with PROLOG; but in fact you can also use this section as an introduction to the larger topic of unification and pattern-matching, which has wide ramifications in computer science.

One other thing you should know. This book comes from the famous Équipe de Logique Mathématique at the University of Paris, a research team that has had an enormous influence on the development of mathematical logic and its links with other branches of mathematics. Read it with confidence.

Preface

This book is based upon several years' experience teaching logic at the UFR of Mathematics of the University of Paris 7, at the beginning graduate level as well as within the DEA of Logic and the Foundations of Computer Science.

As soon as we began to prepare our first lectures, we realized that it was going to be very difficult to introduce our students to general works about logic written in (or even translated into) French. We therefore decided to take advantage of this opportunity to correct the situation. Thus the first versions of the eight chapters that you are about to read were drafted at the same time that their content was being taught. We insist on warmly thanking all the students who contributed thereby to a tangible improvement of the initial presentation.

Our thanks also go to all our colleagues and logician friends, from Paris 7 and elsewhere, who brought us much appreciated help in the form of many comments and moral support of a rare quality. Nearly all of them are co-authors of this work since, to assemble the lists of exercises that accompany each chapter, we have borrowed unashamedly from the invaluable resource that comprises the hundreds and hundreds of pages of written material that were handed out to students over the course of more than twenty-five years during which the University of Paris 7, a pioneer in this matter, has organized courses in logic open to a wide public.

At this point, the reader generally expects a phrase of the following type: 'they are so numerous that we are obviously unable to name them all'. It is true, there are very many to whom we extend our gratitude, but why shouldn't we attempt to name them all?

Thank you therefore to Josette Adda, Marouan Ajlani, Daniel Andler, Gilles Amiot, Fred Appenzeller, Jean-Claude Archer, Jean-Pierre Azra, Jean-Pierre Bénéjam, Chantal Berline, Claude-Laurent Bernard, Georges Blanc, Elisabeth Bouscaren, Albert Burroni, Jean-Pierre Calais, Zoé Chatzidakis, Peter Clote, François Conduché, Jean Coret, Maryvonne Daguenet, Vincent Danos, Max Dickmann, Patrick Dehornoy, Françoise Delon, Florence Duchêne, Jean-Louis Duret, Marie-Christine Ferbus, Jean-Yves Girard, Danièle Gondard, Catherine Gourion, Serge Grigorieff, Ursula Gropp, Philippe Ithier, Bernard Jaulin, Ying Jiang, Anatole Khélif, Georg Kreisel, Jean-Louis Krivine, Ramez Labib-Sami, Daniel Lacombe, Thierry Lacoste, Richard Lassaigne, Yves Legrandgérard, Alain Louveau, François Lucas, Kenneth MacAloon, Gilles Macario-Rat, Sophie Malecki, Jean Malifaud, Pascal Manoury, François Métayer, Marie-Hélène Mourgues, Catherine Muhlrad-Greif, Francis Oger, Michel Parigot, Donald

Pelletier, Marie-Jeanne Perrin, Bruno Poizat, Jean Porte, Claude Précetti, Christophe Raffalli, Laurent Régnier, Jean-Pierre Ressayre, Iégor Reznikoff, Philippe Royer, Paul Rozière, Gabriel Sabbagh, Claire Santoni, Marianne Simonot, Gerald Stahl, Jacques Stern, Anne Strauss, Claude Sureson, Jacques Van de Wiele, Françoise Ville.

We wish also to pay homage to the administrative and technical work accomplished by Mesdames Sylviane Barrier, Gisèle Goeminne, and Claude Orieux.

May those whom we have forgotten forgive us. They are so numerous that we are unable to name them all.

September 1993

The typographical errors in the first printing were so numerous that even Alain Kapur was unable to locate them all. May he be assured of all our encouragement for the onerous task that still awaits him.

We also thank Edouard Dorard and Thierry Joly for their very careful reading.

Contents

Contents of Part II

Notes from the translator

In everyday mathematical language, the English word 'contains' is often used indifferently, sometimes referring to membership of an element in a set, $\in$, and sometimes to the inclusion relation between sets, $\subseteq$. For a reader who is even slightly familiar with the subject, this is not a serious issue since the meaning is nearly always clear from the context. But because this distinction is precisely one of the stumbling blocks encountered by beginning students of logic and set theory, I have chosen to consistently use the word 'contains' when the meaning is $\in$ and the word 'includes' when the meaning is $\subseteq$.

It is perhaps more common in mathematical English to use the phrases 'one-to-one' and 'onto' in place of the more formal-sounding 'injective' and 'surjective'. I have none the less retained 'injective' and 'surjective' as more in keeping with the style of the original; even those who object must admit that 'bijective' has the advantage over 'one-to-one and onto'.

Where the original refers the reader to various standard texts in French for some basic facts of algebra or topology, I have replaced these references with suitable English-language equivalents.

It is useful to distinguish between bold zero and one (**0** and **1**) and plain zero and one (0 and 1). The plain characters are part of the metalanguage and have their usual denotations as integers. The bold characters are used, by convention, to denote the truth values of two-valued logic; they are also used to denote the respective identity elements for the operations of addition and multiplication in a Boolean algebra.

April, 2000.   Donald H. Pelletier

Notes to the reader

The book is divided into two parts. The first consists of Chapters 1 through 4; Chapters 5 through 8 comprise the second. Concepts presented in a given chapter presume knowledge from the preceding chapters (but Chapters 2 and 5 are exceptions to this rule).

Each of the eight chapters is divided into sections, which, in turn, are composed of several subsections that are numbered in an obvious way (see the Contents).

Each chapter concludes with a section devoted to exercises. The solutions to these are grouped together at the end of the corresponding volume.

The solutions, especially for the first few chapters, are rather detailed.

Our reader is assumed to have acquired a certain practice of mathematics and a level of knowledge corresponding, roughly, to classical mathematics as taught in high school and in the first years of university. We will refer freely to what we have called this 'common foundation', especially in the examples and the exercises.

None the less, the course overall assumes no prior knowledge in particular.

Concerning the familiar set-theoretical (meta-)language, we will use the terminology and notations that are most commonly encountered: operations on sets, relations, maps, etc., as well as $\mathbb{N}$, $\mathbb{Z}$, $\mathbb{Z}/n\mathbb{Z}$, $\mathbb{Q}$, $\mathbb{R}$ for the sets we meet every day. We will use $\mathbb{N}^*$ to denote $\mathbb{N} - \{0\}$.

If E and F are sets and if f is a map defined on a subset of E with values in F, the **domain** of f is denoted by $\mathsf{dom}(f)$ (it is the set of elements in E for which f is defined), and its **image** is denoted by $\mathsf{Im}(f)$ (it is the set of elements y in F for which $y = f(x)$ is true for at least one element x in E). If A is a subset of the domain of f, the **restriction** of f to A is the map from A into F, denoted by $f \upharpoonright A$, which, with each element x in A, associates $f(x)$. The image of the map $f \upharpoonright A$ is also called the **direct image of A under** f and is denoted by $f[A]$. If B is a subset of F, the **inverse image of B under** f, denoted by $f^{-1}[B]$, consists of those elements x in E such that $f(x) \in B$. In fact, with any given map f from a set E into a set F, we can associate, in a canonical way, a map from $\wp(E)$ (the set of subsets of E) into $\wp(F)$: this is the 'direct image' map, denoted by $\overline{f}$ which, with any subset A of E, associates $f[A]$, which we could then just as well denote by $\overline{f}(A)$. In the same way, with this given map f, we could associate a map from $\wp(F)$ into $\wp(E)$, called the 'inverse image' map and denoted by $\overline{f}^{-1}$, which, with any subset B of F, associates $f^{-1}[B]$, which we could then just as well denote by $\overline{f}^{-1}(B)$. (See also Exercise 19 from Chapter 2.)

Perhaps it is also useful to present some details concerning the notion of word on an alphabet; this concept will be required at the outset.

Let E be a set, finite or infinite, which we will call the **alphabet.** A **word,** w, on the alphabet E is a finite sequence of elements of E (i.e. a map from the set $\{0, 1, \ldots, n - 1\}$ (where n is an integer) into E); $w = (a_0, a_1, \ldots, a_{n-1})$, or even $a_0 a_1 \ldots a_{n-1}$, represents the word whose domain is $\{0, 1, \ldots, n - 1\}$ and which associates a_i with i (for $0 \leq i \leq n - 1$). The integer n is called the **length** of the word w and is denoted by $\lg[w]$. The set of words on E is denoted by $\mathcal{W}(E)$.

If $n = 0$, we obtain the **empty word.** We will adopt the abuse of language that consists in simply writing a for the word (a) of length 1. The set $\mathcal{W}(E)$ can also support a binary operation called **concatenation:** let $w_1 = (a_0, a_1, \ldots, a_{n-1})$ and $w_2 = (b_0, b_1, \ldots, b_{m-1})$ be two words; we can form the new word $w = (a_0, a_1, \ldots, a_{n-1}, b_0, b_1, \ldots, b_{m-1})$, i.e. the map w defined on $\{0, 1, \ldots, n + m - 1\}$ as follows:

$$
w(i) = \begin{cases} a_i & \text{if } 0 \leq i \leq n - 1; \\ b_{i-n} & \text{if } n \leq i \leq n + m - 1. \end{cases}
$$

This word is called the **concatenation** of w_1 with w_2 and is denoted by $w_1 w_2$. This parenthesis-free notation is justified by the fact that the operation of concatenation is associative.

Given two words w and w_1, we say that w_1 is an **initial segment** of w if there exists a word w_2 such that $w = w_1 w_2$. To put it differently, if $w = (a_0, a_1, \ldots, a_{n-1})$, the initial segments of w are the words of the form $(a_0, a_1, \ldots, a_{p-1})$ where p is an integer less than or equal to n. We say that w_1 is a **final segment** of w if there exists a word w_2 such that $w = w_2 w_1$; so the final segments of $(a_0, a_1, \ldots, a_{n-1})$ are the words of the form $(a_p, a_{p+1}, \ldots, a_{n-1})$ where p is an integer less than or equal to n. In particular, the empty word and w itself are both initial segments and final segments of w. A segment (initial or final) of w is **proper** if it is different from w and the empty word.

When an element b of the alphabet 'appears' in a word $w = a_0 a_1 \ldots a_{n-1}$, we say that it has an **occurrence in** w and the various 'positions' where it appears are called the **occurrences of** b **in** w. We could, of course, be more precise and more formal: we will say that b **has an occurrence in** w if b is equal to one of the a_i for i between 0 and $n - 1$ (i.e. if b belongs to the image of w). An **occurrence of** b **in** w is an integer k, less than $\lg[w]$, such that $b = a_k$. For example, the third occurrence of b in w is the third element of the set $\{k : 0 \leq k \leq n - 1 \text{ and } a_k = b\}$ in increasing order. This formalism will not be used explicitly in the text; the idea sketched at the beginning of this paragraph will be more than adequate for what we have to do.

The following facts are more or less obvious and will be in constant use:

- for all words w_1 and w_2, $\lg[w_1 w_2] = \lg[w_1] + \lg[w_2]$;

- for all words w_1, w_2 and w_3, the equality $w_1 w_2 = w_1 w_3$ implies the equality $w_2 = w_3$ (we call this **left cancellation**);

- for all words w_1, w_2 and w_3, the equality $w_1 w_2 = w_3 w_2$ implies the equality $w_1 = w_3$ (we call this **right cancellation**);

- for all words w_1, w_2, w_3 and w_4, if $w_1 w_2 = w_3 w_4$, then either w_1 is an initial segment of w_3 or else w_3 is an initial segment of w_1. Analogously, under the same assumptions, either w_2 is a final segment of w_4 or else w_4 is a final segment of w_2;

- if w_1 is an initial segment of w_2 and w_2 is an initial segment of w_1, then $w_1 = w_2$.

We will also use the fact that $\mathcal{W}(E)$ is countable if E is either finite or countable (this is a theorem from Chapter 7).

Introduction

There are many who consider that logic, as a branch of mathematics, has a somewhat special status that distinguishes it from all the others. Curiously, both its keenest adversaries and some of its most enthusiastic disciples concur with this conception which places logic near the margin of mathematics, at its border, or even outside it. For the first, logic does not belong to 'real' mathematics; the others, on the contrary, see it as the reigning discipline within mathematics, the one that transcends all the others, that supports the grand structure.

To the reader who has come to meet us in this volume seeking an introduction to mathematical logic, the first advice we would give is to adopt a point of view that is radically different from the one above. The frame of mind to be adopted should be the same as when consulting a treatise in algebra or differential calculus. It is a mathematical text that we are presenting here; in it, we will be doing mathematics, not something else. It seems to us that this is an essential precondition for a proper understanding of the concepts that will be presented.

That does not mean that the question of the place of logic in mathematics is without interest. On the contrary, it is enthralling, but it concerns problems external to mathematics. Any mathematician can (and we will even say must) at certain times reflect on his work, transform himself into an epistemologist, a philosopher or historian of science; but the point must be clearly made that in doing this, he temporarily ceases his mathematical activity. Besides, there is generally no ambiguity: when reading a text in analysis, what the student of mathematics expects to find there are definitions, theorems, and proofs for these theorems. If the author has thought it appropriate to add some comments of a philosophical or historical nature, the reader never has the slightest difficulty separating the concerns contained in these comments from the subject matter itself.

We would like the reader to approach the course that follows in this way and to view logic as a perfectly ordinary branch of mathematics. True, it is not always easy to do this.

The major objection surfaces upon realizing that it is necessary to accept simultaneously the following two ideas:

(1) logic is a branch of mathematics;

(2) the goal of logic is to study mathematics itself.

Faced with this apparent paradox, there are three possible attitudes. First, one may regard it as so serious that to undertake the study of logic is condemned in advance; second, one may deem that the supposed incompatibility between (1) and (2) simply compels the denial of (1), or at least its modification, which leads to the belief that one is not really doing mathematics when one studies logic; the third attitude, finally, consists in dismantling the paradox, becoming convinced that it is not one, and situating mathematical logic in its proper place, within the core of mathematics.

We invite you to follow us in this third path.

Those for whom even the word paradox is too weak will say: 'Wait a minute! Aren't you putting us on when you finally get around, in your Chapter 7, to providing definitions of concepts (intersection, pair, map, ordered set, . . .) that you have been continually using in the six previous chapters? This is certainly paradoxical. You are surely leading us in a vicious circle'.

Well, in fact, no. There is neither a paradox nor a vicious circle.

This text is addressed to readers who have already 'done' some mathematics, who have some prior experience with it, beginning with primary school. We do not ask you to forget all that in order to rebuild everything from scratch. It is the opposite that we ask of you. We wish to exploit the common background that is ours: familiarity with mathematical reasoning (induction, proof by contradiction, . . .), with everyday mathematical objects (sets (yes, even these!), relations, functions, integers, real numbers, polynomials, continuous functions, . . .), and with some concepts that may be less well known (ring, vector space, topological space, . . .). That is what is done in any course in mathematics: we make use of our prior knowledge in the acquisition of new knowledge. We will proceed in exactly this way and we will learn about new objects, possibly about new techniques of proof (but caution: the mathematical reasoning that we habitually employ will never be called into question; on the contrary, this is the only kind contemplated here).

If we simplify a bit, the approach of the mathematician is almost always the same whether the subject matter under study is measure theory, vector spaces, ordered sets, or any other area of so-called classical mathematics. It consists in examining structures, i.e. sets on which relations and functions have been defined, and correspondences among these structures. But, for each of these classical areas, there was a particular motivation that gave birth to it and nurtured its development. The purpose was to provide a mathematical model of some more or less 'concrete' situation, to respond to an expressed desire arising from the world outside mathematics, to furnish a useful mathematical tool (as a banal illustration of this, consider that vector spaces arose, originally, to represent the physical space in which we live).

Logic, too, follows this same approach; its particularity is that the reality it attempts to describe is not one from outside the world of mathematics, but rather the reality that is mathematics itself.

This should not be awkward, provided we remain aware of precisely what is involved. No student of mathematics confuses his physical environment with an oriented three-dimensional euclidean vector space, but the knowledge of this environment assists one's intuition when it comes to proving some property of this mathematical structure. The same applies to logic: in a certain way, we are going to manufacture a copy, a prototype, we dare say a reduced model of the universe of mathematics, with which we are already relatively familiar. More precisely, we will build a whole collection of models, more or less successful (not every vector space resembles our physical space). In addition to a specimen that is truly similar to the original, we will inevitably have created others (at the close of Chapter 6, we should be in position to understand why), often rather different from what we imagined at the outset. The study of this collection teaches us many lessons; notably, it permits those who undertake this study to ask themselves interesting questions about their perceptions and their intuitions of the mathematical world. Be that as it may, we must understand that it is essential not to confuse the original that inspired us with the copy or copies. But the original is indispensable for the production of the copy: our familiarity with the world of mathematics guides us in fabricating the representation of it that we will provide. But at the same time, our undertaking is a mathematical one, within this universe that we are attempting to better comprehend.

So there is no vicious circle. Rather than a circle, imagine a helix (nothing vicious there!), a kind of spiral staircase: we are on the landing of the nth floor, where our mathematical universe is located; call this the 'intuitive level'. Our work takes us down a level, to the $(n-1)$st floor, where we find the prototype, the reduced model; we will then be at the 'formal' level and our passage from one level to the other will be called 'formalization'. What is the value of n? This makes absolutely no difference; there is no first nor last level. Indeed, if our model is well constructed, if in reproducing the mathematical universe it has not omitted any detail, then it will also contain the counterpart of our very own work on formalization; this requires us to consider level $n-2$, and so on. At the beginning of this book, we find ourselves at the intuitive level. The souls that inhabit it will also be called intuitive objects; we will distinguish these from their formal replica by attaching the prefix 'meta' to their names (meta-integers, meta-relations, even meta-universe since the word 'universe' will be given a precise technical meaning in Chapter 7). We will go so far as to say that for any value of n, the nth level in our staircase is intuitive relative to level $n-1$ and is formal relative to level $n+1$. As we descend, i.e. as we progress in our formalization, we could stop for a rest at any moment, and take the opportunity to verify that the formal model, or at least what we can see of it, agrees with the intuitive original. This rest period concerns the meta-intuitive, i.e. level $n+1$.

So we must face the facts: it is no more feasible to build all of mathematics '*ex nihilo*' than to write an English–English dictionary that would be of use to a Martian who knows nothing of our lovely language. We are faced here with a question that had considerable importance in the development of logic at the beginning of the century and about which it is worth saying a few words.

Set theory (it matters little which theory: ZF, Z, or some other), by giving legitimacy to infinite objects and by allowing these to be manipulated just like 'real' objects (the integers, for example), with the same logical rules, spawned a fair amount of resistance among certain mathematicians; all the more because the initial attempts turned out to be contradictory. The mathematical world was then split into two clans. On the one hand, there were those who could not resist the freedom that set theory provided, this 'Cantorian paradise' as Hilbert called it. On the other hand were those for whom only finite objects (the integers, or anything that could be obtained from the integers in a finite number of operations) had any meaning and who, as a consequence, denied the validity of proofs that made use of set theory.

To reconcile these points of view, Hilbert had imagined the following strategy (the well-known 'Hilbert programme'): first, proofs would be regarded as finite sequences of symbols, hence, as finite objects (that is what is done in this book in Chapters 4 and 6); second, an algorithm would be found that would transform a proof that used set theory into a finitary proof, i.e. a proof that would be above all suspicion. If this programme could be realized, we would be able to see, for example, that set theory is consistent: for if not, set theory would permit a proof of $0 = 1$ which could then, with the help of the algorithm suggested above, be transformed into a finitary proof, which is absurd.

This hope was dashed by the second incompleteness theorem of Gödel: surely, any set theory worthy of this name allows the construction of the set of natural numbers and, consequently, its consistency would imply the consistency of Peano's axioms. Gödel's theorem asserts that this cannot be done in a finitary way.

The conclusion is that even finitary mathematics does not provide a foundation for our mathematical edifice, as presently constructed.

The process of formalization involves two essential stages. First, we fix the context (the structures) in which the objects evolve while providing a syntax to express their properties (the languages and the formulas). Here, the important concept is the notion of satisfaction which lies at the heart of the area known as semantics. It would be possible to stop at this point but we can also go further and formalize the reasoning itself; this is the second stage in the formalization. Here, we treat deductions or formal proofs as mathematical objects in their own right. We are then not far from proof theory, which is the branch of logic that specializes in these questions.

This book deliberately assigns priority to the first stage. Despite this, we will not ignore the second, which is where the most famous results from mathematical logic (Gödel's theorems) are situated. Chapter 4 is devoted to the positive results in this area: the equivalence between the syntactic and semantic points of view in the context that we have selected. This equivalence is called 'completeness'. There are several versions of this simply because there are many possible choices for a formal system of deduction. One of these systems is in fashion these days because of its use in computer science: this is the method of resolution. We have chosen to introduce it after first presenting the traditional completeness theorem.

The negative results, the incompleteness theorems, will be treated in Chapter 6, following the study of Peano's arithmetic. This involves, as we explained above, abandoning our possible illusions.

The formalization of reasoning will not occur outside the two chapters that we have just mentioned.

Chapter 1 treats the basic operations on truth values, 'true' and 'false'. The syntax required is very simple (propositional formulas) and the semantics (well-known truth tables) is not very complicated. We are interested in the truth value of propositions, while carefully avoiding any discussion of the nature of the properties expressed by means of these propositions. Our concern with what they express, and with the ways they do it, is the purpose of Chapter 3. We see immediately that the operators considered in the first chapter (the connectives 'and', 'or', 'implies' and so on) do not suffice to express familiar mathematical properties. We have to introduce the quantifiers and we must also provide a way of naming mathematical objects. This leads to formulas that are sequences of symbols obeying rather complicated rules. Following the description of a syntax that is considerably more complex than that for propositional calculus, we define the essential concept: satisfaction of a formula in a structure. We will make extensive use of all this, which is called predicate calculus, in Chapters 4 and 6, to which we referred earlier, as well as in Chapters 7 and 8. You will have concluded that it is only Chapter 5 that does not require prior knowledge of predicate calculus. Indeed, it is devoted to the study of recursive functions, a notion that is absolutely fundamental for anyone with even the slightest interest in computer science. We could perfectly well begin with this chapter provided we refer to Chapter 1 for the process of inductive definition, which is described there in detail and which is used as well for recursive functions.

In Chapter 7 we present axiomatic set theory. It is certainly there that the sense of paradox to which we referred will be most strongly felt since we purport to construct mathematical universes as if we were defining a field or a commutative group. But, once a possible moment of doubt has passed, one will find all that a mathematician should know about the important notions of cardinals and ordinals, the axiom of choice, whose status is generally poorly understood, and, naturally, a list of the axioms of set theory.

Chapter 8 carries us a bit further into an area of which we have so far only caught a glimpse: model theory. Its ambition is to give you a taste for this subject and to stimulate your curiosity to learn more. In any case, it should lead you to suspect that mathematical logic is a rich and varied terrain, where one can create beautiful things, though this can also mean difficult things.

Have we forgotten Chapter 2? Not at all! It is just that it constitutes a singularity in this book. To begin with, it is the only one in which we employ notions from classical mathematics that a student does not normally encounter prior to the upper-level university curriculum (topological spaces, rings and ideals). Moreover, the reader could just as well skip it: the concepts developed there are used only in some of the exercises and in one section of the last chapter. But we have included it for at least three reasons: the first is that Boolean algebras are the "correct" algebraic structures for logic; the second is that it affords us an opportunity to display how perfectly classical mathematics, of a not entirely elementary nature, could be linked in a natural way with the study of logic; the third, finally, is that an exposure to Boolean algebras is generally absent from the mathematical literature offered to students, and is even more rarely proposed to students outside the technical schools. So you should consider Chapter 2, if you will, as a little supplement that you may consult or not, as you wish.

We will probably be criticized for not being fair, either in our choice of the subjects we treat or in the relative importance we accord to each of them. The domain of logic is now so vast that it would have been absolutely impossible to introduce every one of its constituents. So we have made choices: as we have already noted, proof theory is barely scratched; lambda calculus and algorithmic complexity are absent despite the fact that they occupy an increasingly important place in research in logic (because of their applications to the theory of computing which have been decisive). The following are also absent: non-classical logics (intuitionist . . .), second-order logic (in which quantifications range over relations on a structure as well as over its elements), or so-called 'infinitary' logics (which allow formulas of infinite length). These choices are dictated, first of all, by our desire to present a basic course. We do not believe that the apprentice logician should commence anywhere else than with a detailed study of the first-order predicate calculus; this is the context that we have set for ourselves (Chapter 3). Starting from this, we wished to present the three areas (set theory, model theory, recursive function theory and decidability) that seem to us to be the most important. Historically speaking, they certainly are. They also are because the 'grand' theorems of logic are all found there. Finally, it is our opinion that familiarity with these three areas is an indispensable prerequisite for anyone who is interested in any other area of mathematical logic. Having chosen this outline, we still had the freedom to modify the relative importance given to these three axes. In this matter, we cannot deny that we allowed our personal preferences to guide us; it is clear that Chapter 8 could just as well have been devoted to something other than model theory.

These lines were drafted only after the book that follows was written. We think that they should be read only after it has been studied. As we have already pointed out, we can only truly speak about an activity, describe it (formalize it!), once we have acquired a certain familiarity with it.

Until then.

1 Propositional calculus

Propositional calculus is the study of the propositional connectives; these are operators on statements or on formulas. First of all there is negation, which we denote by the symbol ¬ which is placed in front of a formula. The other connectives are placed between two formulas: we will consider conjunction ('and', denoted by ∧), disjunction ('or', denoted by ∨), implication (⟹), and equivalence (⟺). Thus, for example, from two statements A and B, it is possible to form their conjunction: this is another statement that is true if and only if A is true and B is true.

The first thing we do is to construct purely formal objects that we will call propositional formulas, or, more simply in this chapter, formulas. As building blocks we will use propositional variables which intuitively represent elementary propositions, and we assemble them using the connectives mentioned above. Initially, formulas appear as suitably assembled sequences of symbols. In Section 1.1, we will present precise rules for their construction and means for recovering the method by which a given formula was constructed, which makes it possible for the formula to be read. All these formal considerations constitute what we call syntax.

This formal construction is obviously not arbitrary. We will subsequently have to give meaning to these formulas. This is the purpose of Section 1.2. If, for each elementary proposition appearing in a formula F, we know whether it is true or not (we speak of the truth value of the proposition), we must be able to decide whether F itself is true or not. For instance, we will say that A ⟹ B is true in three of the four possible cases: when A and B are both true, when A and B are both false, and when A is false and B is true. Notice here the difference from common usage: for example, in everyday language and even in mathematics texts, the phrase 'A implies B' suggests a causal relationship which, in our context, does not exist at all.

Thereby we arrive at the important notions of this chapter: the concept of tautology (this is a formula that is true regardless of the truth values assigned to the propositional variables) and the notion of logical equivalence (two formulas are logically equivalent if they receive the same truth value regardless of the truth values assigned to the propositional variables).

In Section 1.3, we see that a formula is always logically equivalent to a formula that can be written in a very particular form (disjunctive or conjunctive normal form) while Section 1.4 is devoted to the interpolation theorem and the definability theorem whose full import will be appreciated when they are generalized to the predicate calculus (in Chapter 8). In the last section of this chapter, the compactness theorem is particularly important and it too will be generalized, in Chapter 3. It asserts that if it is impossible to assign truth values to propositional variables in a way that makes all the formulas in some infinite set X true, then there is some finite subset of X for which it is also impossible to do this.

1.1 Syntax

1.1.1 Propositional formulas

The preliminary section called 'Notes to the reader' contains, among other items, some general facts about words on an alphabet. The student who is not familiar with these notions should read that section first.

Let us consider a non-empty set P, finite or infinite, which we will call the set of **propositional variables**. The elements of P will usually be denoted by capital letters of the alphabet, possibly bearing subscripts.

In addition, we allow ourselves the following five symbols:

$$\neg \quad \vee \quad \wedge \quad \Rightarrow \quad \Leftrightarrow$$

which we read respectively as: 'not', 'or', 'and', 'implies' and 'is equivalent to' and which we call **symbols for propositional connectives**. We assume that they are not elements of P.

The symbols $\neg, \vee, \wedge, \Rightarrow, \Leftrightarrow$ are respectively called: the symbol for **negation**, the symbol for **disjunction**, the symbol for **conjunction**, the symbol for **implication** and the symbol for **equivalence**.

In view of the roles that will be assigned to them (see Definition 1.2 below), we say that the symbol $\neg$ is **unary** (or **has one place**) and that the other four symbols for connectives are **binary** (or **have two places**).

Finally, we consider the following two symbols:

$$) \qquad ($$

respectively called the **closing parenthesis** and the **opening parenthesis**, distinct from the symbols for the connectives and also not belonging to P.

Certain finite sequences composed of propositional variables, symbols for propositional connectives, and parentheses will be called **propositional formulas** (or **propositions**). Propositional formulas are thus words formed with the following alphabet:

$$\mathcal{A} = P \cup \{\neg, \vee, \wedge, \Rightarrow, \Leftrightarrow\} \cup \{ \,), (\, \}.$$

Remark 1.1 *From the very first sentences in this chapter, we can already sense one of the difficulties which, unless we are careful, will confront us continually during our apprenticeship with the basic notions of formal logic: certain words and certain symbols are both used in everyday mathematical language (which we will call the **metalanguage**) and also appear in the various formal languages that are the principal objects of our study: for example, the word 'implies' and the symbol ⇒, which, to say the least, arise frequently in all mathematical discourse, are used here to denote a precise mathematical object: a symbol for one of the connectives. We will attempt, insofar as is possible, to eliminate from our meta-language any word or symbol which is used in a formal language. It would none the less be difficult to renounce altogether in our discourse the use of words such as 'and', 'or', 'not' or parentheses. (This very sentence illustrates the fact clearly enough.) This is why we wish to call the reader's attention to this problem right from the beginning and we invite the reader to be constantly alert to the distinction between a formal language and the metalanguage. (The same problem will arise again in Chapter 3 with the symbols for the quantifiers.)*

As stated in 'Notes to the Reader', we will identify, by convention, the elements of $\mathcal{A}$ with the corresponding words of length 1 in $\mathcal{W}(\mathcal{A})$. In particular, P will be considered a subset of $\mathcal{W}(\mathcal{A})$.

Definition 1.2 *The set $\mathcal{F}$ of **propositional formulas constructed from** P is the smallest subset of $\mathcal{W}(\mathcal{A})$ which*

- *includes P;*
- *whenever it contains the word F, it also contains the word $\neg F$;*
- *whenever it contains the words F and G, it also contains the words*

$$(F \wedge G), \ (F \vee G), \ (F \Rightarrow G) \ \text{and} \ (F \Leftrightarrow G).$$

In other words, $\mathcal{F}$ is the smallest subset of $\mathcal{W}(\mathcal{A})$ which includes P and which is closed under the operations:

$$F \mapsto \neg F,$$
$$(F, G) \mapsto (F \wedge G),$$
$$(F, G) \mapsto (F \vee G),$$
$$(F, G) \mapsto (F \Rightarrow G),$$
$$(F, G) \mapsto (F \Leftrightarrow G).$$

Observe that there is at least one subset of $\mathcal{W}(\mathcal{A})$ which has these properties, namely $\mathcal{W}(\mathcal{A})$ itself. The set $\mathcal{F}$ is the intersection of all the subsets of $\mathcal{W}(\mathcal{A})$ that have these properties.

Here are examples of formulas (A, B, and C are elements of P):

$$A$$
$$(A \Rightarrow (B \Leftrightarrow A))$$
$$(\neg A \Rightarrow A)$$
$$\neg(A \Rightarrow A)$$
$$(((A \wedge (\neg B \Rightarrow \neg A)) \wedge (\neg B \vee \neg C)) \Rightarrow (C \Rightarrow \neg A)).$$

And here are words that are not formulas:

$$A \wedge B$$
$$\neg(A)$$
$$(A \Rightarrow B \vee C)$$
$$A \Rightarrow B, C$$
$$(A \wedge B \wedge C)$$
$$\forall A(A \vee \neg A)$$
$$((A \wedge (B \Rightarrow C)) \vee (\neg A \Rightarrow (B \wedge C)) \wedge (\neg A \vee B)).$$

Later we will agree to certain abuses in the writing of formulas: for example, $A \wedge B$ could be accepted in some cases as an abbreviation for the formula $(A \wedge B)$. Obviously, this changes nothing in the definition above; we are simply giving ourselves several ways of representing the same object: if $A \wedge B$ is a permitted way to write the formula $(A \wedge B)$, the length of $A \wedge B$ is none the less equal to 5. Observe in passing that the notion of the length of a formula is already defined since we have defined the length of any word on any alphabet. (See the Notes to the Reader.)

It is possible to give a more explicit description of the set $\mathcal{F}$: to do this, we will define, by induction, a sequence $(\mathcal{F}_n)_{n \in \mathbb{N}}$ of subsets of $\mathcal{W}(\mathcal{A})$. We set

$$\mathcal{F}_0 = P$$

and, for each n,

$$\mathcal{F}_{n+1} = \mathcal{F}_n \cup \{\neg F : F \in \mathcal{F}_n\} \cup \{(F \; \alpha \; G) : F, G \in \mathcal{F}_n, \; \alpha \in \{\wedge, \vee, \Rightarrow, \Leftrightarrow\}\}$$

Observe that the sequence $(\mathcal{F}_n)_{n \in \mathbb{N}}$ is increasing (for $n \le m$, we have $\mathcal{F}_n \subseteq \mathcal{F}_m$).

Theorem 1.3 $\mathcal{F} = \bigcup_{n \in \mathbb{N}} \mathcal{F}_n$.

Proof It is clear that $\bigcup_{n \in \mathbb{N}} \mathcal{F}_n$ includes P and is closed under the operations indicated above (if two words F and G belong to $\mathcal{F}_n$ for a certain integer n, then

$$\neg F, \; (F \wedge G), \; (F \vee G), \; (F \Rightarrow G) \text{ and } (F \Leftrightarrow G)$$

belong to $\mathcal{F}_{n+1}$. It follows that $\bigcup_{n \in \mathbb{N}} \mathcal{F}_n$ includes the smallest set having these properties, namely, $\mathcal{F}$.

To obtain the reverse inclusion, we show by induction that for each integer n, we have $\mathcal{F}_n \subseteq \mathcal{F}$. This is true by definition if $n = 0$, and if we assume (this is the induction hypothesis) $\mathcal{F}_k \subseteq \mathcal{F}$, we also have $\mathcal{F}_{k+1} \subseteq \mathcal{F}$ according to the definition of $\mathcal{F}_{k+1}$ and the closure properties of $\mathcal{F}$.   ∎

We thus have two equivalent definitions of the set of propositional formulas. We often speak of 'definition from above' in the first case and 'definition from below' for the one that follows from the previous theorem.

At several places in this book, we will encounter this type of definition, said to be **inductive** or **by induction** (see, for example, the set of terms or the set of formulas of the predicate calculus in Chapter 3, or again the set of recursive functions in Chapter 5). In each case, we are concerned with defining the smallest subset of a fixed set E that includes a given subset and that is closed under certain operations defined on E (this is definition from above). We always have an equivalent definition from below: this consists in constructing the set that we wish to define one level at a time; the subset given initially is the lowest level and the elements of level $n + 1$ are defined to be the images under the given operations of the elements from the lower levels. The set to be defined is then the union of a sequence of subsets, indexed by the set of natural numbers. In all instances of sets defined by induction that we will meet in the future, as well as in the method of proof by induction described below, we will encounter the notion of height.

Definition 1.4 *The height of a formula $F \in \mathcal{F}$ is the least integer n such that $F \in \mathcal{F}_n$. It is denoted by* $\mathsf{h}[F]$.

For example, if A and B are propositional variables, we have

$$\mathsf{h}[A] = 0; \quad \mathsf{h}[((A \vee B) \wedge (B \Rightarrow A))] = 2; \quad \mathsf{h}[\neg\neg\neg\neg\neg A] = 5.$$

Note that $\mathcal{F}_n$ is the set of formulas of height less than or equal to n, and that $\mathcal{F}_{n+1} - \mathcal{F}_n$ is the set of formulas of height exactly $n + 1$.

It also follows from the definition that for all formulas F and $G \in \mathcal{F}$, we have:

$$\mathsf{h}[\neg F] \leq \mathsf{h}[F] + 1 \quad \text{and} \quad \mathsf{h}[(F \, \alpha \, G)] \leq \sup(\mathsf{h}[F], \mathsf{h}[G]) + 1,$$

where α denotes an arbitrary symbol for a binary connective.

(In fact, we will see, after Theorem 1.12, that we can replace these inequalities by equalities).

1.1.2 Proofs by induction on the set of formulas

Suppose we wish to show that a certain property $\mathcal{X}(F)$ is satisfied by every formula $F \in \mathcal{F}$. To do this we can use an argument by induction (in the usual sense) on the height of F: so we would be led first to show that $\mathcal{X}(F)$ is true for every formula F belonging to $\mathcal{F}_0$, and afterwards that if $\mathcal{X}(F)$ is true for every $F \in \mathcal{F}_n$, then it is also true for every $F \in \mathcal{F}_{n+1}$ (and thus, for any n).

This style of argument is associated with definition 'from below' of the set of formulas.

It is more practical and more natural, however, to take the first definition as our point of departure and proceed as follows. The initial step is the same: we show that $\mathcal{X}(F)$ is satisfied for all formulas belonging to P (that is, to $\mathcal{F}_0$); the induction step consists in proving, on the one hand, that if the formula F satisfies the property $\mathcal{X}$, then so does the formula $\neg F$, and on the other hand, that if F and G satisfy $\mathcal{X}$, then so do the formulas $(F \wedge G)$, $(F \vee G)$, $(F \Rightarrow G)$, and $(F \Leftrightarrow G)$.

As we see, the notion of the height of formulas does not appear explicitly in this style of argument, nor does any other natural number.

Before establishing the correctness of this method of proof (which is the purpose of Lemma 1.7) let us give an initial example of its use:

Theorem 1.5 *The height of a formula is always strictly less than its length.*

Proof In this case, the property $\mathcal{X}(F)$ is: $h[F] < \lg[F]$.

If F is a propositional variable, we have $h[F] = 0$ and $\lg[F] = 1$; the inequality is verified. Let us now pass to the induction step. Suppose that a formula F satisfies $h[F] < \lg[F]$; we then have

$$h[\neg F] \leq h[F] + 1 < \lg[F] + 1 = \lg[\neg F],$$

which shows that $\mathcal{X}(\neg F)$ is true; and now suppose that F and G are formulas that satisfy $h[F] < \lg[F]$ and $h[G] < \lg[G]$; then, where α is any symbol for a binary connective, we have

$$h[(F \alpha G)] \leq \sup(h[F], h[G]) + 1 < \sup(\lg[F], \lg[G]) + 1$$
$$< \lg[F] + \lg[G] + 3 = \lg[(F \alpha G)],$$

which means that $\mathcal{X}(F \alpha G)$ is satisfied and the proof is finished. ∎

As a consequence of this property, observe that there are no formulas of length 0 (which is one of the ways to show that the empty word is not a formula!) and that the only formulas of length 1 are the propositional variables.

The two lemmas that follow allow us to justify the method that we just described and used. The first is a variant of it that we will then easily modify below.

Consider a property $\mathcal{Y}(W)$ of an arbitrary word $W \in \mathcal{W}(\mathcal{A})$ (which need not necessarily be a formula). Here is a sufficient condition that every formula satisfies the property $\mathcal{Y}$:

Lemma 1.6 *Suppose, on the one hand, that $\mathcal{Y}(W)$ is true for every word $W \in P$, and, on the other hand, that for any words W and V, if $\mathcal{Y}(W)$ and $\mathcal{Y}(V)$ are true, then*

$$\mathcal{Y}(\neg F), \ \mathcal{Y}(F \wedge G), \ \mathcal{Y}(F \vee G), \ \mathcal{Y}(F \Rightarrow G), \ and \ \mathcal{Y}(F \Leftrightarrow G)$$

are also true. Under these conditions, $\mathcal{Y}(F)$ is true for every formula F.

Proof Let Z be the set of words that have property $\mathcal{Y}$:

$$Z = \{W \in \mathcal{W}(\mathcal{A}) : \mathcal{Y}(W)\}.$$

The hypotheses of the lemma indicate that Z includes P and that it is closed with respect to the operations:

$$W \mapsto \neg W, \ (W, V) \mapsto (W \wedge V), \ (W, V) \mapsto (W \vee V),$$
$$(W, V) \mapsto (W \Rightarrow V) \text{ and } (W, V) \mapsto (W \Leftrightarrow V).$$

It follows, according to Definition 1.2, that $\mathcal{F}$ is included in Z, which means that every element of $\mathcal{F}$ satisfies the property $\mathcal{Y}$. ∎

Let us now consider the case where we have a property $\mathcal{X}(F)$ which is only defined for formulas and not for arbitrary words. (This is the case, for example, with the property: $h[F] < \lg[F]$, since the notion of height is defined only for elements of $\mathcal{F}$.)

Lemma 1.7 *Suppose, on the one hand, that $\mathcal{X}(F)$ is true for every formula $F \in P$, and, on the other hand that, for all formulas F and G, if $\mathcal{X}(F)$ and $\mathcal{X}(G)$ are true, then*

$$\mathcal{X}(\neg F), \ \mathcal{X}(F \wedge G), \ \mathcal{X}(F \vee G), \ \mathcal{X}(F \Rightarrow G), \ and \ \mathcal{X}(F \Leftrightarrow G)$$

are also true. Under these conditions, $\mathcal{X}(F)$ is true for every formula F.

Proof It suffices to consider the property $\mathcal{Y}(W)$: '$W \in \mathcal{F}$ and $\mathcal{X}(W)$', which is defined for every word $W \in \mathcal{W}(\mathcal{A})$. Since $\mathcal{F}$ includes P and is closed under the operations

$$W \mapsto \neg W, \ (W, V) \mapsto (W \wedge V), \ (W, V) \mapsto (W \vee V),$$
$$(W, V) \mapsto (W \Rightarrow V) \text{ and } (W, V) \mapsto (W \Leftrightarrow V),$$

we see immediately that if the property $\mathcal{X}$ satisfies the stated hypotheses, then the property $\mathcal{Y}$ satisfies those of the preceding lemma. We conclude that $\mathcal{Y}(F)$ is true for every formula F and hence the same holds for $\mathcal{X}(F)$. ∎

1.1.3 The decomposition tree of a formula

Among the first examples of formulas that we gave earlier was the following word W:

$$(((A \wedge (\neg B \Rightarrow \neg A)) \wedge (\neg B \vee \neg C)) \Rightarrow (C \Rightarrow \neg A)).$$

The reader, who justifiably has no intention to take us on faith, should be convinced by what follows that this word is indeed a formula:

By setting

$$W_0 = ((A \wedge (\neg B \Rightarrow \neg A)) \wedge (\neg B \vee \neg C))$$

and

$$W_1 = (C \Rightarrow \neg A),$$

we first observe that W can be written as $(W_0 \Rightarrow W_1)$.

Then, after setting

$$W_{00} = (A \wedge (\neg B \Rightarrow \neg A)),$$
$$W_{01} = (\neg B \vee \neg C),$$
$$W_{10} = C,$$
$$W_{11} = \neg A,$$

we can write $W_0 = (W_{00} \wedge W_{01})$ and $W_1 = (W_{10} \Rightarrow W_{11})$.
Continuing in this way, we will be led successively to set

$$W_{000} = A \qquad W_{001} = (\neg B \Rightarrow \neg A)$$
$$W_{010} = \neg B \qquad W_{011} = \neg C$$
$$W_{110} = A \qquad W_{0010} = \neg B$$
$$W_{0011} = \neg A \qquad W_{0100} = B$$
$$W_{0110} = C \qquad W_{00100} = B$$
$$W_{00110} = A$$

in such a way that

$$W_{00} = (W_{000} \wedge W_{001}), \ \ W_{01} = (W_{010} \vee W_{011}), \ \ W_{11} = \neg W_{110},$$
$$W_{001} = (W_{0010} \Rightarrow W_{0011}), \ \ W_{010} = \neg W_{0100}, \ \ W_{011} = \neg W_{0110},$$
$$W_{0010} = \neg W_{00100} \text{ and } W_{0011} = \neg W_{00110}.$$

This shows that the word W was obtained by starting from propositional variables and by applying, a finite number of times, operations allowed in the definition of the set of formulas. It follows that W is a formula.

We can represent the preceding decomposition in the form of a **tree**:

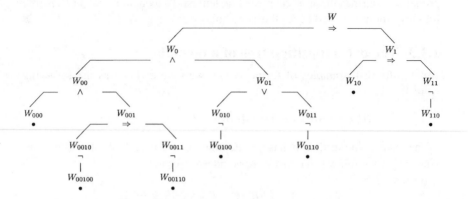

The root of the tree (the formula W) is at the top and the branches 'grow' toward the bottom. Each node of the tree is a word V (which is always a formula if the word at the root of the tree is a formula). Three cases can arise: V is a propositional variable and, in this case, will be an extremity of the tree (the words corresponding to this situation are identified by a black dot in our figure above); V can be written as $\neg V'$ in which case there is a single branch starting from V and ending at the level immediately below at the node V'; or finally, V can be written as $(V' \alpha V'')$ (where α is a symbol for a binary connective) and in this case there are two branches starting from V and ending at the level immediately below at the two nodes V' and V'' (in this case the appropriate symbol for a binary connective has been placed in the figure between the two branches).

The decomposition that we have chosen for our formula shows that it belongs to $\mathcal{F}_5$. Its height is therefore less than or equal to 5. At the moment, nothing allows us to claim that its height is exactly 5. Might we not, in fact, imagine a second way of decomposing this formula which would lead to a shorter tree? All we can say (and this is thanks to Theorem 1.3) is that for every formula $F \in \mathcal{F}$, there is at least one decomposition of the type we have just exhibited. The uniqueness will be established in the next theorem, for which we first require a few lemmas which, with the exception of Lemma 1.10, will be proved by induction on the set of formulas.

1.1.4 The unique decomposition theorem

For each word $W \in \mathcal{W}(\mathcal{A})$, let us agree to denote by $\mathsf{o}[W]$ (respectively: $\mathsf{c}[W]$) the number of opening (respectively: closing) parentheses that occur in W.

Lemma 1.8 *In any formula, the number of opening parentheses is equal to the number of closing parentheses.*

Proof   We argue by induction on the formula F.

- For any formula $F \in P$, we have $\mathsf{o}[F] = \mathsf{c}[F] = 0$.
- For any formula $F \in \mathcal{F}$ such that $\mathsf{o}[F] = \mathsf{c}[F]$, since $\mathsf{o}[\neg F] = \mathsf{o}[F]$ and $\mathsf{c}[\neg F] = \mathsf{c}[F]$, we have $\mathsf{o}[\neg F] = \mathsf{c}[\neg F]$.
- For all formulas F and G belonging to $\mathcal{F}$ such that $\mathsf{o}[F] = \mathsf{c}[F]$ and $\mathsf{o}[G] = \mathsf{c}[G]$, and where α is an arbitrary symbol for a binary connective, we have

$$\mathsf{o}[(F \alpha G)] = \mathsf{o}[F] + \mathsf{o}[G]+1 = \mathsf{c}[F] + \mathsf{c}[G]+1 = \mathsf{c}[(F \alpha G)].$$

Thus, $\mathsf{o}[F] = \mathsf{c}[F]$ for any propositional formula F.   ∎

Lemma 1.9 *For any formula $F \in \mathcal{F}$ and any word $W \in \mathcal{W}(\mathcal{A})$, if W is an initial segment of F, then $\mathsf{o}[W] \geq \mathsf{c}[W]$.*

Proof   The induction is on the formula F.

- If $F \in P$, then for every initial segment W of F, we have $\mathsf{o}[W] = \mathsf{c}[W] = 0$, hence $\mathsf{o}[W] \geq \mathsf{c}[W]$.
- Let F be a formula such that for every initial segment W of F, we have $\mathsf{o}[W] \geq \mathsf{c}[W]$. Consider an initial segment V of $\neg F$: if V is the empty word, then

$o[V] = c[V] = 0$; if V is not the empty word, then there is an initial segment W of F such that $V = \neg W$; we have $o[V] = o[W]$ and $c[V] = c[W]$, and since $o[W] \geq c[W]$ (by the induction hypothesis), we conclude that $o[V] \geq c[V]$.

- Let F and G be two formulas all of whose initial segments have at least as many opening parentheses as closing parentheses, and let α be a symbol for a binary connective. Set $H = (F \alpha G)$. Let V be an initial segment of H. Four cases can arise:

  * either $V = \emptyset$: in this case, $o[V] = c[V] = 0$;

  * or $V = (W$ (where W is an initial segment of F):
    then $o[V] = o[W] + 1$ and $c[V] = c[W]$, and since $o[W] \geq c[W]$ (by the induction hypothesis), we conclude that $o[V] \geq c[V]$;

  * or $V = (F \alpha K$ (where K is an initial segment of G):
    then $o[V] = o[F] + o[K] + 1$ and $c[V] = c[F] + c[K]$; but $o[F] = c[F]$ (by Lemma 1.8) and $o[K] \geq c[K]$ (by the induction hypothesis), which allows us to conclude once more that $o[V] \geq c[V]$;

  * or $V = H$: in which case $o[V] \geq c[V]$ (by Lemma 1.8).

So we see that in all cases, $o[V] \geq c[V]$.   ■

Lemma 1.10 *For any formula $F \in \mathcal{F}$ whose first symbol is an opening parenthesis and for every word $W \in \mathcal{W}(\mathcal{A})$ which is a proper initial segment of F, we have*

$$o[W] > c[W].$$ (*strict inequality*)

Proof For once, the proof is not by induction!

Consider a formula F which can be written as $F = (G \alpha H)$ where G and H are arbitrary formulas and α is a symbol for a binary connective. Let W be a proper initial segment of F. There are two possible cases.

- either $W = (K$ (where K is an (arbitrary) initial segment of G);
  in this case, $o[W] = o[K] + 1$ and $c[W] = c[K]$, and since $o[K] \geq c[K]$ (by Lemma 1.9), we conclude that $o[W] > c[W]$;

- or $W = (G \alpha L$ (where L is an initial segment of H);
  in this case, $o[W] = o[G] + o[L] + 1$ and $c[W] = c[G] + c[L]$; but $o[G] = c[G]$ (by Lemma 1.8) and $o[L] \geq c[L]$ (by Lemma 1.9), which leads once again to $o[W] > c[W]$.   ■

Lemma 1.11 *For any formula $F \in \mathcal{F}$ and for any word $W \in \mathcal{W}(\mathcal{A})$, if W is a proper initial segment of F then W is not a formula.*

Proof Here too, the induction is on the formula F.

- A propositional variable does not have any proper initial segments.

- If F is a formula none of whose proper initial segments is a formula, and if V is a proper initial segment of $\neg F$, then either $V = \neg$ and is not a formula (the only formulas of length 1 are the elements of P), or else $V = \neg W$ where

W is a proper initial segment of F; in this case, W is not a formula (by the induction hypothesis) nor is $V = \neg W$. Observe that, contrary to what we might have been tempted to believe, the fact that

'if W is not a formula, then neither is $\neg W$'

is not a trivial application of the definition of the set of formulas, but requires a proof. Here it is: if $\neg W$ is a formula, examination of its first symbol shows that it is neither a propositional variable nor a formula of the type $(H \; \alpha \; K)$; therefore (by Theorem 1.3) there exists at least one formula G such that $\neg W = \neg G$; if the words $\neg W$ and $\neg G$ are identical, then so also are the words W and G, which proves that W is a formula.

- Let F and G be two arbitrary formulas, α a symbol for a binary connective, and V a proper initial segment of $(F \; \alpha \; G)$. We have $\mathsf{o}[V] > \mathsf{c}[V]$ (by Lemma 1.10). We conclude that V is not a formula (by Lemma 1.8). Note that it was not necessary, in this part of the argument by induction, to assume that the proper initial segments of F and G are not formulas. ∎

Theorem 1.12 (*Unique decomposition*) *For any formula $F \in \mathcal{F}$, one and only one of the following three cases can arise:*
Case 1: $F \in P$.
Case 2: there is a unique formula G such that $F = \neg G$.
Case 3: there is a unique symbol for a binary connective α and a unique pair of formulas $(G, H) \in \mathcal{F}^2$ such that $F = (G \; \alpha \; H)$.

Proof It is obvious that these three cases are mutually exclusive: we are in case 1, in case 2, or in case 3 (still subject to proving uniqueness in each of these cases) according as the first symbol of F is an element of P, the symbol $\neg$, or the symbol ((these are, by virtue of Theorem 1.3, the only possibilities).

What we know already (Theorem 1.3) is this: either $F \in P$, or else there is at least one formula G such that $F = \neg G$, or else there is at least one symbol for a binary connective α and formulas G and H such that $F = (G \; \alpha \; H)$.

So it only remains for us to prove the uniqueness of the decomposition in cases 2 and 3.

This is more or less obvious for case 2: if $F = \neg G = \neg G'$, then $G = G'$.

As for case 3, suppose that there exist formulas $G, H, K,$ and L and symbols for binary connectives α and β such that $F = (G \; \alpha \; H) = (K \; \beta \; L)$. We conclude that the two words $G \; \alpha \; H$ and $K \; \beta \; L$ are equal, which shows that one of the two formulas G and K is an initial segment of the other. By Lemma 1.11, this cannot be a proper initial segment. Since the empty word is not a formula, we conclude that $G = K$. From this, it follows that the words $\alpha \; H$ and $\beta \; L$ are equal. The symbols α and β are therefore identical, as well as the formulas H and L. ∎

As the first application of the unique decomposition theorem, we have the uniqueness of the decomposition tree of a formula, as described above.

We may also conclude from it (as announced at the end of Subsection 1.1.1) that for all formulas F and G belonging to $\mathcal{F}$, we have

$$\mathsf{h}[\neg F] = \mathsf{h}[F] + 1 \quad \text{and} \quad \mathsf{h}[(F \ \alpha \ G)] = \sup(\mathsf{h}[F], \mathsf{h}[G]) + 1$$

for any symbol for a binary connective α.

For example, let us prove the second equality (the other one is treated in an exactly analogous fashion): let H denote the formula $(F \ \alpha \ G)$. Since this is not an element of P, there exists a (unique) integer n such that $\mathsf{h}[H] = n + 1$. This means that $H \in \mathcal{F}_{n+1}$ and $H \notin \mathcal{F}_n$. By the definition of $\mathcal{F}_{n+1}$ and because H begins with an opening parenthesis, we conclude that there exist two formulas H_1 and $H_2 \in \mathcal{F}_n$ and a symbol for a binary connective β such that $H = (H_1 \ \beta \ H_2)$. The unique decomposition theorem then shows that $\beta = \alpha$, $H_1 = F$, and $H_2 = G$. Consequently, F and G belong to $\mathcal{F}_n$. If there were some integer $m < n$ such that F and G belonged to $\mathcal{F}_m$, the formula $(F \ \alpha \ G)$ would belong to $\mathcal{F}_{m+1}$, hence also to $\mathcal{F}_n$, which is false. It follows that at least one of the formulas F and G has height n, and so: $\mathsf{h}[(F \ \alpha \ G)] = \sup(\mathsf{h}[F], \mathsf{h}[G]) + 1$.

1.1.5 Definitions by induction on the set of formulas

Just as we can give proofs by induction on the set of formulas, we can provide **definitions by induction** of functions or of relations whose domain is the set of formulas. The principle is as follows: given an arbitrary set E, to define a mapping ϕ from $\mathcal{F}$ into E, it suffices to give, first of all, the values of ϕ on P, and then to give rules which allow us, for all formulas F and G, to determine the values of

$$\phi(\neg F), \ \ \phi((F \wedge G)), \ \ \phi((F \vee G)), \ \ \phi((F \Rightarrow G)) \text{ and } \phi((F \Leftrightarrow G))$$

from the values of $\phi(F)$ and $\phi(G)$. Let us be more precise:

Lemma 1.13 *Let ϕ_0 be a mapping from P into E, f a mapping from E into E, and $g, h, i,$ and j four mappings from E^2 into E. Then there exists a unique mapping ϕ from $\mathcal{F}$ into E satisfying the following conditions:*

- *the restriction of ϕ to P is ϕ_0;*
- *for any formula $F \in \mathcal{F}$, $\phi(\neg F) = f(\phi(F))$;*
- *for all formulas F and $G \in \mathcal{F}$,*

$$\phi((F \wedge G)) = g(\phi(F), \phi(G)) \qquad \phi((F \vee G)) = h(\phi(F), \phi(G))$$
$$\phi((F \Rightarrow G)) = i(\phi(F), \phi(G)) \text{ and } \phi((F \Leftrightarrow G)) = j(\phi(F), \phi(G))$$

Proof The uniqueness of ϕ is easily proved by induction on the set of formulas using the unique decomposition theorem. The existence of ϕ, which is intuitively clear, is proved with an elementary argument from set theory that we will not present here. ∎

Here is an initial example of definition by induction, for the concept of sub-formula of a propositional formula:

Definition 1.14 *With each formula $F \in \mathcal{F}$ we associate a subset $\mathsf{sf}(F)$ of $\mathcal{F}$, called the **set of sub-formulas of** F, which is defined by induction according to the following conditions:*

- *if $F \in P$,*

$$\mathsf{sf}(F) = \{F\};$$

- *if $F = \neg G$,*

$$\mathsf{sf}(F) = \mathsf{sf}(G) \cup \{F\};$$

- *if $F = (G \; \alpha \; H)$ where $\alpha \in \{\wedge, \vee, \Rightarrow, \Leftrightarrow\}$,*

$$\mathsf{sf}(F) = \mathsf{sf}(G) \cup \mathsf{sf}(H) \cup \{F\}.$$

It is easy to verify that the sub-formulas of a formula are exactly those that appear as nodes in its decomposition tree.

1.1.6 Substitutions in a propositional formula

Let F be a formula in $\mathcal{F}$ and let $A_1, A_2, \ldots, A_n$ be propositional variables from P that are pairwise distinct (this hypothesis is essential). We will use the notation $F[A_1, A_2, \ldots, A_n]$ for F when we wish to emphasize that the elements of P that occur at least once in F are among $A_1, A_2, \ldots, A_n$. For example, the formula $F = (A \Rightarrow (B \vee A))$ could be written as $F[A, B]$, but also as $F[A, B, C, D]$ if it is useful to do so in a given context.

If we are given a formula $F[A_1, A_2, \ldots, A_n, B_1, B_2, \ldots, B_m]$ and n formulas $G_1, G_2, \ldots, G_n$, consider the word obtained by substituting the formula G_1 (respectively: $G_2, \ldots, G_n$) for the variable A_1, (respectively: $A_2, \ldots, A_n$) at each occurrence of these in F. This word will be denoted by $F_{G_1/A_1, G_2/A_2, \ldots, G_n/A_n}$

(read this as 'F sub G_1 replaces A_1, G_2 replaces A_2, et cetera, G_n replaces A_n'),

but we will also denote it by $F[G_1, G_2, \ldots, G_n, B_1, B_2, \ldots, B_m]$ despite the fact that this might cause some delicate problems.

For example, if $F = F[A, B]$ is the formula $(A \Rightarrow (B \vee A))$ and G is the formula $(B \Rightarrow A)$, then $F_{G/A}$ is the word $((B \Rightarrow A) \Rightarrow (B \vee (B \Rightarrow A)))$ which we could denote by $F[G, B]$ or equally well by $F[(B \Rightarrow A), B]$. If we then consider a propositional variable C (distinct from A and B) and the formula $H = C$, then $F_{H/A}$ is the word $(C \Rightarrow (B \vee C))$, which could be written, according to our conventions, as $F[C, B]$. A nasty ambiguity then arises, since it is unclear how to determine from the equalities

$$F[A, B] = (A \Rightarrow (B \vee A)) \qquad \text{and} \qquad F[C, B] = (C \Rightarrow (B \vee C))$$

which of these two formulas is the formula F.

Nevertheless, the notation $F[G_1, G_2, \ldots, G_n, B_1, B_2, \ldots, B_m]$ is extremely practical and, most of the time, perfectly clear. That is why we will permit ourselves to use it, despite the danger we have pointed out, by limiting its use to circumstances where there can be no ambiguity.

In fact we could give a definition of $F_{G_1/A_1, G_2/A_2, \ldots, G_n/A_n}$ by induction on the formula F (where $G_1, G_2, \ldots, G_n \in \mathcal{F}$ and $A_1, A_2, \ldots, A_n \in P$ remain fixed):

- if $F \in P$, then

$$F_{G_1/A_1, G_2/A_2, \ldots, G_n/A_n} = \begin{cases} G_k & \text{if } F = A_k \ (1 \leq k \leq n); \\ F & \text{if } F \notin \{A_1, A_2, \ldots, A_n\}. \end{cases}$$

- if $F = \neg G$, then

$$F_{G_1/A_1, G_2/A_2, \ldots, G_n/A_n} = \neg G_{G_1/A_1, G_2/A_2, \ldots, G_n/A_n};$$

- if $F = (G \ \alpha \ H)$, then

$$F_{G_1/A_1, G_2/A_2, \ldots, G_n/A_n} = (G_{G_1/A_1, G_2/A_2, \ldots, G_n/A_n} \ \alpha \ H_{G_1/A_1, G_2/A_2, \ldots, G_n/A_n})$$

for all formulas G and H and symbol for a binary connective α.

In the examples that we provided, we were able to observe that the word obtained after substituting formulas for propositional variables in a formula was, in every case, itself a formula. There is nothing surprising in this:

Theorem 1.15 *Given an integer n, formulas F, G_1, G_2, \ldots, G_n, and propositional variables A_1, A_2, \ldots, A_n, the word $F_{G_1/A_1, G_2/A_2, \ldots, G_n/A_n}$ is a formula.*

Proof Letting $G_1, G_2, \ldots, G_n \in \mathcal{F}$ and $A_1, A_2, \ldots, A_n \in P$ remain fixed, we prove this by induction on the formula F.

- if $F \in P$, $F_{G_1/A_1, G_2/A_2, \ldots, G_n/A_n}$ is equal to G_k if $F = A_k$ $(1 \leq k \leq n)$ and to F if $F \notin \{A_1, A_2, \ldots, A_n\}$; in both cases this is a formula.

- if $F = \neg G$, and if we suppose that $G_{G_1/A_1, G_2/A_2, \ldots, G_n/A_n}$ is a formula, then $F_{G_1/A_1, G_2/A_2, \ldots, G_n/A_n}$, which is the word $\neg \, G_{G_1/A_1, G_2/A_2, \ldots, G_n/A_n}$, is again a formula.

- if $F = (G \ \alpha \ H)$ (where α is a symbol for a binary connective) and if we suppose that the words $G_{G_1/A_1, G_2/A_2, \ldots, G_n/A_n}$ and $H_{G_1/A_1, G_2/A_2, \ldots, G_n/A_n}$ are formulas, then $F_{G_1/A_1, G_2/A_2, \ldots, G_n/A_n}$, which is the word

$$(G_{G_1/A_1, G_2/A_2, \ldots, G_n/A_n} \ \alpha \ H_{G_1/A_1, G_2/A_2, \ldots, G_n/A_n}),$$

is also a formula. ∎

Remark 1.16 *It behooves us to insist on the fact that the formula*

$$F_{G_1/A_1, G_2/A_2, \ldots, G_n/A_n}$$

is the result of simultaneously substituting the formulas G_1, G_2, $\ldots$, G_n for the variables A_1, A_2, $\ldots$, A_n in the formula F. A priori we would obtain a different formula if we performed these substitutions one after another; moreover, the result we obtain might depend on the order in which these substitutions were performed.

Let us take an example.
Set

$$F = (A_1 \wedge A_2), \quad G_1 = (A_1 \vee A_2) \text{ and } G_2 = (A_1 \Rightarrow A_2).$$

We then have:

$$F_{G_1/A_1, G_2/A_2} = ((A_1 \vee A_2) \wedge (A_1 \Rightarrow A_2));$$

whereas

$$[F_{G_1/A_1}]_{G_2/A_2} = ((A_1 \vee (A_1 \Rightarrow A_2)) \wedge (A_1 \Rightarrow A_2));$$

and

$$[F_{G_2/A_2}]_{G_1/A_1} = ((A_1 \vee A_2) \wedge ((A_1 \vee A_2) \Rightarrow A_2)).$$

We may also, in a given formula F, substitute a formula G for a sub-formula H of F. The word that results from this operation is once again a formula. Although, in practice, this type of substitution is very frequent, we will not introduce a special notation for it and will not enter into details. Let us be satisfied with an example. Suppose that

$$F = (((A \wedge B) \Rightarrow (\neg B \wedge (A \Rightarrow C))) \vee (B \Leftrightarrow (B \Rightarrow (A \vee C)))),$$
$$G = (A \Leftrightarrow (B \vee C)) \text{ and}$$
$$H = (\neg B \wedge (A \Rightarrow C)).$$

Then, by substituting G for the sub-formula H in the formula F, we obtain the formula

$$(((A \wedge B) \Rightarrow (A \Leftrightarrow (B \vee C))) \vee (B \Leftrightarrow (B \Rightarrow (A \vee C)))).$$

1.2 Semantics

1.2.1 Assignments of truth values and truth tables

Definition 1.17 *An **assignment of truth values** to P is a mapping from P into the set $\{0, 1\}$.*

Instead of 'assignment of truth values', some speak of a 'valuation', others of an 'evaluation', or 'distribution of truth values'.

An assignment of truth values to P is therefore an element of the set $\{\mathbf{0}, \mathbf{1}\}^P$.

An assignment of truth values $\delta \in \{\mathbf{0}, \mathbf{1}\}^P$ attributes, to each propositional variable A, a value $\delta(A)$ which is $\mathbf{0}$ or $\mathbf{1}$ (intuitively, false or true). Once this is done, we will see that it is then possible, in one and only one way, to extend δ to the set of all propositional formulas while respecting the rules that agree, more or less, with our intuition as suggested by the names we have given to the various symbols for propositional connectives. Why 'more or less'? Because, although it is doubtful that anyone would be surprised that a formula F will receive the value $\mathbf{1}$ if and only if the formula $\neg F$ receives the value $\mathbf{0}$, the decision to attribute the value $\mathbf{1}$ to the formula $(F \Rightarrow G)$ when the formulas F and G each have the value $\mathbf{0}$ will perhaps give rise to more uneasiness (at least at first sight). One way to dissipate this uneasiness is to ask ourselves under what circumstances the formula $(F \Rightarrow G)$ could be considered false: we would probably agree that this would only happen in the case where F were true without G being true, which leads us to attribute the value $\mathbf{1}$ to $(F \Rightarrow G)$ in the other three possible cases. The difficulty no doubt arises from the fact that, in mathematical arguments, we have the impression that we practically never have to consider situations of the type 'false implies false' or 'false implies true'. But this impression is misleading. No one will contest, for example, that the statement 'for every natural number n, n divisible by 4 implies n is even' is true. But an inevitable consequence of this is that the following two statements are true:

'1 divisible by 4 implies 1 is even';

'2 divisible by 4 implies 2 is even'.

The situations 'false implies false' and 'false implies true' were already present in our initial statement; let us simply say that we don't really care.

Another remark is called for here. If we took a poll among mathematicians asking them whether the statement 'n divisible by 3 implies n odd' is true or false, the second response would win overwhelmingly. For any mathematician would think to have read or heard the statement 'every natural number divisible by 3 is odd' and would legitimately respond: it's false. This is because standard mathematical usage considers that statements having the form of an implication are automatically accompanied by a universal quantifier that is taken for granted.

There is no shortage of examples (the statement

$$\forall \epsilon > 0 \, \exists \delta > 0 \, (|x - y| < \delta \Rightarrow |f(x) - f(y)| < \epsilon)$$

is often taken as the definition of uniform continuity of a function f; in this context, the quantifiers $\forall x$ and $\forall y$, which should be placed between '$\exists \delta > 0$' and the first opening parenthesis, are frequently omitted because they are considered to be understood; this could explain the difficulty that certain students have when asked to describe a function that is not uniformly continuous...). As for the question in our poll, it makes no sense as we have stated it, until we know what the integer n is.

And it suffices to replace n by 3, by 4 or by 5 for the statement to be true (it will be, respectively, of type 'true implies true', 'false implies false', and 'false implies true').

There is another difficulty concerning implication, which is that mathematicians generally see in it a notion of causality, which propositional calculus absolutely does not take into account. If P_1 and P_2 are two true statements, propositional logic imposes the value true on the statement 'P_1 implies P_2'. But a mathematician will refuse, more often than not, to affirm that 'P_1 implies P_2' is true when the statements P_1 and P_2 have 'nothing to do' with one another. Is it true that Rolle's theorem implies the Pythagorean theorem? Those who do not dismiss this question as absurd will generally respond no, because to say yes suggests that they are in position to provide a proof of the Pythagorean theorem in which Rolle's theorem is actually involved.

Though the conflicts between intuition or mathematical usage and the definitions that we are about to give arise especially with implication, the other connectives may also make a modest contribution to this (disjunction is often interpreted as exclusive (A or B but not both) whereas our $\vee$ will not be).

In propositional calculus, these kinds of questions are not to be considered. We will be content to perform very simple operations on two objects: **0** and **1**, and our only reference will be to the definitions of these operations, i.e. to what we will later call the truth tables.

Let it be perfectly clear that the intuition we referred to above is exclusively mathematical intuition. Our concern is not at all to invoke 'everyday' logic (the one that is known as 'common sense'). Mathematicians make no pretence of possessing a universal mode of reasoning. It is hard to resist applying mathematical reasoning to situations outside mathematics, seduced as we are by the rigour of this reasoning, when we discover it. But the result is not what we had hoped: we soon face the fact that human problems do not allow themselves to be resolved by mathematical logic. As for the supposed pedagogical virtues of giving 'real-life examples', they are the opposite of what some may expect from them. This type of approach does not, in any determining way, make the apprenticeship of the rules of mathematical logic any easier, but it is very useful for teaching us prudence, and even humility: to learn mathematical reasoning, let us study mathematics. In fact, we can ask ourselves whether, to illustrate that a formula involving implication is equivalent to its contrapositive, it is more convincing to take the very celebrated example 'if it is raining I take my umbrella' compared to 'if I do not take my umbrella it is not raining' or the example 'if n is prime then n is odd' compared to 'if n is even then n is not prime'. Indeed, it suffices to perform the two experiments to conclude that the example of the umbrella immediately provokes, justifiably, a slew of objections. It is not uninteresting to also note that the contrapositive of 'if it is raining I take my umbrella' is most often stated in the form: 'I do not take my umbrella, therefore it is not raining', which more closely resembles an 'argued conjunction' than an implication. The application of mathematical logic to 'everyday life' has produced a choice collection of hilarious examples that the students of Daniel Lacombe know well and which have attained a certain popularity among logicians:

- A father threatens his son: 'if you are not quiet you will get a slap!', and he proceeds to administer the slap even though the child had immediately become quiet; from the point of view of mathematical logic, this man is not at fault: the truth table for $\Rightarrow$ shows that, by becoming quiet, the child makes the implication true regardless of the truth value of 'you will get a slap'... (a good father should have said 'you will get a slap if and only if you are not quiet').

- In view of tautology no. 17 from Section 1.2.3, what should we think about the equivalence between 'if you are hungry, there is some meat in the fridge' and its contrapositive: 'if there is no meat in the fridge, you are not hungry'?

- When a contest offers as its first prize 'a new car or a cheque for \$100 000', why shouldn't the winner claim both the car and the cheque, relying on the truth table for disjunction?

As we see, all this no doubt has an amusing side, but it does not help us at all in resolving exercises from mathematics in general or mathematical logic in particular. We will therefore leave our umbrella in the closet and remain in the world of mathematics, where there is already enough to do.

Theorem 1.18 *For any assignment of truth values* $\delta \in \{0, 1\}^P$, *there exists a unique map* $\overline{\delta} : \mathcal{F} \Rightarrow \{0, 1\}$ *which agrees with* δ *on* P *(i.e. it extends* δ) *and which satisfies the following properties:*

(1) *for any formula F,*

$$\overline{\delta}(\neg F) = 1 \text{ if and only if } \overline{\delta}(F) = 0;$$

(2) *for all formulas F and G,*

$$\overline{\delta}(F \wedge G) = 1 \text{ if and only if } \overline{\delta}(F) = \overline{\delta}(G) = 1;$$

(3) *for all formulas F and G,*

$$\overline{\delta}(F \vee G) = 0 \text{ if and only if } \overline{\delta}(F) = \overline{\delta}(G) = 0;$$

(4) *for all formulas F and G,*

$$\overline{\delta}(F \Rightarrow G) = 0 \text{ if and only if } \overline{\delta}(F) = 1 \text{ and } \overline{\delta}(G) = 0;$$

(5) *for all formulas F and G,*

$$\overline{\delta}(F \Leftrightarrow G) = 1 \text{ if and only if } \overline{\delta}(F) = \overline{\delta}(G).$$

Proof To simplify the notation, let us observe right away that conditions (1) to (5) can be expressed using the operations of addition and multiplication on the two-element field $\mathbb{Z}/2\mathbb{Z}$, with which we can naturally identify the set $\{0, 1\}$. These conditions then become equivalent to:

for all formulas F and G:

$$\text{(i')} \quad \overline{\delta}(\neg F) = \mathbf{1} + \overline{\delta}(F);$$
$$\text{(ii')} \quad \overline{\delta}((F \wedge G)) = \overline{\delta}(F)\,\overline{\delta}(G);$$
$$\text{(iii')} \quad \overline{\delta}((F \vee G)) = \overline{\delta}(F) + \overline{\delta}(G) + \overline{\delta}(F)\,\overline{\delta}(G)$$
$$\text{(iv')} \quad \overline{\delta}((F \Rightarrow G)) = \mathbf{1} + \overline{\delta}(F) + \overline{\delta}(F)\,\overline{\delta}(G);$$
$$\text{(v')} \quad \overline{\delta}((F \Leftrightarrow G)) = \mathbf{1} + \overline{\delta}(F) + \overline{\delta}(G).$$

(The proof is immediate.)

We see that the function $\overline{\delta}$ is defined by induction on the set of formulas, which is what guarantees its existence and uniqueness by Lemma 1.13; here, the functions f, g, h, i, and j are defined on $\mathbb{Z}/2\mathbb{Z}$ by: for all x and y,

$$f(x) = 1 + x, \quad g(x, y) = xy, \quad h(x, y) = x + y + xy,$$
$$i(x, y) = 1 + x + xy \text{ and } j(x, y) = 1 + x + y. \quad \blacksquare$$

We should point out that identifying $\{\mathbf{0}, \mathbf{1}\}$ with $\mathbb{Z}/2\mathbb{Z}$ is extremely practical and will be used in what follows.

We can recapitulate conditions (i') to (v') above in tables which we call the **truth tables** for negation, for conjunction, for disjunction, for implication and for equivalence:

F	$\neg F$
0	**1**
1	**0**

F	G	$(F \wedge G)$
0	**0**	**0**
0	**1**	**0**
1	**0**	**0**
1	**1**	**1**

F	G	$(F \vee G)$
0	**0**	**0**
0	**1**	**1**
1	**0**	**1**
1	**1**	**1**

F	G	$(F \Rightarrow G)$
0	**0**	**1**
0	**1**	**1**
1	**0**	**0**
1	**1**	**1**

F	G	$(F \Leftrightarrow G)$
0	**0**	**1**
0	**1**	**0**
1	**0**	**0**
1	**1**	**1**

In practice, we will not really make the distinction between an assignment of truth values and its extension to the set of formulas. We will speak of the 'truth value of the formula F under the assignment δ' and we will eventually forget the bar over the δ which would have indicated that we were dealing with the extension.

If F is a formula and δ is an assignment of truth values, we will say that F **is satisfied by** δ, or that δ **satisfies** F, when $\overline{\delta}(F) = \mathbf{1}$.

Given a formula F and an assignment of truth values δ, the definition of the extension $\overline{\delta}$ clearly points to a method for calculating $\overline{\delta}(F)$: this consists in calculating the values taken by $\overline{\delta}$ on the various sub-formulas of F, beginning with the sub-formulas of height 1 (the values for those of height 0 being precisely what is given), and applying the tables above as many times as necessary. For example, if F is the formula $((A \Rightarrow B) \Rightarrow (B \vee (A \Leftrightarrow C)))$ and if δ is an assignment of truth values for which $\delta(A) = \delta(B) = \mathbf{0}$ and $\delta(C) = \mathbf{1}$, then we have successively:

$$\overline{\delta}((A \Rightarrow B)) = \mathbf{1} \ ; \ \overline{\delta}((A \Leftrightarrow C)) = \mathbf{0} \ ; \ \overline{\delta}((B \vee (A \Leftrightarrow C))) = \mathbf{0} \text{ and } \overline{\delta}(F) = \mathbf{0}.$$

Of course, it can happen that calculating the values of $\overline{\delta}$ for all the sub-formulas of F is unnecessary: to see this, consider the formula

$$G = (A \Rightarrow (((B \wedge \neg A) \vee (\neg C \wedge A)) \Leftrightarrow (A \vee (A \Rightarrow \neg B))))$$

and an assignment of truth values λ for which $\lambda(A) = \mathbf{0}$; we may conclude that $\lambda(G) = \mathbf{1}$ without bothering with the truth value of the sub-formula

$$(((B \wedge \neg A) \vee (\neg C \wedge A)) \Leftrightarrow (A \vee (A \Rightarrow \neg B))).$$

In the examples that we have just examined, to calculate the truth value of a formula, we only used the values taken by the assignment of truth values under consideration on the variables that actually occur in the formula. It is clear that this is always the case.

Lemma 1.19 *For any formula $F[A_1, A_2, \ldots, A_n]$ (involving no propositional variables other than A_1, A_2, $\ldots$, A_n) and any assignments of truth values λ and $\mu \in \{0, 1\}^P$, if λ and μ agree on $\{A_1, A_2, \ldots, A\}$, then $\overline{\lambda}(F) = \overline{\mu}(F)$.*

Proof The proof involves no difficulties. It is done by induction on the formulas.
∎

Let $G[A_1, A_2, \ldots, A_n]$ be a formula. To discover the set of truth values of G (corresponding to the set of all possible assignments), we see that it is sufficient to 'forget' momentarily the variables in P that do not occur in G, and to suppose that the set of propositional variables is just $\{A_1, A_2, \ldots, A_n\}$. There are then only a finite number of assignments of truth values to consider: this is the number of mappings from $\{A_1, A_2, \ldots, A_n\}$ into $\{0, 1\}$, namely 2^n (recall that the notation $G[A_1, A_2, \ldots, A_n]$ presumes that the variables A_i are pairwise distinct). We can identify each mapping δ from $\{A_1, A_2, \ldots, A_n\}$ into $\{0, 1\}$ with the n-tuple $(\delta(A_1), \delta(A_2), \ldots, \delta(A_n)) \in \{0, 1\}^n$ and place the set of truth values taken by G into a tableau in which each row would correspond to one of the 2^n n-tuples and would contain the corresponding truth value of G. Such a tableau, which could also contain the truth values of the sub-formulas of G, will be called the **truth table** of the formula G. Ultimately, this is nothing more than the table of values of a certain mapping from $\{0, 1\}^n$ into $\{0, 1\}$.

Let us return to the example given just above:

$$G = (A \Rightarrow (((B \wedge \neg A) \vee (\neg C \wedge A)) \Leftrightarrow (A \vee (A \Rightarrow \neg B)))).$$

Set

$$H = (B \wedge \neg A), \ I = (\neg C \wedge A), \ J = (A \Rightarrow \neg B),$$
$$K = (H \vee I), \quad L = (A \vee J), \quad M = (K \Leftrightarrow L).$$

Then we have $G = (A \Rightarrow M)$. Here is the truth table for G:

A	B	C	¬A	¬B	¬C	H	I	J	K	L	M	G
0	0	0	1	1	1	0	0	1	0	1	0	1
0	0	1	1	1	0	0	0	1	0	1	0	1
0	1	0	1	0	1	1	0	1	1	1	1	1
0	1	1	1	0	0	1	0	1	1	1	1	1
1	0	0	0	1	1	0	1	1	1	1	1	1
1	0	1	0	1	0	0	0	1	0	1	0	0
1	1	0	0	0	1	0	1	0	1	1	1	1
1	1	1	0	0	0	0	0	0	0	1	0	0

We should note that, with our conventions concerning the notation

$$G[A_1, A_2, \ldots, A_n],$$

we do not have uniqueness of the truth table for a formula (for example, the first four columns of the table above could be considered as the truth table of the formula $\neg A$). There is, nonetheless, a 'minimal' table for every formula, the one which involves only those propositional variables that occur at least once in the formula.

However, even restricting ourselves to this notion of minimal table, there can still be, for the same formula, many tables which differ in the order in which the n-tuples from $\{0, 1\}^n$ are presented.

It is reasonable to choose, once and for all, a particular order (among the $2^n!$ that are possible) and to adopt it systematically. We have chosen the lexicographical order (the 'dictionary' order): in the table, the n-tuple $(a_1, a_2, \ldots, a_n)$ will be placed ahead of $(b_1, b_2, \ldots, b_n)$ if, for the first subscript $j \in \{1, 2, \ldots, n\}$ for which $a_j \neq b_j$, we have $a_j < b_j$.

In view of these remarks, we will allow ourselves to speak about 'the' truth table of a formula.

1.2.2 Tautologies and logically equivalent formulas

Definition 1.20

- A **tautology** is a formula that assumes the value **1** under every assignment of truth values.

- *the notation for 'F is a tautology' is:* $\vdash^* F$;
  whereas $\nvdash^* F$ *signifies: 'F is not a tautology'.*

- *Given two formulas F and G, F is* **logically equivalent** *to G if and only if the formula* $(F \Leftrightarrow G)$ *is a tautology.*

The notation for '*F* is logically equivalent to *G*' is: $F \sim G$.

Remark 1.21 *The next two properties follow immediately from these definitions:*

- *For all formulas F and G, we have* $F \sim G$ *if and only if for every assignment of truth values* $\delta \in \{\mathbf{0}, \mathbf{1}\}^P$, $\overline{\delta}(F) = \overline{\delta}(G)$.
- *The binary relation* $\sim$ *is an equivalence relation on* $\mathcal{F}$.

The equivalence class of the formula F for the relation $\sim$ is denoted by $\mathsf{cl}(F)$.

A tautology is therefore a formula whose truth table contains only **1**s in its last column, in other words, a formula that is 'always true'. Two logically equivalent formulas are two formulas that are satisfied by exactly the same assignments of truth values, which thus have the same truth table. Any formula logically equivalent to a tautology is a tautology. Therefore the tautologies constitute one of the equivalence classes for the relation $\sim$, denoted by **1**. The formulas whose negations are tautologies (some call these **antilogies**, others **antitautologies**) constitute another equivalence class, distinct from **1**, denoted by **0**: these are the formulas that are 'always false', which is to say that their truth tables contain only **0**s in the last column.

When we do semantics, we argue 'up to equivalence'. This will be justified by the study of the set of equivalence classes for the equivalence relation $\sim$, which we will do a bit further on and will complete in Chapter 2.

Let us now examine the effect of substitutions on the truth values of formulas:

Theorem 1.22 *Given an assignment of truth values, δ, a natural number, n, formulas F, G_1, G_2, ..., G_n, and pairwise distinct propositional variables A_1, A_2, ..., A_n, let λ be the assignment of truth values defined by*

$$for\ all\ X \in P,\ \lambda(X) = \begin{cases} \delta(X) & \text{if } X \notin \{A_1, A_2, \ldots, A_n\}; \\ \overline{\delta}(G_i) & \text{if } X = A_i \quad (1 \leq i \leq n). \end{cases}$$

We then have

$$\overline{\delta}(F_{G_1/A_1, G_2/A_2, \ldots, G_n/A_n}) = \overline{\lambda}(F).$$

Proof We argue by induction on the formula F:

- if F is an element of P, then:

  either $F \notin \{A_1, A_2, \ldots, A_n\}$; in this case, $F_{G_1/A_1, G_2/A_2, \ldots, G_n/A_n} = F$ and

  $$\overline{\delta}(F_{G_1/A_1, G_2/A_2, \ldots, G_n/A_n}) = \overline{\delta}(F) = \delta(F) = \lambda(F) = \overline{\lambda}(F);$$

or else $F = A_i$ $(1 \le i \le n)$; in this case, $F_{G_1/A_1, G_2/A_2, \ldots, G_n/A_n} = G_i$ and

$$\overline{\delta}(F_{G_1/A_1, G_2/A_2, \ldots, G_n/A_n}) = \overline{\delta}(G_i) = \lambda(A_i) = \lambda(F) = \overline{\lambda}(F)$$

by definition of λ.

- If $F = \neg G$, and if we suppose that $\overline{\delta}(G_{G_1/A_1, G_2/A_2, \ldots, G_n/A_n}) = \overline{\lambda}(G)$ (the induction hypothesis), then

$$
\begin{aligned}
\overline{\delta}(F_{G_1/A_1, G_2/A_2, \ldots, G_n/A_n}) &= \overline{\delta}(\neg G_{G_1/A_1, G_2/A_2, \ldots, G_n/A_n}) \\
&= 1 + \overline{\delta}(G_{G_1/A_1, G_2/A_2, \ldots, G_n/A_n}) \\
&= 1 + \overline{\lambda}(G) = \overline{\lambda}(\neg G) = \overline{\lambda}(F).
\end{aligned}
$$

- If $F = (G \wedge H)$, and if we suppose (the induction hypothesis)

$$\overline{\delta}(G_{G_1/A_1, G_2/A_2, \ldots, G_n/A_n}) = \overline{\lambda}(G) \text{ and } \overline{\delta}(H_{G_1/A_1, G_2/A_2, \ldots, G_n/A_n}) = \overline{\lambda}(H),$$

then

$$
\begin{aligned}
\overline{\delta}(F_{G_1/A_1, G_2/A_2, \ldots, G_n/A_n}) &= \overline{\delta}((G_{G_1/A_1, G_2/A_2, \ldots, G_n/A_n} \\
&\quad \wedge H_{G_1/A_1, G_2/A_2, \ldots, G_n/A_n})) \\
&= \overline{\delta}(G_{G_1/A_1, G_2/A_2, \ldots, G_n/A_n}) \\
&\quad \times \overline{\delta}(H_{G_1/A_1, G_2/A_2, \ldots, G_n/A_n}) \\
&= \overline{\lambda}(G)\, \overline{\lambda}(H) = \overline{\lambda}((G \wedge H)) = \overline{\lambda}(F).
\end{aligned}
$$

- The cases $F = (G \vee H)$, $F = (G \Rightarrow H)$, and $F = (G \Leftrightarrow H)$ are treated in a similar fashion without the slightest difficulty; indeed, we could really not bother with them at all (for this, see Remark 1.33 later on). ∎

The next corollary follows immediately from the theorem:

Corollary 1.23 *For all formulas $F, G_1, G_2, \ldots, G_n$ and pairwise distinct propositional variables $A_1, A_2, \ldots, A_n$, if F is a tautology, then so is the formula*

$$F_{G_1/A_1, G_2/A_2, \ldots, G_n/A_n}.$$

Proof Given an arbitrary assignment of truth values δ, by defining the assignment λ as in the previous theorem, we have

$$\overline{\delta}(F_{G_1/A_1, G_2/A_2, \ldots, G_n/A_n}) = \overline{\lambda}(F) = \mathbf{1},$$

since F is a tautology. ∎

Another type of substitution also allows us to preserve logical equivalence of formulas:

Theorem 1.24 *Consider a formula F, a sub-formula G of F and a formula H that is logically equivalent to G. Then the formula F', obtained from F by substituting H for the sub-formula G, is logically equivalent to F.*

Proof   We argue by induction on the formula F.

- If $F \in P$, then, necessarily, $G = F$ and $F' = H$. We certainly have $F' \sim F$.

- If $F = \neg F_1$, then either $G = F$, $F' = H$, and we have $F' \sim F$, or else G is a sub-formula of F_1 and, by the induction hypothesis, the formula F'_1, which results from substituting H for G in F_1, is logically equivalent to F_1. Then the formula F' is the formula $\neg F'_1$; it is therefore logically equivalent to F since, for any assignment of truth values δ, we have

$$\overline{\delta}(F') = 1 + \overline{\delta}(F'_1) = 1 + \overline{\delta}(F_1) = \overline{\delta}(\neg F_1) = \overline{\delta}(F).$$

- If $F = (F_1 \wedge F_2)$, then there are three possibilities. Either $G = F$, $F' = H$, and we have $F' \sim F$. Or else G is a sub-formula of F_1, and, by the induction hypothesis, the formula F'_1, which results from substituting H for G in F_1, is logically equivalent to F_1. Then the formula F' is the formula $(F'_1 \wedge F_2)$; it is logically equivalent to F because, for any assignment δ, we have

$$\overline{\delta}(F') = \overline{\delta}(F'_1)\,\overline{\delta}(F_2) = \overline{\delta}(F_1)\,\overline{\delta}(F_2) = \overline{\delta}((F_1 \wedge F_2)) = \overline{\delta}(F).$$

The argument is strictly similar in the third case, when G is a sub-formula of F_2.

The cases $F = (F_1 \vee F_2)$, $F = (F_1 \Rightarrow F_2)$, and $F = (F_1 \Leftrightarrow F_2)$ are treated in an analogous fashion using relations (iii') to (v') from Theorem 1.18.   ■

In practice, to show that a formula is a tautology, or that two formulas are logically equivalent, we have several methods available. First of all, we could use truth tables, but this is no longer viable once the number of variables exceeds 3 or 4. In certain cases, we could have recourse to what might be called 'economical truth tables': this consists in discussing the values taken by a restricted number of variables; in a way, we are treating several lines of the truth table in a single step. Let us take an example: we will show that the following formula F is a tautology:

$$((A \Rightarrow ((B \vee \neg C) \wedge \neg(A \Rightarrow D))) \vee ((D \wedge \neg E) \vee (A \vee C))).$$

By setting

$$H = (A \Rightarrow ((B \vee \neg C) \wedge \neg(A \Rightarrow D))) \text{ and}$$
$$K = ((D \wedge \neg E) \vee (A \vee C)),$$

we have $F = (H \vee K)$. Next, consider an assignment of truth values δ. If $\delta(A) = \mathbf{0}$, then we see that $\overline{\delta}(H) = \mathbf{1}$, thus also $\overline{\delta}(F) = \mathbf{1}$. If $\delta(A) = \mathbf{1}$, then $\overline{\delta}(A \vee C) = \mathbf{1}$, hence $\overline{\delta}(K) = \mathbf{1}$ and $\overline{\delta}(F) = \mathbf{1}$.

Just as well, we could invoke Corollary 1.23 and Theorem 1.24 by making use of certain 'basic' tautologies (see the list in Section 1.2.3). For example, to show that the formula $G = ((\neg A \vee B) \vee \neg(A \Rightarrow B))$ is a tautology, we first use the fact that the formulas $(\neg A \vee B)$ and $(A \Rightarrow B)$ are logically equivalent, which shows that G is logically equivalent to $(\,(A \Rightarrow B) \vee \neg(A \Rightarrow B))$ (by Theorem 1.24), then observe that this latter formula is obtained by substituting the formula $(A \Rightarrow B)$ for the variable A in the tautology $(A \vee \neg A)$, and is therefore itself a tautology (by Corollary 1.23).

Needless to say, we will only rarely be led to present such an argument with this degree of detail.

There are also purely syntactical methods which can be used to prove that a formula is a tautology (see Chapter 4).

Finally, Exercise 14 shows how we can reduce all this to a simple calculation involving polynomials.

1.2.3 Some tautologies

Here is a list of common tautologies (which are just so many exercises for the reader, with no solutions given!):

(A, B and C denote propositional variables (but we may, by Corollary 1.23, substitute arbitrary formulas in their place); $\top$ denotes an arbitrary tautology and $\bot$ the negation of $\top$, which is to say a formula that always takes the value **0**.)

(1) $((A \wedge A) \Leftrightarrow A)$
(2) $((A \vee A) \Leftrightarrow A)$
(3) $((A \wedge B) \Leftrightarrow (B \wedge A))$
(4) $((A \vee B) \Leftrightarrow (B \vee A))$
(5) $((A \wedge (B \wedge C)) \Leftrightarrow ((A \wedge B) \wedge C))$
(6) $((A \vee (B \vee C)) \Leftrightarrow ((A \vee B) \vee C))$
(7) $((A \wedge (B \vee C)) \Leftrightarrow ((A \wedge B) \vee (A \wedge C)))$
(8) $((A \vee (B \wedge C)) \Leftrightarrow ((A \vee B) \wedge (A \vee C)))$
(9) $((A \wedge (A \vee B)) \Leftrightarrow A)$
(10) $((A \vee (A \wedge B)) \Leftrightarrow A)$
(11) $(\neg(A \vee B) \Leftrightarrow (\neg A \wedge \neg B))$
(12) $(\neg(A \wedge B) \Leftrightarrow (\neg A \vee \neg B))$
(13) $((A \wedge \top) \Leftrightarrow A)$
(14) $((A \vee \bot) \Leftrightarrow A)$
(15) $((A \wedge \bot) \Leftrightarrow \bot)$
(16) $((A \vee \top) \Leftrightarrow \top)$
(17) $((A \Rightarrow B) \Leftrightarrow (\neg B \Rightarrow \neg A))$

These tautologies reflect important properties. Numbers (1) and (2) express the **idempotence** of conjunction and disjunction, (3) and (4) their **commutativity**, (5) and (6) their **associativity**, (7) and (8) the **distributivity** of each over the other. But be careful! All this is taking place up to logical equivalence (which is to say that these properties are really properties of operations on the set $\mathcal{F}/\sim$ of equivalence classes for the relation $\sim$ on $\mathcal{F}$: for more details, refer to Exercise 1 from Chapter 2). Numbers (9) and (10) are called the **absorption laws.** Numbers (11) and (12) express **de Morgan's laws** (see Chapter 2 as well). Tautology number (13) (respectively, number (14)) expresses that the class of tautologies **1** (respectively, the class of antilogies **0**) is the **identity element** for conjunction (respectively, for disjunction). Number (15) (respectively, number (16)) expresses that the class **0** (respectively, the class **1**) is the **zero element** for conjunction (respectively, for disjunction). The formula $(\neg B \Rightarrow \neg A)$ is called the **contrapositive** of $(A \Rightarrow B)$ and tautology number (17) expresses that every implicative formula is logically equivalent to its contrapositive.

We will now continue our list with additional common tautologies:

(18) $(A \vee \neg A)$

(19) $(A \Rightarrow A)$

(20) $(A \Leftrightarrow A)$

(21) $(\neg\neg A \Rightarrow A)$

(22) $(A \Rightarrow (A \vee B))$

(23) $((A \wedge B) \Rightarrow A)$

(24) $(((A \Rightarrow B) \wedge A) \Rightarrow B)$

(25) $(((A \Rightarrow B) \wedge \neg B) \Rightarrow \neg A)$

(26) $((\neg A \Rightarrow A) \Rightarrow A)$

(27) $((\neg A \Rightarrow A) \Leftrightarrow A)$

(28) $(\neg A \Rightarrow (A \Rightarrow B))$

(29) $(A \vee (A \Rightarrow B))$

(30) $(A \Rightarrow (B \Rightarrow A))$

(31) $(((A \Rightarrow B) \wedge (B \Rightarrow C)) \Rightarrow (A \Rightarrow C))$

(32) $((A \Rightarrow B) \vee (C \Rightarrow A))$

(33) $((A \Rightarrow B) \vee (\neg A \Rightarrow B))$

(34) $((A \Rightarrow B) \vee (A \Rightarrow \neg B))$

(35) $((A \Rightarrow B) \Rightarrow ((B \Rightarrow C) \Rightarrow (A \Rightarrow C)))$

(36) $(\neg A \Rightarrow (\neg B \Leftrightarrow (B \Rightarrow A)))$

(37) $((A \Rightarrow B) \Rightarrow (((A \Rightarrow C) \Rightarrow B) \Rightarrow B))$

Moreover, in the list below, formulas that are all on the same line are pairwise logically equivalent.

(38) $(A \Rightarrow B), (\neg A \vee B), (\neg B \Rightarrow \neg A), ((A \wedge B) \Leftrightarrow A), ((A \vee B) \Leftrightarrow B)$

(39) $\neg(A \Rightarrow B), (A \wedge \neg B)$

(40) $(A \Leftrightarrow B), ((A \wedge B) \vee (\neg A \wedge \neg B)), ((\neg A \vee B) \wedge (\neg B \vee A))$

(41) $(A \Leftrightarrow B), ((A \Rightarrow B) \wedge (B \Rightarrow A)), (\neg A \Leftrightarrow \neg B), (B \Leftrightarrow A)$

(42) $(A \Leftrightarrow B), ((A \vee B) \Rightarrow (A \wedge B))$

(43) $\neg(A \Leftrightarrow B), (A \Leftrightarrow \neg B), (\neg A \Leftrightarrow B)$

(44) $A, \neg\neg A, (A \wedge A), (A \vee A), (A \vee (A \wedge B)), (A \wedge (A \vee B))$

(45) $A, (\neg A \Rightarrow A), (A \Rightarrow B) \Rightarrow A), ((B \Rightarrow A) \wedge (\neg B \Rightarrow A))$

(46) $A, (A \wedge \top), (A \vee \top), (A \Leftrightarrow \top), (\top \Rightarrow A)$

(47) $\neg A, (A \Rightarrow \neg A), ((A \Rightarrow B) \wedge (A \Rightarrow \neg B))$

(48) $\neg A, (A \Rightarrow \bot), (A \Leftrightarrow \bot)$

(49) $\perp, (A \wedge \perp), (A \Leftrightarrow \neg A)$

(50) $\top, (A \vee \top), (A \Rightarrow \top), (\perp \Rightarrow A)$

(51) $(A \wedge B), (B \wedge A), (A \wedge (\neg A \vee B)), \neg (A \Rightarrow \neg B)$

(52) $(A \vee B), (B \vee A), (A \vee (\neg A \wedge B)), (\neg A \Rightarrow B), ((A \Rightarrow B) \Rightarrow B)$

(53) $(A \Rightarrow (B \Rightarrow C)), ((A \wedge B) \Rightarrow C), (B \Rightarrow (A \Rightarrow C)), ((A \Rightarrow B) \Rightarrow$
      $(A \Rightarrow C))$

(54) $(A \Rightarrow (B \wedge C)), ((A \Rightarrow B) \wedge (A \Rightarrow C))$

(55) $(A \Rightarrow (B \vee C)), ((A \Rightarrow B) \vee (A \Rightarrow C))$

(56) $((A \wedge B) \Rightarrow C), ((A \Rightarrow C) \vee (B \Rightarrow C))$

(57) $((A \vee B) \Rightarrow C), ((A \Rightarrow C) \wedge (B \Rightarrow C))$

(58) $(A \Leftrightarrow (B \Leftrightarrow C)), ((A \Leftrightarrow B) \Leftrightarrow C).$

We should take notice, from lines (54) to (57), that implication does not distribute over conjunction nor over disjunction. We see however that it does distribute from the left ((54) and (55)), which is to say when $\wedge$ and $\vee$ occur to the right of $\Rightarrow$. In the case when one or the other is located to the left of $\Rightarrow$, we have a kind of artificial distributivity, the $\wedge$ (respectively, the $\vee$) being transformed into $\vee$ (respectively, into $\wedge$) after its 'distribution' ((56) and (57)). It behoves us to be vigilant in all cases when manipulating this type of formula.

From now on, we will admit the following abuses of notation:

- In general, in writing a formula, we will allow ourselves to omit the outer-most parentheses. This convention supposes that these parentheses automatically reappear as soon as this formula occurs as a (strict) sub-formula of another formula: for example, we will accept the formula $F = A \Leftrightarrow B$, and the formula $F \Rightarrow \neg C$, but the latter will obviously be written $(A \Leftrightarrow B) \Rightarrow \neg C$ and not as $A \Leftrightarrow B \Rightarrow \neg C$.

- For all formulas F, G, and H,
  the formula $((F \wedge G) \wedge H)$ will be written $(F \wedge G \wedge H)$,
  the formula $((F \vee G) \vee H)$ will be written $(F \vee G \vee H)$.

We could also, by applying the previous convention concerning the omission of parentheses, write $F \wedge G \wedge H$ or $F \vee G \vee H$.

- More generally, for any non-zero natural number k, if $F_1, F_2, \ldots, F_k$ are formulas, we will let $F_1 \wedge F_2 \wedge \cdots \wedge F_k$ represent the formula

$$((\ldots (F_1 \wedge F_2) \wedge F_3) \wedge \cdots \wedge F_k)$$

(which begins with $k-1$ occurrences of the open parenthesis symbol). Of course we make the analogous convention for disjunction.

- If $I = \{i_1, i_2, \ldots, i_k\}$ is a non-empty finite set of indices and if $F_{i_1}, F_{i_2}, \ldots, F_{i_k}$ are formulas, the formula $F_{i_1} \wedge F_{i_2} \wedge \cdots \wedge F_{i_k}$ will also be written:

$$\bigwedge_{j \in I} F_j$$

(to be read as 'the conjunction of the F_j for j belonging to I').

We will notice that with this notation, there is an ambiguity relating to the order of the indices in the set I, which needs to be fixed for this manner of writing to have a meaning. But in fact, as long as we are concerned with semantics, the choice of this ordering has no importance whatever in view of the commutativity of conjunction.

In the same way, the formula $F_{i_1} \vee F_{i_2} \vee \cdots \vee F_{i_k}$ will be abbreviated:

$$\bigvee_{j \in I} F_j$$

(to be read as 'the disjunction of the F_j for j belonging to I').

Naturally, we will also have variants, such as $\bigvee_{1 \leq k \leq n} G_k$ or $\bigwedge_{F \in X} F$ (where X is a non-empty finite set of formulas), whose meaning is clear.

In fact, our decision, for example, that the expression $A \vee B \vee C$ represents the formula $((A \vee B) \vee C)$ is based on an arbitrary choice. We could just as well have opted for the formula $(A \vee (B \vee C))$ which is logically equivalent to the first one. It is the associativity of conjunction and disjunction (numbers (5) and (6) from Section 1.2.3) that led us suppress these parentheses knowing that, whatever method is used to reintroduce them, we obtain a formula from the same equivalence class. (In Exercise 16, we will understand why it would be imprudent to allow the analogous abuses of notation in the case of $\Leftrightarrow$, although this appears, according to number (58) from Section 1.2.3, to allow it just as well as $\wedge$ and $\vee$).

1.3 Normal forms and complete sets of connectives

1.3.1 Operations on $\{0, 1\}$ and formulas

Up to and including Section 1.3.3, we will assume that the set P of propositional variables is a finite set of n elements ($n \geq 1$):

$$P = \{A_1, A_2, \ldots, A_n\}.$$

This allows us to consider that every formula $F \in \mathcal{F}$ has its variables among $A_1, A_2, \ldots, A_n$ and to write $F = F[A_1, A_2, \ldots, A_n]$.

Notation:

- For every n-tuple $(\varepsilon_1, \varepsilon_2, \ldots, \varepsilon_n) \in \{0, 1\}^n$, $\delta_{\varepsilon_1, \varepsilon_2, \ldots, \varepsilon_n}$ denotes the distribution of truth values defined by $\delta_{\varepsilon_1, \varepsilon_2, \ldots, \varepsilon_n}(A_i) = \varepsilon_i$ for each $i \in \{1, 2, \ldots, n\}$.
- For each propositional variable A and for each element $\varepsilon \in \{0, 1\}$ we let εA denote the formula that is equal to A if $\varepsilon = 1$ and to $\neg A$ if $\varepsilon = 0$.

For each formula F, we let $\Delta(F)$ denote the set of distributions of truth values that satisfy F:

$$\Delta(F) = \{\delta \in \{0, 1\}^P : \overline{\delta}(F) = 1\}.$$

For each formula F, we define a mapping φ_F from $\{0, 1\}^P$ into $\{0, 1\}$ by

$$\varphi_F(\varepsilon_1, \varepsilon_2, \ldots, \varepsilon_n) = \overline{\delta_{\varepsilon_1, \varepsilon_2, \ldots, \varepsilon_n}}(F).$$

The mapping φ_F is thus none other than the one defined by the truth table of F. We will allow ourselves the slight abuse of language involved in saying that φ_F is the truth table of F.

Notice that two formulas F and G are logically equivalent if and only if $\varphi_F = \varphi_G$. What this means precisely is that the mapping $F \mapsto \varphi_F$ (from $\mathcal{F}$ into $\{0, 1\}^{(\{0,1\}^n)}$) is compatible with the relation $\sim$. We also see that this mapping is not injective (for example, for any formula F, we have: $\varphi_{\neg\neg F} = \varphi_F$), but that the mapping that it induces, from $\mathcal{F}/\sim$ into $\{0, 1\}^{(\{0,1\}^n)}$ (the mapping $\mathrm{cl}(F) \mapsto \varphi_F$) is injective (recall that $\mathrm{cl}(F)$ denotes the equivalence class of the formula F under the equivalence relation $\sim$). This shows that the number of equivalence classes for the relation $\sim$ on $\mathcal{F}$ is at most equal to the number of mappings from $\{0, 1\}^n$ into $\{0, 1\}$, which is to say 2^{2^n}.

It remains to discover if there are exactly 2^{2^n} equivalence classes of formulas or if there are fewer. In other words, is the mapping $F \mapsto \varphi_F$ surjective? Or again, can the table of an arbitrary mapping from $\{0, 1\}^n$ into $\{0, 1\}$ be viewed as the truth table of some formula?

The answer to these questions is positive, as we will see with the next theorem. The proof of the theorem will furnish us with an explicit method for finding such a formula, knowing only its truth table.

Lemma 1.25 *For any n-tuple $(\varepsilon_1, \varepsilon_2, \ldots, \varepsilon_n) \in \{0, 1\}^n$, the formula*

$$\bigwedge_{1 \leq k \leq n} \varepsilon_k A_k$$

is satisfied by the distribution of truth values $\delta_{\varepsilon_1, \varepsilon_2, \ldots, \varepsilon_n}$ and by no other.

In our notation, this would be written: $\Delta(\bigwedge_{1 \leq k \leq n} \varepsilon_k A_k) = \{\delta_{\varepsilon_1, \varepsilon_2, \ldots, \varepsilon_n}\}$.

Proof For any distribution of truth values λ, we have $\overline{\lambda}(\bigwedge_{1 \leq k \leq n} \varepsilon_k A_k) = 1$ if and only if for every $k \in \{1, 2, \ldots, n\}$, $\overline{\lambda}(\varepsilon_k A_k) = 1$, which, in view of the definition of $\delta_{\varepsilon_1, \varepsilon_2, \ldots, \varepsilon_n}$, is equivalent to:

for every $k \in \{1, 2, \ldots, n\}$, $\lambda(A_k) = \delta_{\varepsilon_1, \varepsilon_2, \ldots, \varepsilon_n}(A_k)$,

in other words, to $\lambda = \delta_{\varepsilon_1, \varepsilon_2, \ldots, \varepsilon_n}$. ∎

Lemma 1.26 *Let X be a non-empty subset of $\{0, 1\}^n$ and let F_X be the formula*

$$\bigvee_{(\varepsilon_1, \varepsilon_2, \ldots, \varepsilon_n) \in X} \left(\bigwedge_{1 \leq i \leq n} \varepsilon_i A_i \right).$$

Then the formula F_X is satisfied by those distributions of truth values $\delta_{\varepsilon_1, \varepsilon_2, \ldots, \varepsilon_n}$ for which $(\varepsilon_1, \varepsilon_2, \ldots, \varepsilon_n) \in X$ and only by these.

With our notations,

$$\Delta\left(\bigvee_{(\varepsilon_1,\varepsilon_2,\ldots,\varepsilon_n)\in X}\left(\bigwedge_{1\leq i\leq n}\varepsilon_i A_i\right)\right)=\{\delta_{\varepsilon_1,\varepsilon_2,\ldots,\varepsilon_n}:(\varepsilon_1,\varepsilon_2,\ldots,\varepsilon_n)\in X\}.$$

Proof For any distribution of truth values λ, we have $\lambda(F_X)=\mathbf{1}$ if and only if there exists an n-tuple $(\varepsilon_1,\varepsilon_2,\ldots,\varepsilon_n)\in X$ such that $\overline{\lambda}(\bigwedge_{1\leq i\leq n}\varepsilon_i A_i)=\mathbf{1}$, which, according to Lemma 1.25, is equivalent to: there exists an n-tuple $(\varepsilon_1,\varepsilon_2,\ldots,\varepsilon_n)\in X$ such that $\lambda=\delta_{\varepsilon_1,\varepsilon_2,\ldots,\varepsilon_n}$, or equivalently, to

$$\lambda\in\{\delta_{\varepsilon_1,\varepsilon_2,\ldots,\varepsilon_n}:(\varepsilon_1,\varepsilon_2,\ldots,\varepsilon_n)\in X\}.$$

∎

Theorem 1.27 *For any mapping φ from $\{\mathbf{0},\mathbf{1}\}^n$ into $\{\mathbf{0},\mathbf{1}\}$, there exists at least one formula F such that $\varphi_F=\varphi$.*

(In other words, every mapping from $\{\mathbf{0},\mathbf{1}\}^n$ into $\{\mathbf{0},\mathbf{1}\}$ is a truth table).

Proof Fix a mapping φ from $\{\mathbf{0},\mathbf{1}\}^n$ into $\{\mathbf{0},\mathbf{1}\}$.

- If it assumes only the value $\mathbf{0}$, then it is a truth table, for example, of the formula $F=(A_1\wedge\neg A_1)$.

- In the opposite case, the set

$$X=\varphi^{-1}(\{\mathbf{1}\})=\{(\varepsilon_1,\varepsilon_2,\ldots,\varepsilon_n)\in\{\mathbf{0},\mathbf{1}\}^n:\varphi(\varepsilon_1,\varepsilon_2,\ldots,\varepsilon_n)=\mathbf{1}\}$$

is non-empty and, by virtue of Lemma 1.26, the formula

$$F_X=\bigvee_{(\varepsilon_1,\varepsilon_2,\ldots,\varepsilon_n)\in X}\left(\bigwedge_{1\leq i\leq n}\varepsilon_i A_i\right)$$

is satisfied by those distributions of truth values $\delta_{\varepsilon_1,\varepsilon_2,\ldots,\varepsilon_n}$ for which $\varphi(\varepsilon_1,\varepsilon_2,\ldots,\varepsilon_n)=\mathbf{1}$ and only by these.

In other words, for any n-tuple $(\varepsilon_1,\varepsilon_2,\ldots,\varepsilon_n)\in\{\mathbf{0},\mathbf{1}\}^n$, we have

$$\overline{\delta_{\varepsilon_1,\varepsilon_2,\ldots,\varepsilon_n}}(F)=\mathbf{1}\text{ if and only if }\varphi(\varepsilon_1,\varepsilon_2,\ldots,\varepsilon_n)=\mathbf{1}.$$

This means precisely that φ is the function φ_{F_X}, the truth table of the formula F.

∎

Thus we see that there are 2^{2^n} equivalence classes of formulas on a set of n propositional variables, corresponding to the 2^{2^n} possible truth tables.

Mappings from $\{\mathbf{0},\mathbf{1}\}^n$ into $\{\mathbf{0},\mathbf{1}\}$ are sometimes called n-**place propositional connectives**. We see that it is harmless to identify such an object with the class of formulas which is naturally associated to it.

In the cases $n=1$ and $n=2$ which we will examine in detail (and which lead respectively to 4 and to 16 truth tables), we will rediscover, among the common names for these one- or two-place connectives, names that were already used to denote symbols for propositional connectives. Thus, for example, $\vee$ simultaneously denotes the symbol for the propositional connective and the equivalence class of the formula $(A_1 \vee A_2)$ as well as the corresponding mapping from $\{0, 1\}^2$ into $\{0, 1\}$.

Tables 1.1 and 1.2 present all the two-place and one-place propositional connectives (φ_1 to φ_{16} and ψ_1 to ψ_4). The first columns give the values of each mapping at each point of $\{0, 1\}^2$ or of $\{0, 1\}$. The column that follows gives a formula belonging to the corresponding equivalence class. Finally, the last column displays the symbol in common use, if any, that represents the connective or its usual name.

Table 1.1   The two-place connectives

Values of φ_i					Example of a formula whose truth table is φ_i	Usual denotation for φ_i	
ϵ_1	**0**	**0**	**1**	**1**			
ϵ_2	**0**	**1**	**0**	**1**		Symbol	Name
$\varphi_1(\epsilon_1, \epsilon_2)$	**0**	**0**	**0**	**0**	$(A_1 \wedge \neg A_1)$	**0**	FALSE
$\varphi_2(\epsilon_1, \epsilon_2)$	**0**	**0**	**0**	**1**	$(A_1 \wedge A_2)$	$\wedge$	AND
$\varphi_3(\epsilon_1, \epsilon_2)$	**0**	**0**	**1**	**0**	$\neg(A_1 \Rightarrow A_2)$	$\not\Rightarrow$	DOES NOT IMPLY
$\varphi_4(\epsilon_1, \epsilon_2)$	**0**	**0**	**1**	**1**	A_1		
$\varphi_5(\epsilon_1, \epsilon_2)$	**0**	**1**	**0**	**0**	$\neg(A_2 \Rightarrow A_1)$	$\not\Leftarrow$	
$\varphi_6(\epsilon_1, \epsilon_2)$	**0**	**1**	**0**	**1**	A_2		
$\varphi_7(\epsilon_1, \epsilon_2)$	**0**	**1**	**1**	**0**	$\neg(A_1 \Leftrightarrow A_2)$	$\not\Leftrightarrow$	NOT EQUIVALENT
$\varphi_8(\epsilon_1, \epsilon_2)$	**0**	**1**	**1**	**1**	$(A_1 \vee A_2)$	$\vee$	OR
$\varphi_9(\epsilon_1, \epsilon_2)$	**1**	**0**	**0**	**0**	$\neg(A_1 \vee A_2)$	$\curlyvee$	SHEFFER'S 'OR'
$\varphi_{10}(\epsilon_1, \epsilon_2)$	**1**	**0**	**0**	**1**	$(A_1 \Leftrightarrow A_2)$	$\Leftrightarrow$	IS EQUIVALENT TO
$\varphi_{11}(\epsilon_1, \epsilon_2)$	**1**	**0**	**1**	**0**	$\neg A_2$		
$\varphi_{12}(\epsilon_1, \epsilon_2)$	**1**	**0**	**1**	**1**	$(A_2 \Rightarrow A_1)$	$\Leftarrow$	
$\varphi_{13}(\epsilon_1, \epsilon_2)$	**1**	**1**	**0**	**0**	$\neg A_1$		
$\varphi_{14}(\epsilon_1, \epsilon_2)$	**1**	**1**	**0**	**1**	$(A_1 \Rightarrow A_2)$	$\Rightarrow$	IMPLIES
$\varphi_{15}(\epsilon_1, \epsilon_2)$	**1**	**1**	**1**	**0**	$\neg(A_1 \wedge A_2)$	$\curlywedge$	SHEFFER'S 'AND'
$\varphi_{16}(\epsilon_1, \epsilon_2)$	**1**	**1**	**1**	**1**	$(A_1 \vee \neg A_1)$	**1**	TRUE

Table 1.2   The one-place connectives

Values of ψ_i			Example of a formula whose truth table is ψ_i	Usual designation of ψ_i
ϵ_1	**0**	**1**		
$\psi_1(\epsilon_1)$	**0**	**0**	$(A_1 \wedge \neg A_1)$	**0** (FALSE)
$\psi_2(\epsilon_1)$	**0**	**1**	A_1	IDENTITY
$\psi_3(\epsilon_1)$	**1**	**0**	$\neg A_1$	$\neg$ (NOT)
$\psi_4(\epsilon_1)$	**1**	**1**	$(A_1 \vee \neg A_1)$	**1** (TRUE)

1.3.2 Normal forms

There are important consequences of Theorem 1.27. Before examining them, we need some definitions:

Definition 1.28

(1) *A formula F is in **disjunctive normal form** (DNF) if and only if there exist*
   (a) an integer $m \geq 1$,
   (b) integers $k_1, k_2, \ldots, k_m \geq 1$,
   (c) for every $i \in \{1, 2, \ldots, m\}$, k_i propositional variables: $B_{i1}, B_{i2}, \ldots, B_{ik_i}$ and k_i elements $\varepsilon_{i1}, \varepsilon_{i2}, \ldots, \varepsilon_{ik_i}$ in $\{0, 1\}$, such that

$$F = \bigvee_{1 \leq i \leq m} (\varepsilon_{i1} B_{i1} \wedge \varepsilon_{i2} B_{i2} \wedge \cdots \wedge \varepsilon_{ik_i} B_{ik_i}).$$

(2) *A formula F is in **canonical disjunctive normal form** (CDNF) if and only if there exists a non-empty subset X of $\{0, 1\}^n$ such that*

$$F = \bigvee_{(\varepsilon_1, \varepsilon_2, \ldots, \varepsilon_n) \in X} \left(\bigwedge_{1 \leq i \leq n} \varepsilon_i A_i \right)$$

(3) *By interchanging the symbols for disjunction and conjunction in parts (1) and (2), we obtain respectively the definitions of a formula being in **conjunctive normal form** (CNF) and a formula being in **canonical conjunctive normal form** (CCNF).*

These definitions call for a few remarks. First of all, we see that to be in canonical disjunctive normal form is a special case of being in disjunctive normal form (the case where each k_i is equal to n, where for each $i \in \{1, 2, \ldots, n\}$ and $j \in \{1, 2, \ldots, m\}$, $B_{ij} = A_j$ and where the m n-tuples $\varepsilon_{i1}, \varepsilon_{i2}, \ldots, \varepsilon_{ik_i}$ are pairwise distinct; note, in passing, that this forces m to be at most equal to 2^n).

Furthermore, by examining the proof of Theorem 1.27, we see that, given a mapping φ from $\{0, 1\}^n$ into $\{0, 1\}$ distinct from the zero mapping, there exists a formula F in canonical disjunctive normal form such that $\varphi_F = \varphi$. (The formula F_X that we have been considering is certainly in CDNF.) As well, we conclude a kind of uniqueness for canonical disjunctive (or conjunctive) normal forms, in the sense that two canonical disjunctive (or conjunctive) normal forms which are logically equivalent can differ only in the 'order of their factors'. More precisely, if the formulas:

$$\bigvee_{(\varepsilon_1, \varepsilon_2, \ldots, \varepsilon_n) \in X} \left(\bigwedge_{1 \leq i \leq n} \varepsilon_i A_i \right) \text{ and } \bigvee_{(\eta_1, \eta_2, \ldots, \eta_n) \in Y} \left(\bigwedge_{1 \leq i \leq n} \eta_i A_i \right)$$

are logically equivalent, then the subsets X and Y of $\{0, 1\}^n$ are identical. The analogous fact is obviously true for conjunctive normal forms.

These remarks lead us to the following **normal form theorem**:

Theorem 1.29 *Every formula is logically equivalent to at least one formula in disjunctive normal form and to at least one formula in conjunctive normal form. Any formula that does not belong to the class* **0** *is logically equivalent to a unique formula in CDNF; every formula that does not belong to the class* **1** *is logically equivalent to a unique formula in CCNF, where uniqueness is understood to be 'up to the order of the factors'.*

Proof Let F be a formula.

- If F is a tautology, it is logically equivalent to $A_1 \vee \neg A_1$, which is both a DNF and a CNF.

- If $\neg F$ is a tautology, F is logically equivalent to $A_1 \wedge \neg A_1$, which is DNF and CNF.

- In the other cases, we have just observed that F is logically equivalent to a formula in CDNF. But this is also true for $\neg F$, which means that there is a non-empty subset X of $\{0, 1\}^n$ such that

$$\neg F \quad \sim \quad \bigvee_{(\varepsilon_1, \varepsilon_2, \ldots, \varepsilon_n) \in X} \left(\bigwedge_{1 \leq i \leq n} \varepsilon_i A_i \right).$$

Therefore we have

$$F \sim \neg \neg F$$

$$\sim \neg \left(\bigvee_{(\varepsilon_1, \varepsilon_2, \ldots, \varepsilon_n) \in X} \left(\bigwedge_{1 \leq i \leq n} \varepsilon_i A_i \right) \right)$$

$$\sim \left(\bigwedge_{(\varepsilon_1, \varepsilon_2, \ldots, \varepsilon_n) \in X} \left(\bigvee_{1 \leq i \leq n} \neg \varepsilon_i A_i \right) \right)$$

(by de Morgan's laws). This last formula, once we delete any double negations, is in CCNF.

The second part of the theorem clearly follows from the first and from the remarks that preceded it. ■

We can therefore speak of '*the* CDNF' of a formula (provided it is not an antilogy) and of '*the* CCNF' of a formula (provided it is not a tautology).

The normal form theorem also furnishes us with a practical method for obtaining the CDNF and the CCNF of a formula (when they exist) once we know its truth table. Thus, for example, the formula

$$G = (A \Rightarrow (((B \wedge \neg A) \vee (\neg C \wedge A)) \Leftrightarrow (A \vee (A \Rightarrow \neg B)))),$$

whose truth table was given in Section 1.2.1, is satisfied by the distributions $(\mathbf{0,0,0})$, $(\mathbf{0,0,1})$, $(\mathbf{0,1,0})$, $(\mathbf{0,1,1})$, $(\mathbf{1,0,0})$, $(\mathbf{1,0,0})$ and $(\mathbf{1,1,0})$, while $\neg G$ is satisfied by $(\mathbf{1,0,1})$

and $(\mathbf{1,1,1})$. From this we conclude that the CDNF of G is

$$(\neg A \wedge \neg B \wedge \neg C) \vee (\neg A \wedge \neg B \wedge C) \vee (\neg A \wedge B \wedge \neg C) \vee$$
$$(\neg A \wedge B \wedge C) \vee (A \wedge \neg B \wedge \neg C) \vee (A \wedge B \wedge \neg C),$$

and the CDNF of $\neg G$ is

$$(A \wedge \neg B \wedge C) \vee (A \wedge B \wedge C),$$

and finally the CCNF of G is

$$(\neg A \vee B \vee \neg C) \wedge (\neg A \vee \neg B \vee \neg C).$$

Remark 1.30 *We should mention that formulas of the type $\varepsilon_i A_i$ are sometimes called **literals** (mostly by computer scientists), that formulas of the type $\bigvee_{k \in J} \eta_k B_k$ (i.e. a disjunction of literals) are often called **clauses** and that conjunctive normal forms are then called **clausal forms**. We will encounter this terminology in Chapter 4.*

1.3.3 Complete sets of connectives

In a formula that is in disjunctive normal form, the only symbols for connectives that can occur are $\neg$, $\wedge$ and $\vee$. So we may conclude from Theorem 1.29 that every formula is equivalent to at least one formula in which these are the only connectives that may appear.

This property can be restated in terms of propositional connectives, that is, in terms of operations on $\{\mathbf{0}, \mathbf{1}\}$:

Lemma 1.31 *For every integer $m \geq 1$, every mapping from $\{\mathbf{0}, \mathbf{1}\}^m$ into $\{\mathbf{0}, \mathbf{1}\}$ can be obtained by composition of the mappings $\neg$ (from $\{\mathbf{0}, \mathbf{1}\}$ into $\{\mathbf{0}, \mathbf{1}\}$), together with $\wedge$ and $\vee$ (from $\{\mathbf{0}, \mathbf{1}\}^2$ into $\{\mathbf{0}, \mathbf{1}\}$).*

Proof Let m be a non-zero natural number and φ be a mapping from $\{\mathbf{0}, \mathbf{1}\}^m$ into $\{\mathbf{0}, \mathbf{1}\}$. Choose a formula F which has φ as its truth table and which is written with no symbols for connectives other than $\neg$, $\wedge$ and $\vee$ (for example, a formula in DNF). The decomposition tree of F then gives us a composition of mappings taken from among the mappings $\neg$, $\wedge$ and $\vee$ which coincides with the function φ. Without going into uselessly heavy detail, let us be satisfied with an example.

The mapping φ from $\{\mathbf{0}, \mathbf{1}\}^3$ into $\{\mathbf{0}, \mathbf{1}\}$ which assumes the value $\mathbf{0}$ for $(\mathbf{1,0,1})$ and $(\mathbf{1,1,1})$ and the value $\mathbf{1}$ for the six other triples in $\{\mathbf{0}, \mathbf{1}\}^3$ is, as we have already shown above, the truth table of the formula

$$(\neg A \vee B \vee \neg C) \wedge (\neg A \vee \neg B \vee \neg C).$$

(Here we chose the CCNF, which is much shorter than the CDNF and is also written using only $\neg$, $\wedge$ and $\vee$).

The truly correct way to write this formula is

$$(((\neg A \lor B) \lor \neg C) \land ((\neg A \lor \neg B) \lor \neg C)).$$

We conclude that for any elements x, y and z from $\{\mathbf{0}, \mathbf{1}\}$, we have

$$\varphi(x, y, z) = \land(\lor(\lor(\neg x, y), \neg z), \lor(\lor(\neg x, \neg y), \neg z)).$$

(In this expression, $\neg$, $\land$ and $\lor$ are on this occasion denoting operations on $\{\mathbf{0}, \mathbf{1}\}$).

We see that the operations $\neg$, $\land$ and $\lor$ generate all possible operations (with any number of places) on $\{\mathbf{0}, \mathbf{1}\}$. ∎

We express the property that has just been exhibited by saying that $\{\neg, \land, \lor\}$ is a **complete set of connectives**.

Definition 1.32 *A set of connectives is called* **complete** *if it generates, under composition, the set of all propositional connectives. A complete set of connectives is called* **minimal** *when no proper subset is a complete set of connectives.*

The set $\{\neg, \land, \lor\}$ is not a minimal complete set. Actually, with every formula F which involves no connectives other than $\neg$, $\land$ and $\lor$ we can associate a logically equivalent formula that involves only the symbols for connectives $\neg$ and $\lor$: it suffices to substitute, for each sub-formula of F of the form $(H \land K)$, the logically equivalent formula $\neg(\neg H \lor \neg K)$, repeating the operation as many times as necessary to eliminate all the $\land$. This shows that $\{\neg, \lor\}$ is a complete set of connectives that is a proper subset of $\{\neg, \land, \lor\}$.

The set $\{\neg, \lor\}$ is a minimal complete set. To be sure, it is sufficient to show that $\{\neg\}$ and $\{\lor\}$ are not complete sets.

Formulas in which no symbol for a connective other than $\neg$ occur are the formulas of the type $\neg\neg \ldots \neg A$ (a propositional variable preceded by a finite number, possibly zero, of occurrences of the symbol $\neg$). A formula of this type is logically equivalent either to A or to $\neg A$, and it is clear that there are formulas (for example, $(A \lor B)$) that are not logically equivalent to any formula of this type. Thus, $\{\neg\}$ is not complete. As for $\{\lor\}$, note that a formula in which the only symbol for a connective that occurs is $\lor$ will be satisfied by the distribution of truth values δ_1 defined by $\delta_1(X) = \mathbf{1}$ for every propositional variable X. This can be proved without difficulty by induction (Exercise 20). From this we conclude that the formula $\neg A_1$, which takes the value $\mathbf{0}$ for δ_1, cannot be logically equivalent to any formula which uses only $\lor$ as a symbol for a connective. Thus $\{\lor\}$ is not complete.

In Exercise 15, we will see that each of the connectives known as Sheffer strokes ($\curlyvee$ and $\curlywedge$) has the property of being, by itself alone, a complete set of connectives. We will also show that, among the one- or two-place connectives, these are the only ones with this property.

Remark 1.33 *Suppose we wish to show, by induction, that a certain property $\mathcal{X}(F)$ is true for every formula $F \in \mathcal{F}$, and suppose that this property is compatible with the relation $\sim$ (which is to say that any formula that is logically equivalent to* *a formula having property $\mathcal{X}$ also has the property). We can then exploit the fact that $\{\neg, \vee\}$ is a complete set by limiting ourselves, in the proof by induction, to the induction steps relating to $\neg$ and to $\vee$. If we prove that $\mathcal{X}(F)$ is true when F is an element of P, and that, whenever $\mathcal{X}(F)$ and $\mathcal{X}(G)$ are true, then $\mathcal{X}(\neg F)$ and $\mathcal{X}((F \vee G))$ are also true, this will guarantee that the property $\mathcal{X}$ is true for all formulas in which no symbols for connectives other than $\neg$ and $\vee$ occur. Now let H be an arbitrary formula from $\mathcal{F}$. Since $\{\neg, \vee\}$ is complete, H is logically equivalent to at least one formula K which can be written using only these connectives. So $\mathcal{X}(K)$ is true, and since $\mathcal{X}$ is compatible with $\sim$, $\mathcal{X}(H)$ is also true. Of course, this remark applies just as well to any other complete set of connectives.*

To give an example, note that in the proof of Theorem 1.22 we could have legitimately dispensed with the steps relating to $\vee$, $\Rightarrow$ and $\Leftrightarrow$ thanks to the remark that was just made (for $\{\neg, \wedge\}$ is a complete set), to the compatibility of the property in question with $\sim$ (which is obvious), and subject to verifying that the completeness of the set $\{\neg, \vee\}$ can be proved without recourse to Theorem 1.22 (otherwise we would be running in circles!).

1.4 The interpolation lemma

1.4.1 Interpolation lemma

Lemma 1.34 *Let F and G be two formulas having no propositional variable in common. The following two properties are equivalent:*

(1) *The formula $(F \Rightarrow G)$ is a tautology.*

(2) *At least one of the formulas $\neg F$ or G is a tautology.*

Proof It is clear first of all that the second property implies the first: for any distribution of truth values δ, we have $\delta(G) = \mathbf{1}$ if G is a tautology and $\delta(F) = \mathbf{0}$ if $\neg F$ is one. In both cases, $\delta((F \Rightarrow G)) = \mathbf{1}$.

Now suppose that property (2) is false. Then we can choose a distribution of truth values λ such that $\lambda(\neg F) = \mathbf{0}$, which is to say $\lambda(F) = \mathbf{1}$, and a distribution of truth values μ such that $\mu(G) = \mathbf{0}$. Now define a distribution of truth values δ by setting, for each propositional variable X,

$$\delta(X) = \begin{cases} \lambda(X) & \text{if } X \text{ occurs at least once in } F; \\ \mu(X) & \text{if } X \text{ does not occur in } F. \end{cases}$$

Since, by hypothesis, any variable that occurs in G cannot occur in F, we see that δ coincides with λ on the set of variables of F and with μ on the set of variables of G. We conclude (Lemma 1.19) that $\delta(F) = \lambda(F) = \mathbf{1}$ and that $\delta(G) = \mu(G) = \mathbf{0}$, and, consequently, that $\delta((F \Rightarrow G)) = \mathbf{0}$. So property (1) fails. ∎

The following result is known as the **interpolation lemma**:

Theorem 1.35 *Let n be a non-zero integer, $A_1, A_2, \ldots, A_n$ pairwise distinct propositional variables, and F and G two formulas that have (at most) the*

propositional variables A_1, A_2, ..., A_n in common. The following two properties are equivalent:

(1) The formula $(F \Rightarrow G)$ is a tautology.

(2) There is at least one formula H, containing no propositional variables other
than A_1, A_2, ..., A_n, such that the formulas

$$(F \Rightarrow H) \text{ and } (H \Rightarrow G)$$

are tautologies.

(Such a formula H is called an **interpolant between F and G**).

Proof Suppose $\vdash^* (F \Rightarrow H)$ and $\vdash^* (H \Rightarrow G)$ and consider an arbitrary distribution of truth values δ. If $\delta(H) = 0$, then $\delta(F) = 0$ (because $\delta((F \Rightarrow H)) = 1$);
if $\delta(H) = 1$, then $\delta(G) = 1$ (because $\delta((H \Rightarrow G)) = 1$). In both cases,
$\delta((F \Rightarrow G)) = 1$, which proves property (1).

To show the converse, we will assume $\vdash^* (F \Rightarrow G)$ and argue by induction on
the number of propositional variables which have at least one occurrence in F but
have none in G.

- If this number is zero, then by setting $H = F$ we clearly obtain a formula which
contains no propositional variables other than A_1, A_2, ..., A_n and is such that
$\vdash^* (F \Rightarrow H)$ and $\vdash^* (H \Rightarrow G)$.

- Suppose (the induction hypothesis) that property (2) is true for formulas F
that contain at most m variables that do not occur in G and let us examine
the case in which there are $m + 1$. Let B_1, B_2, ..., B_m, B_{m+1} denote the
variables of F that do not occur in G. According to our conventions, we thus
have $F = F[A_1, A_2, ..., A_n, B_1, B_2, ..., B_m, B_{m+1}]$. Set

$$F_1 = F[A_1, A_2, ..., A_n, B_1, B_2, ..., B_m, A_1] = F_{A_1/B_{m+1}}$$
$$F_0 = F[A_1, A_2, ..., A_n, B_1, B_2, ..., B_m, \neg A_1] = F_{\neg A_1/B_{m+1}}.$$

Notice that because B_{m+1} does not occur in G, the result of substituting the
formula A_1 for the variable B_{m+1} in the formula $(F \Rightarrow G)$ is the formula
$(F_1 \Rightarrow G)$, and the result of substituting the formula $\neg A_1$ for the variable B_{m+1}
in the formula $(F \Rightarrow G)$ is the formula $(F_0 \Rightarrow G)$. Invoking Corollary 1.23
and our hypothesis, we conclude that $(F_1 \Rightarrow G)$ and $(F_0 \Rightarrow G)$ are tautologies,
and hence so are the formulas

$$((F_1 \Rightarrow G) \wedge (F_0 \Rightarrow G)) \text{ and } ((F_1 \vee F_0) \Rightarrow G)$$

(see number (57) in our list in Section 1.2.3).

The variables in the formula $(F_1 \vee F_0)$ are among $A_1, A_2, ..., A_n, B_1, B_2, ..., B_m$.
So we can apply our induction hypothesis and find a formula H that is an interpolant

between $(F_1 \vee F_0)$ and G, which is to say that its variables are among $A_1, A_2, \ldots, A_n$ and it is such that

$$\vdash^* ((F_1 \vee F_0) \Rightarrow H) \text{ and } \vdash^* (H \Rightarrow G).$$

We will now show that $(F \Rightarrow (F_1 \vee F_0))$ is also a tautology. This will conclude the proof since we can then conclude (invoking tautology number (31)) $\vdash^* (F \Rightarrow H)$, which will make H an interpolant between F and G.

So let δ be an assignment of truth values which satisfies F. We have (by Theorem 1.22):

   either $\delta(A_1) = \delta(B_{m+1})$, and in this case $\delta(F_1) = \delta(F) = \mathbf{1}$,
   or else $\delta(A_1) \neq \delta(B_{m+1})$, and so $\delta(F_0) = \delta(F) = \mathbf{1}$.

In all cases, $\delta((F_1 \vee F_0)) = \mathbf{1}$. And so $\vdash^* (F \Rightarrow (F_1 \vee F_0))$. ∎

1.4.2 The definability theorem

Here is a corollary of the interpolation lemma, the **definability theorem**:

Theorem 1.36 *Let A, B, A_1, A_2, $\ldots$, A_k be pairwise distinct propositional variables and $F = F[A, A_1, A_2, \ldots, A_k]$ be a formula (whose variables are therefore among A, A_1, A_2, $\ldots$, A_k). We assume that the formula*

$$((F[A, A_1, A_2, \ldots, A_k] \wedge F[B, A_1, A_2, \ldots, A_k]) \Rightarrow (A \Leftrightarrow B))$$

is a tautology. Then there exists a formula $G = G[A_1, A_2, \ldots, A_k]$, whose only variables are among A_1, A_2, $\ldots$, A_k and is such that the formula

$$(F[A, A_1, A_2, \ldots, A_k] \Rightarrow (A \Leftrightarrow G[A_1, A_2, \ldots, A_k]))$$

is a tautology.

Intuitively, the hypothesis is saying that the formula $F[A, A_1, A_2, \ldots, A_k]$ determines the value of A as a function of the values of $A_1, A_2, \ldots, A_k$, in the sense that distributions of truth values that satisfy F and that assign the same value to $A_1, A_2, \ldots, A_k$ must also assign the same value to A; the conclusion is that the value thereby assigned to A is the value taken by a certain formula $G[A_1, A_2, \ldots, A_k]$ which does not depend on A and which could be called a '**definition of A modulo F**'. Exercise 18 suggests a proof of the theorem directly inspired by this intuition. Here, we are content to apply the preceding lemma.

Proof Taking into consideration numbers (41), (53), and (54) from our list in Section 1.2.3, the hypothesis leads successively to the following tautologies:

$$\vdash^* ((F[A, A_1, A_2, \ldots, A_k] \wedge F[B, A_1, A_2, \ldots, A_k]) \Rightarrow (A \Rightarrow B)),$$
$$\vdash^* (((F[A, A_1, A_2, \ldots, A_k] \wedge F[B, A_1, A_2, \ldots, A_k]) \wedge A) \Rightarrow B),$$
$$\vdash^* (((F[A, A_1, A_2, \ldots, A_k] \wedge A) \wedge F[B, A_1, A_2, \ldots, A_k]) \Rightarrow B),$$
$$\vdash^* ((F[A, A_1, A_2, \ldots, A_k] \wedge A) \Rightarrow (F[B, A_1, A_2, \ldots, A_k] \Rightarrow B)).$$

THE COMPACTNESS THEOREM 45

The interpolation lemma then guarantees the existence of an interpolant

$G[A_1, A_2, \ldots, A_k]$ between $(F[A, A_1, A_2, \ldots, A_k] \wedge A)$
and $(F[B, A_1, A_2, \ldots, A_k]) \Rightarrow B)$.

So, in particular,

$$\vdash^* ((F[A, A_1, A_2, \ldots, A_k] \wedge A) \Rightarrow G)$$

and hence

$$\vdash^* (F[A, A_1, A_2, \ldots, A_k] \Rightarrow (A \Rightarrow G)). \qquad (\star)$$

On the other hand,

$$\vdash^* (G \Rightarrow (F[B, A_1, A_2, \ldots, A_k] \Rightarrow B)),$$

and so

$$\vdash^* ((G \wedge F[B, A_1, A_2, \ldots, A_k]) \Rightarrow B)$$
$$\vdash^* ((F[B, A_1, A_2, \ldots, A_k] \wedge G) \Rightarrow B),$$
$$\vdash^* (F[B, A_1, A_2, \ldots, A_k] \Rightarrow (G \Rightarrow B)).$$

The result of substituting A for B in this latter formula is again a tautology (Corollary 1.23):

$$\vdash^* (F[A, A_1, A_2, \ldots, A_k] \Rightarrow (G \Rightarrow A)). \qquad (\star\star)$$

Properties $(\star)$ and $(\star\star)$ together with tautologies (41) and (54) from Section 1.2.3 finally give us

$$\vdash^* (F[A, A_1, A_2, \ldots, A_k] \Rightarrow (A \Leftrightarrow G)). \qquad \blacksquare$$

1.5 The compactness theorem

1.5.1 Satisfaction of a set of formulas

Definition 1.37 *Let $\mathcal{A}$ and $\mathcal{B}$ be two sets of formulas of the propositional calculus on the set of propositional variables P, let G be a formula and let δ be a distribution of truth values on P.*

- *$\mathcal{A}$ is **satisfied** by δ (or δ **satisfies** $\mathcal{A}$) if and only if δ satisfies all the formulas belonging to $\mathcal{A}$.*

- *$\mathcal{A}$ is **satisfiable** (or **consistent**, or **non-contradictory**) if and only if there exists at least one distribution of truth values that satisfies $\mathcal{A}$.*

- *$\mathcal{A}$ is **finitely satisfiable** if and only if every finite subset of $\mathcal{A}$ is satisfiable.*

- *$\mathcal{A}$ is **contradictory** if and only if it is not satisfiable.*

- G is a **consequence of** A (which we denote by: $A \vdash^* G$) if and only if every distribution of truth values that satisfies A satisfies G.
  (The notation for 'G is not a consequence of A' is: $A \nvdash^* G$).

- A and B are **equivalent** if and only if every formula of A is a consequence of B and every formula of B is a consequence of A.

For example, consider pairwise distinct propositional variables $A, B, A_1,$ $A_2, \ldots, A_m, \ldots$: the set $\{A, B, (\neg A \vee B)\}$ is satisfiable; $\{A, \neg B, (A \Rightarrow B)\}$ is contradictory; the empty set is satisfied by any distribution of truth values whatever (if this were not true, we could find a distribution of truth values δ and a formula $F \in \emptyset$ such that $\delta(F) = \mathbf{0}$; but such a feat is clearly impossible...). We have

$$\{A, \ B\} \vdash^* (A \wedge B) \text{ and } \{A, \ (A \Rightarrow B)\} \vdash^* B.$$

The sets $\{A, B\}$ and $\{(A \wedge B)\}$ are equivalent, as are the sets

$$\{A_1, A_2, \ldots, A_m, \ldots\} \text{ and } \{A_1, A_1 \wedge A_2, \ldots, A_1 \wedge A_2 \wedge \ldots \wedge A_m, \ldots\}.$$

The following lemma lists a certain number of properties that follow from these definitions. Nearly all of them are immediate consequences. It will profit the beginning reader to carefully prove all of these. We will content ourselves with proving the three properties marked by two bullets ($\bullet\bullet$) rather than one.

Lemma 1.38 *For all sets of formulas A and B, integers m and $p \geq 1$, and formulas $G, H, F_1, F_2, \ldots, F_m$ and $G_1, G_2, \ldots, G_p$, the following properties are verified:*

- $A \vdash^* G$ if and only if $A \cup \{\neg G\}$ is contradictory.
- If A is satisfiable and if $B \subseteq A$, then B is satisfiable.
- If A is satisfiable, then A is finitely satisfiable.
- If A is contradictory and if $A \subseteq B$, then B is contradictory.
- If $A \vdash^* G$ and if $A \subseteq B$, then $B \vdash^* G$.
- $A \cup \{G\} \vdash^* H$ if and only if $A \vdash^* (G \Rightarrow H)$.
- $A \vdash^* (G \wedge H)$ if and only if $A \vdash^* G$ and $A \vdash^* H$.
- $\{F_1, F_2, \ldots, F_m\} \vdash^* G$ if and only if $\vdash^* ((F_1 \wedge F_2 \wedge \ldots \wedge F_m) \Rightarrow G)$.
- $\bullet$ G is a tautology if and only if G is a consequence of the empty set.
- G is a tautology if and only if G is a consequence of any set of formulas whatever.
- A is contradictory if and only if $A \vdash^* (G \wedge \neg G)$.
- A is contradictory if and only if every formula is a consequence of A.
- A is contradictory if and only if every antilogy is a consequence of A.
- A is contradictory if and only if there exists at least one antilogy that is a consequence of A.

- $\{F_1, F_2, \ldots, F_m\}$ *is contradictory if and only if* $(\neg F_1 \vee \neg F_2 \vee \ldots \vee \neg F_m)$ *is a tautology.*

- $\mathcal{A}$ *and* $\mathcal{B}$ *are equivalent if and only if they are satisfied by the same assignments of truth values.*

- *When we replace each formula in* $\mathcal{A}$ *by a logically equivalent formula, we obtain a set that is equivalent to* $\mathcal{A}$*.*

- *If* $\mathcal{A}$ *is contradictory, then* $\mathcal{B}$ *is equivalent to* $\mathcal{A}$ *if and only if* $\mathcal{B}$ *is contradictory.*

- • $\mathcal{A}$ *is equivalent to the empty set if and only if every formula belonging to* $\mathcal{A}$ *is a tautology.*

- *The empty set is satisfiable.*

- *The set* $\mathcal{F}$ *of all formulas is contradictory.*

- *The sets* $\{G\}$ *and* $\{H\}$ *are equivalent if and only if the formulas G and H are logically equivalent.*

- *The sets* $\{F_1, F_2, \ldots, F_m\}$ *and* $\{G_1, G_2, \ldots, G_p\}$ *are equivalent if and only if the formula* $((F_1 \wedge F_2 \wedge \cdots \wedge F_m) \Leftrightarrow (G_1 \wedge G_2 \wedge \ldots \wedge G_p))$ *is a tautology.*

- *Every finite set of formulas is equivalent to a set consisting of a single formula.*

- • *When the set P is infinite, and only in this case, there exist sets of formulas that are not equivalent to any finite set of formulas.*

- *The binary relation 'is equivalent to' is an equivalence relation on the set of subsets of* $\mathcal{F}$*.*

Proof •• $\vdash^* G$ if and only if $\emptyset \vdash^* G$: because the empty set is satisfied by every distribution of truth values, G is a consequence of the empty set if and only if every distribution of truth values satisfies G, in other words: if and only if G is a tautology. Observe, as a result, that the notation $\vdash^* G$ for 'G is a tautology' seems natural.

•• $\mathcal{A}$ is equivalent to $\emptyset$ if and only if every element of $\mathcal{A}$ is a tautology: it is clear, first of all, that every formula belonging to $\emptyset$ is a consequence of $\mathcal{A}$, and this holds for any set $\mathcal{A}$ whatever (otherwise, there would be a formula belonging to $\emptyset$ which would not be a consequence of $\mathcal{A}$, and this is clearly impossible); so what we have to prove is that every formula in $\mathcal{A}$ is a consequence of $\emptyset$ if and only if every formula in $\mathcal{A}$ is a tautology; but this is precisely the preceding property.

•• P is infinite if and only if there exists a set of formulas that is not equivalent to any finite set: if P is finite and has n elements, there are 2^{2^n} classes of logically equivalent formulas; choose a representative from each class. We can then, given an arbitrary set of formulas $\mathcal{X}$, replace each formula of $\mathcal{X}$ by the representative chosen from its equivalence class; the resulting set is equivalent to $\mathcal{X}$ and is finite since it can contain at most 2^{2^n} elements. If P is infinite, consider the infinite set of formulas $\mathcal{Y} = \{A_1, A_2, \ldots, A_m, \ldots\}$ (where the A_i are pairwise distinct propositional variables); if $\mathcal{Y}$ were equivalent to a finite set of formulas $\mathcal{Z}$, then $\mathcal{Z}$

would be satisfied, as is $\mathcal{Y}$, by the constant distribution δ_1 equal to $\mathbf{1}$, and we could choose at least one integer k such that the variable A_k does not occur in any of the formulas of $\mathcal{Z}$ (which are finite in number); so the distribution λ which takes the value $\mathbf{1}$ everywhere except at A_k, where its value is $\mathbf{0}$, would still satisfy $\mathcal{Z}$ (Lemma 1.19) but would obviously not satisfy $\mathcal{Y}$, thereby yielding a contradiction: thus we have a set $\mathcal{Y}$ which is not equivalent to any finite set. ∎

1.5.2 The compactness theorem for propositional calculus

We have arrived at what is incontestably the major theorem of this chapter. We will see several applications of it in the exercises.

It can be stated in several equivalent forms:

The compactness theorem, version 1:

Theorem 1.39 *For any set $\mathcal{A}$ of formulas of the propositional calculus, $\mathcal{A}$ is satisfiable if and only if $\mathcal{A}$ is finitely satisfiable.*

The compactness theorem, version 2:

Theorem 1.40 *For any set $\mathcal{A}$ of formulas of the propositional calculus, $\mathcal{A}$ is contradictory if and only if at least one finite subset of $\mathcal{A}$ is contradictory.*

The compactness theorem, version 3:

Theorem 1.41 *For any set $\mathcal{A}$ of formulas of the propositional calculus and for any formula F, F is a consequence of $\mathcal{A}$ if and only if F is a consequence of at least one finite subset of $\mathcal{A}$.*

Proof The proof that the three versions are equivalent is a simple exercise that uses the elementary properties stated in Lemma 1.38. We also observe that the 'only if' direction of version 1 and the 'if' direction of versions 2 and 3 are obvious.

We will now prove the 'if' direction of version 1.

Here is a first proof that is valid for the case in which the set of propositional variables, P, is countably infinite:

$$P = \{A_0, A_1, A_2, \dots, A_m, \dots\}.$$

(For the case when P is finite, the theorem is more or less obvious (there is only a finite number of equivalence classes of formulas), but we can always invoke the situation of the present proof by extending P to a countable set.)

So consider a set $\mathcal{A}$ of formulas that is finitely satisfiable. We must prove the existence of a distribution of truth values that satisfies all of the formulas in $\mathcal{A}$. To do this, we will define, by induction, a sequence $(\varepsilon_n)_{n \in \mathbb{N}}$ of elements of $\{\mathbf{0}, \mathbf{1}\}$ such that the distribution of truth values δ_0 defined by:

$$\text{for every } n \in \mathbb{N}, \ \delta_0(A_n) = \varepsilon_n,$$

satisfies $\mathcal{A}$.

To define ε_0, we distinguish two cases:

- Case 0_0: for every finite subset $\mathcal{B} \subset \mathcal{A}$, there exists at least one distribution of truth values $\delta \in \{\mathbf{0}, \mathbf{1}\}^P$ which satisfies $\mathcal{B}$ and is such that $\delta(A_0) = \mathbf{0}$.

In this case, we set $\varepsilon_0 = \mathbf{0}$.

- Case 1_0: this is the contrary case: we can choose a finite subset $\mathcal{B}_0 \subseteq \mathcal{A}$ such that, for every distribution of truth values $\delta \in \{\mathbf{0}, \mathbf{1}\}^P$ which satisfies $\mathcal{B}_0$, we have $\delta(A_0) = \mathbf{1}$.

In this case, we set $\varepsilon_0 = \mathbf{1}$.

In case 1_0, the following claim is verified:

For every finite subset $\mathcal{B} \subseteq \mathcal{A}$, there exists at least one distribution of truth values $\delta \in \{\mathbf{0}, \mathbf{1}\}^P$ which satisfies $\mathcal{B}$ and is such that $\delta(A_0) = \mathbf{1}$.

To see this, given a finite subset $\mathcal{B} \subseteq \mathcal{A}$, note that $\mathcal{B} \cup \mathcal{B}_0$ is a finite subset of $\mathcal{A}$ that is satisfiable according to the initial hypothesis. Choose a distribution of truth values δ that satisfies it. Then δ satisfies $\mathcal{B}_0$ (which is a subset of $\mathcal{B} \cup \mathcal{B}_0$!), and, by the choice of $\mathcal{B}_0$, we have $\delta(A_0) = \mathbf{1}$. But since δ also satisfies $\mathcal{B}$, the claim is established.

Thus, from our definition of ε_0, we may conclude the following property (R_0):

(R_0) | For every finite subset $\mathcal{B} \subseteq \mathcal{A}$, there exists at least one assignment of truth values $\delta \in \{\mathbf{0}, \mathbf{1}\}^P$ which satisfies $\mathcal{B}$ and is such that $\delta(A_0) = \varepsilon_0$.

Suppose (the induction hypothesis) that $\varepsilon_0, \varepsilon_1, \ldots, \varepsilon_n$ (elements of $\{\mathbf{0}, \mathbf{1}\}$) have been defined in such a way that the following property (R_n) is satisfied:

(R_n) | For every finite subset $\mathcal{B} \subseteq \mathcal{A}$, there exists at least one assignment of truth values $\delta \in \{\mathbf{0}, \mathbf{1}\}^P$ which satisfies $\mathcal{B}$ and is such that $\delta(A_0) = \varepsilon_0, \ \delta(A_1) = \varepsilon_1, \ \ldots, \delta(A_{n-1}) = \varepsilon_{n-1}, \ \text{and } \delta(A_n) = \varepsilon_n$.

We then define ε_{n+1} by distinguishing two cases:

- Case 0_{n+1}: For every finite subset $\mathcal{B} \subseteq \mathcal{A}$, there exists at least one distribution of truth values $\delta \in \{\mathbf{0}, \mathbf{1}\}^P$ which satisfies $\mathcal{B}$ and is such that $\delta(A_0) = \varepsilon_0, \delta(A_1) = \varepsilon_1, \ldots, \delta(A_n) = \varepsilon_n$ and $\delta(A_{n+1}) = \mathbf{0}$.

In this case we set $\varepsilon_{n+1} = \mathbf{0}$.

- Case 1_{n+1}: this is the contrary case : we can choose a finite subset $\mathcal{B}_{n+1} \subseteq \mathcal{A}$ such that, for every distribution of truth values $\delta \in \{\mathbf{0}, \mathbf{1}\}^P$ which satisfies $\mathcal{B}_{n+1}$ and which is such that $\delta(A_0) = \varepsilon_0, \delta(A_1) = \varepsilon_1, \ldots, \delta(A_n) = \varepsilon_n$, we have $\delta(A_{n+1}) = \mathbf{1}$.

In this case, we set $\varepsilon_{n+1} = \mathbf{1}$.

Let us show that property (R_{n+1}) is then satisfied. This amounts to proving, for case 1_{n+1}, that for every finite subset $\mathcal{B} \subseteq \mathcal{A}$, there exists at least one distribution of truth values $\delta \in \{0, 1\}^P$ which satisfies B and is such that $\delta(A_0) = \varepsilon_0$, $\delta(A_1) = \varepsilon_1, \ldots, \delta(A_n) = \varepsilon_n$ and $\delta(A_{n+1}) = 1$.

So consider a finite subset $B \subseteq A$. Then $B \cup B_{n+1}$ is a finite subset of A; and, according to property (R_n), we can choose a distribution of truth values δ which satisfies it and is such that $\delta(A_0) = \varepsilon_0, \delta(A_1) = \varepsilon_1, \ldots, \delta(A_n) = \varepsilon_n$. So δ satisfies B_{n+1} and, because of the way this set was chosen, we may conclude that $\delta(A_{n+1}) = 1$. Since δ satisfies B, our objective is achieved.

The sequence $(\varepsilon_n)_{n \in \mathbb{N}}$ is thus defined; and, for every integer n, property (R_n) is satisfied.

As anticipated, set $\delta_0(A_n) = \varepsilon_n$ for every n.

Let F be formula belonging to A, and let k be a natural number such that all the propositional variables that occur in F are among $\{A_1, A_2, \ldots, A_k\}$ (F being a finite string of symbols, such an integer necessarily exists). Property (R_k) and the fact that $\{F\}$ is a finite subset of A show that we can find a distribution of truth values $\delta \in \{0, 1\}^P$ which satisfies F and is such that $\delta(A_0) = \varepsilon_0, \delta(A_1) = \varepsilon_1, \ldots, \delta(A_k) = \varepsilon_k$. We see that δ and δ_0 agree on the set $\{A_1, A_2, \ldots, A_k\}$, which allows us to conclude (Lemma 1.19) that $\delta_0(F) = \delta(F) = 1$.

The conclusion is that δ_0 satisfies all the formulas in A. ∎

Let us get to the proof of the theorem in the general case: we no longer make any particular assumptions about the set P.

We have to invoke **Zorn's lemma** (see Part 2, Chapter 7).

Proof Once more, we are given a finitely satisfiable set of formulas, A.

Let $\mathcal{E}$ denote the set of mappings whose domain is a subset of P, which take values in $\{0, 1\}$ and which, for every finite subset $B \subset A$, have an extension to the whole of P which is a distribution of truth values that satisfies B.

Formally:

$$\mathcal{E} = \{\varphi \in \bigcup_{X \subseteq P} \{0, 1\}^X :$$

$$(\forall B \in \wp_F(A)) \, (\exists \delta \in \{0, 1\}^P) \, (\delta \restriction X = \varphi \text{ and } (\forall F \in B) \, (\delta(F) = 1))\}.$$

Note that this set is not empty, for it contains the empty mapping (those who find this object perplexing can find additional comments concerning it in Part 2, Chapter 7). To see this, note that by our hypothesis, there exists a distribution of truth values δ on P that satisfies B. As δ is obviously an extension of the empty mapping, the latter satisfies the condition for membership in $\mathcal{E}$.

It is interesting to observe that this is the only point in the proof where we use the hypothesis that A is finitely satisfiable.

Define the binary relation $\leq$ on $\mathcal{E}$ by

$$\varphi \leq \psi \text{ if and only if } \psi \text{ is an extension of } \varphi$$

(in other words, $\mathsf{dom}(\varphi) \subseteq \mathsf{dom}(\psi)$ and for every $A \in \mathsf{dom}(\varphi), \varphi(A) = \psi(A)$). It is very easy to verify that $\leq$ is an order relation on $\mathcal{E}$.

We will prove that the ordered set $(\mathcal{E}, \leq)$ is **inductive**, which is to say that every subset of $\mathcal{E}$ that is totally ordered by $\leq$ has an upper bound in $\mathcal{E}$. This is the same (see Part 2, Chapter 7) as showing that $\mathcal{E}$ is non-empty and that every non-empty subset of $\mathcal{E}$ that is totally ordered by $\leq$ has an upper bound in $\mathcal{E}$. This will permit us (by Zorn's lemma) to assert the existence of a maximal element in $\mathcal{E}$ for the order $\leq$.

We have already observed that $\mathcal{E}$ is non-empty. Consider a non-empty subset $\mathcal{C} \subseteq \mathcal{E}$ that is totally ordered by $\leq$. We define a mapping λ as follows:

- The domain of λ is the union of the domains of the elements of $\mathcal{C}$.
- For every $A \in \mathsf{dom}(\lambda)$ and for every $\varphi \in \mathcal{C}$, if $A \in \mathsf{dom}(\varphi)$, then $\lambda(A) = \varphi(A)$.

This definition makes sense because, if φ and ψ are elements of $\mathcal{E}$ such that $A \in \mathsf{dom}(\varphi)$ and $A \in \mathsf{dom}(\psi)$, then we have either $\varphi \leq \psi$ or $\psi \leq \varphi$, and in both cases $\varphi(A) = \psi(A)$; so the value of the mapping λ at the point A can be legitimately defined as the value at A taken by an arbitrary mapping that belongs to the subset $\mathcal{C}$ and is defined at A; thus λ is the natural common extension of all the elements of $\mathcal{C}$.

Let us show that λ is an element of $\mathcal{E}$. For this, given a finite subset $\mathcal{B} \subseteq \mathcal{A}$, we must find a distribution of truth values $\mu \in \{0, 1\}^P$ which extends λ and which satisfies $\mathcal{B}$. Since $\mathcal{B}$ is finite, there are at most a finite number of propositional variables appearing in the formulas of $\mathcal{B}$.

Let $A_1, A_2, \ldots, A_n$ be the propositional variables that occur in at least one formula from $\mathcal{B}$ and which belong to the domain of λ, in other words, to the union of the domains of the elements of $\mathcal{C}$. Then there exist in $\mathcal{C}$ elements $\varphi_1, \varphi_2, \ldots, \varphi_n$ such that $A_1 \in \mathsf{dom}(\varphi_1)$, $A_2 \in \mathsf{dom}(\varphi_2)$, $\ldots$, $A_n \in \mathsf{dom}(\varphi_n)$. Because $\mathcal{C}$ is totally ordered by $\leq$, one of the φ_i is an extension of all the others: call it φ_0. Thus we have $\varphi_0 \in \mathcal{C}$ and $\{A_1, A_2, \ldots, A_n\} \subseteq \mathsf{dom}(\varphi_0)$. Being an element of $\mathcal{E}$, φ_0 has an extension ψ_0 to P that satisfies $\mathcal{B}$. Let us define the mapping μ from P into $\{0, 1\}$ as follows:

$$\mu(A) = \begin{cases} \lambda(A) & \text{if } A \in \mathsf{dom}(\lambda); \\ \psi_0(A) & \text{if } A \notin \mathsf{dom}(\lambda). \end{cases}$$

- μ is an extension of λ: it agrees with λ on $\mathsf{dom}(\lambda)$.
- μ satisfies $\mathcal{B}$: to see this, we have on the one hand that for every variable $A \in \mathsf{dom}(\varphi_0)$,

$$\mu(A) = \lambda(A) = \varphi_0(A) = \psi_0(A);$$

we conclude from this that μ agrees with ψ_0 on $\{A_1, A_2, \ldots, A_n\}$; on the other hand, if A is a variable that occurs in some formula of $\mathcal{B}$ without belonging to the set $\{A_1, A_2, \ldots, A_n\}$, we have $A \notin \mathsf{dom}(\lambda)$, so $\mu(A) = \psi_0(A)$; thus we see that μ takes the same value as ψ_0 on all propositional variables that are involved in the set $\mathcal{B}$; and since ψ_0 satisfies $\mathcal{B}$, so does μ (Lemma 1.19).

So we have found a distribution of truth values that extends λ and that satisfies $\mathcal{B}$; thus $\lambda \in \mathcal{E}$ and $\mathcal{E}$ is seen to be an inductive ordered set. Zorn's lemma then allows us to choose an element γ from $\mathcal{E}$ that is maximal for the order $\leq$.

Suppose that the domain of γ is not the whole of P and consider a propositional variable A that does not belong to the domain of γ. We will define an extension γ' of γ to the set $\mathsf{dom}(\gamma) \cup \{A\}$ in the following way:

- $\gamma' \lceil_{\mathsf{dom}(\gamma)} = \gamma$;
- $\gamma'(A) = \mathbf{0}$ if for every finite subset $\mathcal{B}$ of $\mathcal{A}$ there exists a distribution of truth values δ on P that satisfies $\mathcal{B}$, that extends γ, and is such that $\delta(A) = \mathbf{0}$;
- $\gamma'(A) = \mathbf{1}$ otherwise.

We then make the following claim: if $\gamma'(A) = \mathbf{1}$, then for every finite subset $\mathcal{B} \subseteq \mathcal{A}$, there is a distribution of truth values δ on P that satisfies $\mathcal{B}$, that extends γ, and is such that $\delta(A) = \mathbf{1}$.

To see this, note that if $\gamma'(A) \neq \mathbf{0}$, we can find a finite subset $\mathcal{B}_0 \subseteq \mathcal{A}$ such that for every distribution of truth values δ that satisfies $\mathcal{B}_0$ and extends γ, we have $\delta(A) = \mathbf{1}$. Now let $\mathcal{B}$ be an arbitrary finite subset of $\mathcal{A}$. The set $\mathcal{B} \cup \mathcal{B}_0$ is a finite subset of $\mathcal{A}$ so there is (by definition of the set $\mathcal{E}$ to which γ belongs) a distribution of truth values δ that extends γ and satisfies $\mathcal{B} \cup \mathcal{B}_0$; δ satisfies $\mathcal{B}_0$ and extends γ: so $\delta(A) = \mathbf{1}$. So we have truly found an extension of γ to P that satisfies $\mathcal{B}$ (since $\mathcal{B} \subseteq \mathcal{B} \cup \mathcal{B}_0$) and takes the value $\mathbf{1}$ at the point A.

So we see that whatever the value of $\gamma'(A)$, there exists, for every finite subset $\mathcal{B}$ of $\mathcal{A}$, an extension δ of γ to P that satisfies $\mathcal{B}$ and is such that $\delta(A) = \gamma'(A)$. But this simply amounts to saying that δ is in fact an extension of γ'. Consequently, for every finite subset $\mathcal{B} \subseteq \mathcal{A}$, γ' can be extended to a distribution of truth values that satisfies $\mathcal{B}$. This means that γ' belongs to $\mathcal{E}$; and so γ' is in $\mathcal{E}$ and is strictly greater than γ for the order $\leq$ ($\mathsf{dom}(\gamma) \subsetneq \mathsf{dom}(\gamma')$), which contradicts the fact that γ is a maximal element of $\mathcal{E}$.

So the assumption we made about the domain of γ was absurd.

It follows that $\mathsf{dom}(\gamma) = P$. Thus we see that γ is a distribution of truth values on P and that any extension of γ to P is equal to γ. So by definition of $\mathcal{E}$, every finite subset $\mathcal{B}$ of $\mathcal{A}$ is satisfied by γ. In particular, this is true for every singleton subset, which means that every formula $F \in \mathcal{A}$ is satisfied by γ. So $\mathcal{A}$ is satisfiable. ∎

In Chapter 2, we will give two other proofs of this theorem.

EXERCISES FOR CHAPTER 1

1. Given two natural numbers m and n, what is the length of a formula of the propositional calculus that has n occurrences of symbols for binary connectives and m occurrences of the symbol for negation?

2. Consider formulas of the propositional calculus on a set of propositional variables P. Given a natural number n, determine the possible different lengths of a formula whose height is n.

3. The set of pseudoformulas constructed from a set of propositional variables P is defined to be the smallest set of words on the alphabet $P \cup \{\neg, \wedge, \vee, \Rightarrow, \Leftrightarrow, (\}$ that satisfies the following conditions:

   • every element of P is a pseudoformula;

   • if F is a pseudoformula, then so is $\neg F$;

   • if F and G are pseudoformulas, then so are the words:

$$(F \wedge G, \qquad (F \vee G, \qquad (F \Rightarrow G, \qquad (F \Leftrightarrow G.$$

   Thus the pseudoformulas are the words obtained from the usual formulas by suppressing all closing parentheses.

   (a) Show that there is a unique readability theorem for pseudoformulas analogous to the one for the usual formulas.

   (b) Would the analogous result be true if we suppressed of all the opening parentheses rather than the closing parentheses?

4. Let $\mathcal{F}$ denote the set of all formulas constructed from a given set of propositional variables P. Let $\mathcal{F}^*$ denote the subset of $\mathcal{F}$ consisting of those formulas in which the symbols $\wedge$, $\vee$, and $\Leftrightarrow$ do not occur.

   (a) Give an inductive definition of the set $\mathcal{F}^*$.

   (b) Let μ be a map from P into $\mathcal{F}^*$. Show that there exists a unique extension $\widehat{\mu}$ of μ to $\mathcal{F}^*$ such that for all formulas F and G belonging to $\mathcal{F}^*$:

$$\widehat{\mu}(\neg F) = \neg \widehat{\mu}(F), \text{ and}$$
$$\widehat{\mu}(F \Rightarrow G) = \neg(\widehat{\mu}(G) \Rightarrow \widehat{\mu}(F)).$$

   (c) Show that for all formulas F and G belonging to $\mathcal{F}^*$, if F is a sub-formula of G, then $\widehat{\mu}(F)$ is a sub-formula of $\widehat{\mu}(G)$.

   (d) Define $\mu_0 : P \to \mathcal{F}^*$ as follows: for all A belonging to P, $\mu_0(A) = \neg A$. Write down the formulas $\widehat{\mu_0}((A \Rightarrow B))$ and $\widehat{\mu_0}((\neg A \Rightarrow B))$. Show that for every formula F in $\mathcal{F}^*$, $\widehat{\mu_0}(F)$ is logically equivalent to F.

5. (a) Show that the following two formulas are tautologies. (They use the propositional variables A_1, A_2, A_3, B_1, B_2, and B_3 and their written form involves some abuses of language.)

$$F_2 = (((A_1 \Rightarrow B_1) \wedge (A_2 \Rightarrow B_2) \wedge \neg(B_1 \wedge B_2) \wedge (A_1 \vee A_2))$$
$$\Rightarrow ((B_1 \Rightarrow A_1) \wedge (B_2 \Rightarrow A_2)));$$

$$F_3 = (((A_1 \Rightarrow B_1) \wedge (A_2 \Rightarrow B_2) \wedge (A_3 \Rightarrow B_3)$$
$$\wedge \neg(B_1 \wedge B_2) \wedge \neg(B_1 \wedge B_3) \wedge \neg(B_2 \wedge B_3) \wedge (A_1 \vee A_2 \vee A_3))$$
$$\Rightarrow (B_2 \Rightarrow A_2) \wedge (B_3 \Rightarrow A_3))).$$

(b) Write a tautology F_n that generalizes F_2 and F_3 using the $2n$ variables A_1, $B_1 \ A_2, B_2, \ldots, A_n, B_n$.

6. E is the formula $((B \wedge C) \Rightarrow (A \Leftrightarrow (\neg B \vee C)))$ in which A, B, and C are propositional variables.

(a) Find a formula that is logically equivalent to E and that is written using only the connectives $\Rightarrow$ and $\Leftrightarrow$.

(b) Find a disjunctive normal form for E that is as reduced as possible.

(c) How many terms (elementary conjunctions) are there in the canonical disjunctive normal form of E?

(d) Show that the formulas

$$(C \Rightarrow (B \Rightarrow (A \Leftrightarrow (B \Rightarrow C)))) \text{ and } (C \Rightarrow (B \Rightarrow A))$$

are logically equivalent.

7. What are the assignments of truth values on $P = \{A_1, A_2, \ldots, A_n\}$ that satisfy:

(a) the formula $F = ((A_1 \Rightarrow A_2) \wedge (A_2 \Rightarrow A_3) \wedge \cdots \wedge (A_{n-1} \Rightarrow A_n))$?

(b) the formula $G = (F \wedge (A_n \Rightarrow A_1))$?

(c) the formula $H = \bigwedge_{\substack{1 \le i \le n \\ 1 \le j \le n \\ i \ne j}} (A_i \Rightarrow \neg A_j)$?

Write down disjunctive normal forms for F, G, and H.

8. Consider the set of propositional variables $P = \{A_1, A_2, \ldots, A_n\}$.

(a) Show that the following formula is a tautology:

$$\left(\bigvee_{1 \le i < j \le n} (A_i \wedge A_j) \right) \Leftrightarrow \bigwedge_{1 \le i \le n} \left(\bigvee_{j \ne i} A_j \right).$$

(b) Which assignments of truth values on P make the following formula false:

$$\left(\bigvee_{1\le i\le n} A_i\right) \Leftrightarrow \bigwedge_{1\le i\le n}\left(\bigvee_{j\ne i} A_j\right) ?$$

(c) Show that the preceding formula is logically equivalent to

$$\bigwedge_{1\le i\le n}\left(A_i \Rightarrow \bigvee_{j\ne i} A_j\right).$$

9. A safe has n locks and can be opened only when all n of the locks are open. Five people, a, b, c, d, and e are to receive keys to some of the locks. Each key can be duplicated any number of times. Find the smallest value of n, and a corresponding distribution of keys to the five people, so that the safe can be opened if and only if at least one of the following situations applies:

- a and b are present together;
- a, c, and d are present together;
- b, d, and e are present together.

10. Consider a set of 15 propositional variables:

$$P = \{A_0, A_1, \ldots, A_{14}\}.$$

The subscripts are viewed as elements of the additive group $\langle \mathbb{Z}/15\mathbb{Z}, + \rangle$ and the operations ($+$ and $-$) on the subscripts are those of this group.

Find all assignments of truth values to P that satisfy the following set of formulas

$$\{A_0\} \cup \{(A_i \Rightarrow A_{-i}) : 0 \le i \le 14\} \cup \{((A_i \wedge A_j) \Rightarrow A_{i+j}) :$$
$$0 \le i \le 14, 0 \le j \le 14\}.$$

11. Consider distinct propositional variables A and B and a symbol α for a binary connective. A formula that is neither a tautology nor an antilogy will be called **neutral**. For each of the formulas

$$F_\alpha = (A \; \alpha \; (B \; \alpha \; A)), \text{ and}$$
$$G_\alpha = ((B \; \alpha \; A) \; \alpha \; \neg(A \; \alpha \; B)),$$

determine whether it is a tautology, an antilogy or neutral when

(a) $\alpha = \wedge$ (b) $\alpha = \vee$ (c) $\alpha = \Rightarrow$
(d) $\alpha = \Leftrightarrow$ (e) $\alpha = \not\Rightarrow$ (f) $\alpha = \curlyvee$

Of course, in cases (e) and (f), α is not a symbol for a connective, but it is used with the obvious conventions; for example, $(B \curlywedge A)$ is the formula $\neg(B \wedge A)$.

12. (a) Show that there exists a unique 3-place connective φ such that for all t belonging to $\{0, 1\}$, we have

$$\varphi(t, \neg t, t) = \varphi(t, 0, 0) = 1 \text{ and}$$
$$\varphi(t, t, \neg t) = \varphi(t, 1, 1) = 0.$$

(b) Find a disjunctive normal form for the connective defined in (a) that is as reduced as possible.

(c) In each of the following cases, give an example of a formula F of the propositional calculus on $\{A, B, C\}$ which, for all assignments of truth values $\delta \in \{0, 1\}^{\{A,B,C\}}$, satisfies the condition indicated:

(1) $\overline{\delta}(F) = \varphi(\delta(A), \delta(A), \delta(A))$
(2) $\overline{\delta}(F) = \varphi(\delta(A), \delta(B), \delta(B))$
(3) $\overline{\delta}(F) = \varphi(\delta(A), \delta(A), \delta(B))$
(4) $\overline{\delta}(F) = \varphi(\delta(A), \delta(B), \delta(A))$
(5) $\overline{\delta}(F) = \varphi(\delta(A), \varphi(\delta(B), \delta(B), \delta(B)), \delta(A))$
(6) $\overline{\delta}(F) = \varphi(\delta(A), \delta(B), \delta(B)) \Rightarrow \varphi(\delta(A), \delta(B), \delta(A))$.

[Notice that in (6), $\Rightarrow$ is not a symbol for a binary connective (since this is not a formula in the formal language) but rather denotes the corresponding binary operation on $\{0, 1\}$. An analogous remark applies to the use of $\neg$ in the conditions imposed on φ in part (a).]

(d) Can the connective $\vee$ be obtained by composition from the connective φ?

(e) Is $\{\varphi\}$ a complete set of connectives?

13. When we add two numbers that when written in the binary number system (numbers in base 2) use at most two digits, say ab and cd, we obtain a number with at most three digits, pqr. For example, $11 + 01 = 100$. Using the standard connectives, write formulas that express p, q, and r as functions of a, b, c, and d.

14. Consider a set of propositional variables P.
   We identify the set $\{0, 1\}$ with the field $\langle \mathbb{Z}/2\mathbb{Z}, +, \times, 0, 1 \rangle$.

(a) Express the usual connectives using the operations $+$ and $\times$.

(b) Express the operations $+$ and $\times$ using the usual connectives.

(c) Show that with each propositional formula $F[A_1, A_2, \ldots, A_n]$ we can associate a polynomial in n unknowns $P_F \in \mathbb{Z}/2\mathbb{Z}[X_1, X_2, \ldots, X_n]$ such that for all assignments of truth values $\delta \in \{0, 1\}^P$, we have

$$\overline{\delta}(F) = \widetilde{P_F}(\delta(A_1), \delta(A_2), \ldots, \delta(A_n)),$$

where $\widetilde{P_F}$ denotes the polynomial function (from $\{0, 1\}^n$ into $\{0, 1\}$) associated with the polynomial P_F.
   For a given formula F, is the polynomial P_F unique?

(d) From the preceding, deduce a procedure for determining whether two for-mulas are logically equivalent or whether a formula is a tautology.

15. We propose to slightly modify the notion of propositional calculus that we have defined by adding to its syntax the 'constants true and false'.

We still have a set of propositional variables P, the parentheses and the five symbols for connectives, and, to complete the alphabet from which the formulas will be built, we add two new symbols $\top$ (the constant, true) and $\bot$ (the constant, false) which we may consider, if we wish, to be symbols for 0-**ary** connectives. The only modification to the definition of the set of formulas is to admit two new formulas of height 0:

$$\top \text{ and } \bot.$$

From the semantical point of view, we must augment the definition of the extension $\bar{\delta}$ of an assignment of truth values δ (which remains a map from P into $\{\mathbf{0}, \mathbf{1}\}$) in the following way:

$$\bar{\delta}(\top) = \mathbf{1} \text{ and } \bar{\delta}(\bot) = \mathbf{0}.$$

All other definitions are unchanged.

The formula $\top$ belongs to $\mathbf{1}$, the class of tautologies, and the formula $\bot$ to the class of antilogies, $\mathbf{0}$. This justifies the use we made of these symbols in our earlier list of tautologies.

(a) Show that, in this new context, the interpolation lemma is true, even without the hypothesis that the formulas F and G have at least one variable in common.

(b) Show that any formula that is written with the unique variable A together with the symbols for connectives $\wedge$, $\vee$, $\top$, and $\bot$ (to the exclusion of all others) is logically equivalent to one of the three formulas $\top$, $\bot$, A.

(c) Show that any formula that is written with the two variables A and B together with the symbols for connectives $\neg$, $\Leftrightarrow$, $\top$, and $\bot$ (to the exclusion of all others) is logically equivalent to one of the eight formulas

$$\top, \ \bot, \ A, \ B, \ \neg A, \ \neg B, \ (A \Leftrightarrow B), \ \neg(A \Leftrightarrow B).$$

(d) Show that the following systems of connectives are complete:

$$\{\Rightarrow, \mathbf{0}\}; \ \{\mathbf{0}, \Leftrightarrow, \vee\}; \ \{\mathbf{0}, \Leftrightarrow, \wedge\}; \ \{\curlyvee\}; \ \{\curlywedge\}.$$

(e) Show that the following systems of connectives are not complete:

$$\{\mathbf{1}, \Rightarrow, \wedge, \vee\}; \ \{\mathbf{0}, \mathbf{1}, \wedge, \vee\}; \ \{\mathbf{0}, \mathbf{1}, \neg, \Leftrightarrow\}.$$

(f) Show that among the zero-place or one-place or two-place connectives, the only ones that, by themselves alone, constitute a complete set of connectives are the Sheffer strokes, $\curlyvee$ and $\curlywedge$.

16. (a) Show that the formula $(A \Leftrightarrow (B \Leftrightarrow C))$ is logically equivalent to $((A \Leftrightarrow B) \Leftrightarrow C)$ but not to $((A \Leftrightarrow B) \wedge (B \Leftrightarrow C))$. The first of these observations might have led us to adopt a simplified way of writing the formula as $A \Leftrightarrow B \Leftrightarrow C$ as we did for conjunction and disjunction. Explain why the second of these observations motivates us to avoid doing so.

(b) Consider a natural number $n \geq 2$ and a set of pairwise distinct propositional variables $\mathcal{B} = \{B_1, B_2, \ldots, B_n\}$. Let $\mathcal{G}(\mathcal{B})$ denote the set of formulas that can be written using one occurrence of each of the n propositional variables $B_1, B_2, \ldots, B_n$, $n - 1$ occurrences of the opening parenthesis, $n - 1$ occurrences of the closing parenthesis and $n - 1$ occurrences of the symbol $\Leftrightarrow$. Show that all the formulas in $\mathcal{G}(\mathcal{B})$ are (pairwise) logically equivalent and are satisfied by a given assignment of truth values δ if and only if the number of variables B_i $(1 \leq i \leq n)$ assigned the value false by δ is even.

(c) For any formula $G \in \mathcal{G}(\mathcal{B})$, $\widetilde{G}$ denotes the formula obtained from G by replacing all occurrences of $\Leftrightarrow$ with $\nLeftrightarrow$. Show that $\widetilde{G}$ is logically equivalent to G if n is odd and to $\neg G$ if n is even.

(d) Let E be a set. For every natural number $k \geq 2$ and for all subsets X_1, $X_2, \ldots, X_k$ of E, we define the **symmetric difference** of $X_1, X_2, \ldots, X_k$, denoted by $X_1 \Delta X_2 \Delta \ldots \Delta X_k$, by induction in the following way:

$$X_1 \Delta X_2 = \{x \in E : x \in X_1 \nLeftrightarrow x \in X_2\};$$
$$X_1 \Delta X_2 \Delta \ldots \Delta X_{k+1} = (X_1 \Delta X_2 \Delta \ldots \Delta X_k) \Delta X_{k+1}.$$

Show that for every natural number $k \geq 2$ and for all subsets $X_1, X_2, \ldots, X_k$ of E, $X_1 \Delta X_2 \Delta \ldots \Delta X_k$ is the set of elements of E that belong to an odd number of the subsets X_i.

17. Consider propositional variables $A, B, A_1, A_2, \ldots, A_n$.

(a) Prove the converse of the definability theorem:
for any formula $F[A_1, A_2, \ldots, A_n, A]$, if there exists a formula $G[A_1, A_2, \ldots, A_n]$ such that the formula

$$(F[A_1, A_2, \ldots, A_n, A] \Rightarrow (G[A_1, A_2, \ldots, A_n] \Leftrightarrow A))$$

is a tautology (we say in this case that G is a definition of A modulo F), then the formula

$$((F[A_1, A_2, \ldots, A_n, A] \wedge F[A_1, A_2, \ldots, A_n, B]) \Rightarrow (A \Leftrightarrow B))$$

is also a tautology.

(b) In each of the five following cases, with the given formula of the form $F[A_1, A_2, \ldots, A_n, A]$, associate a formula $G[A_1, A_2, \ldots, A_n]$ that is a definition of A modulo F.

    1. $F = A_1 \Leftrightarrow A$;
    2. $F = (A_1 \Rightarrow A) \wedge (A \Rightarrow A_2) \wedge (A_1 \Leftrightarrow A_2)$;
    3. $F = A_1 \wedge A_2 \wedge A$;
    4. $F = (A_1 \Rightarrow A) \wedge (A \vee A_2) \wedge \neg(A \wedge A_2) \wedge (A_2 \Rightarrow A_1)$;
    5. $F = (A_1 \Rightarrow A) \wedge (A_2 \Rightarrow A) \wedge (A_3 \Rightarrow A) \wedge (\neg A_1 \Leftrightarrow (A_2 \Leftrightarrow \neg A_3))$.

18. This exercise suggests another proof of the definability theorem.

Let $F[A_1, A_2, \ldots, A_n, A]$ be a propositional formula such that the formula

$$(F[A_1, A_2, \ldots, A_n, A] \wedge F[A_1, A_2, \ldots, A_n, B]) \Rightarrow (A \Leftrightarrow B)$$

is a tautology. Recall that φ_F denotes the map from $\{0, 1\}^{n+1}$ into $\{0, 1\}$ associated with F (its 'truth table').

We define a map ψ from $\{0, 1\}^n$ into $\{0, 1\}$ as follows: for all elements ε_1, $\varepsilon_2, \ldots, \varepsilon_n$ in $\{0, 1\}$

$$\psi(\varepsilon_1, \varepsilon_2, \ldots, \varepsilon_n) = \begin{cases} 0 & \text{if } \varphi_F(\varepsilon_1, \varepsilon_2, \ldots, \varepsilon_n, 0) = 1 \\ 1 & \text{otherwise.} \end{cases}$$

(a) Show that if $\varphi_F(\varepsilon_1, \varepsilon_2, \ldots, \varepsilon_n, 1) = 1$, then $\psi(\varepsilon_1, \varepsilon_2, \ldots, \varepsilon_n) = 1$.

(b) Let $G = G[A_1, A_2, \ldots, A_n]$ be a formula that has ψ as its truth table (i.e. is such that $\varphi_G = \psi$). Show that G is a definition of A modulo F, i.e. that the formula

$$(F[A_1, A_2, \ldots, A_n, A] \Rightarrow (G[A_1, A_2, \ldots, A_n] \Leftrightarrow A))$$

is a tautology.

19. Consider a set with five propositional variables, $P = \{A, B, C, D, E\}$.

(a) How many formulas, up to logical equivalence, are satisfied by exactly seventeen assignments of truth values?

(b) How many formulas, up to logical equivalence, are consequences of the formula $(A \wedge B)$?

20. Consider a set of propositional variables P.

Let δ_1 denote the assignment of truth values on P defined by

$$\delta_1(A) = 1 \text{ for every element } A \in P.$$

(a) Show that for every formula F, there exists at least one formula G that does not contain the symbol $\neg$ and is such that F is logically equivalent either to G or to $\neg G$.

(b) Show that for every formula F, the following three properties are equivalent:

    (i) F is logically equivalent to at least one formula in which the only connectives that may appear are $\wedge$, $\vee$, and $\Rightarrow$.

    (ii) F is logically equivalent to at least one formula which does not contain the symbol $\neg$.

    (iii) $\overline{\delta_1}(F) = \mathbf{1}$.

21. Consider a finite set of propositional variables $P = \{A_1, A_2, \ldots, A_n\}$.

On the set of assignments of truth values on P, we define a binary relation $\ll$ by the condition

$$\lambda \ll \mu \text{ if and only if for all } i \in \{1, 2, \ldots, n\}, \ \lambda(A_i) \leq \mu(A_i).$$

(a) Show that $\ll$ is an order relation. Is it a total ordering?

(b) A formula F is called **increasing** if and only if for all assignments of truth values λ and μ to P, if $\lambda \ll \mu$, then $\overline{\lambda}(F) \leq \overline{\mu}(F)$.

Is the negation of a formula that is not increasing necessarily an increasing formula?

(c) Show that for every formula F, F is increasing if and only if:

    (i) F is a tautology, or

    (ii) $\neg F$ is a tautology, or

    (iii) there exists a formula G that is logically equivalent to F and in which none of the three connectives $\neg$, $\Rightarrow$, and $\Leftrightarrow$ occurs.

22. A set $\mathcal{A}$ of formulas of the propositional calculus is called **independent** if and only if for every formula $F \in \mathcal{A}$, F is not a consequence of $\mathcal{A} - \{F\}$.

(a) Which of the following sets of formulas are independent?

$$\{(A \Rightarrow B), (B \Rightarrow C), (C \Rightarrow A)\}$$
$$\{(A \Rightarrow B), (B \Rightarrow C), (A \Rightarrow C)\}$$
$$\{(A \vee B), (A \Rightarrow C), (B \Rightarrow C), (\neg A \Rightarrow (B \vee C))\};$$
$$\{A, B, (A \Rightarrow C), (C \Rightarrow B)\};$$
$$\{(A \Rightarrow (B \vee C)), (C \Rightarrow \neg B), (B \Rightarrow (A \vee C)), ((B \wedge C) \Leftrightarrow B),$$
$$(A \Rightarrow C), (B \Rightarrow A)\};$$
$$\{((A \Rightarrow B) \Rightarrow C), (A \Rightarrow C), (B \Rightarrow C), (C \Rightarrow (B \Rightarrow A)),$$
$$((A \Rightarrow B) \Rightarrow (A \Leftrightarrow B))\}.$$

For each of them, if it is not independent, find one (and, if possible, several) equivalent independent subsets.

(b) Is the empty set independent? Provide a necessary and sufficient condition for the independence of a set consisting of a single formula.

(c) Show that every finite set of formulas has at least one independent equivalent subset.

(d) Show that for a set of formulas to be independent, it is necessary and sufficient that all its finite subsets be independent.

(e) Does the infinite set $\{A_1, A_1 \wedge A_2, A_1 \wedge A_2 \wedge A_3, \ldots, A_1 \wedge A_2 \wedge \cdots \wedge A_n, \ldots\}$ have an equivalent independent subset? (The A_i are propositional variables.) Does there exist any independent set that is equivalent to it?

(f) Show that for any countable set of formulas $\{F_1, F_2, \ldots, F_n, \ldots\}$, there exists at least one equivalent independent set.

23. Given a set E, a **graph** on E is a binary relation G that is symmetric and **antireflexive** (which means that for every element $x \in E$, $(x, x) \notin G$).

   If k is a non-zero natural number and if G is a graph on E, we say that G is k-**colourable** if and only if there exists a map f from E into $\{1, 2, \ldots, k\}$ such that for all $(x, y) \in G$, $f(x) \neq f(y)$.

(a) For every pair $(x, i) \in E \times \{1, 2, \ldots, k\}$, let $A_{x,i}$ be a propositional variable. Define a set $\mathcal{A}(E, G, k)$ of formulas of the propositional calculus on the set of variables $A_{x,i}$ that is satisfiable if and only if the graph is k-colourable.

(b) Show that for a graph to be k-colourable, it is necessary and sufficient that all of its finite restrictions be k-colourable.

24. An abelian group $\langle G, \cdot, 1 \rangle$ is called **orderable** if and only if there exists a total order relation $\leq$ on G that is **compatible** with the group operation, which means that for all elements x, y, and z of G, if $x \leq y$, then $x \cdot z \leq y \cdot z$.

   An abelian group $\langle G, \cdot, 1 \rangle$ is called **torsion-free** if and only if for any element x of G different from 1 and for any non-zero natural number n, x^n is different from 1. (x^k is defined by induction: $x^1 = x$ and for all integers $k \geq 1$, $x^{k+1} = x \cdot x^k$.)

   An abelian group $\langle G, \cdot, 1 \rangle$ is said to be **of finite type** if and only if it is generated by a finite subset of G (which means that there is a finite subset $X \subseteq G$ such that the smallest subgroup of G that includes X is G itself).

   We will use the following theorem from algebra (for example, see Theorem 5.09 in The Theory of Groups by I.D. Macdonald, Oxford University Press, 1968) *For every torsion-free abelian group of finite type $\langle G, \cdot, 1 \rangle$ that does not reduce to the single element 1, there is a non-zero natural number p such that $\langle G, \cdot, 1 \rangle$ is isomorphic to the group $\langle \mathbb{Z}^p, +, 0 \rangle$.*

(a) Let $\langle G, \cdot, 1 \rangle$ be an abelian group. Take $\{A_{x,y} : (x, y) \in G^2\}$ as the set of propositional variables and write down a set $\mathcal{A}(G)$ of formulas of propositional calculus that is satisfiable if and only if the group G is orderable.

(b) Show that for an abelian group to be orderable, it is necessary and sufficient that all of its subgroups of finite type be orderable.

(c) Show that for an abelian group to be orderable, it is necessary and sufficient that it be torsion-free.

25. Consider two sets E and F and a binary relation $R \subseteq E \times F$.

For every element $x \in E$, let R_x denote the set of elements of F that are related to x under R:

$$R_x = \{y \in F : (x, y) \in R\}.$$

For every subset $A \subseteq E$, the following set is called the **image of A under** R:

$$R_A = \bigcup_{x \in A} R_x.$$

We make the following two hypotheses:

(I) For every subset A of E, the cardinality of R_A is greater than or equal to the cardinality of A.

(II) For every element $x \in E$, the set R_x is finite.

The purpose of this exercise is to prove the following property:

(III) There exists an injective map f from E into F such that for every element $x \in E$, $f(x) \in R_x$ (i.e. an injective map from E into F that is included in R).

(a) Suppose that E is finite. Without using hypothesis II, prove III by induction on the cardinality of E by examining two cases:

1. there is at least one subset A of E such that $A \neq \emptyset$, $A \neq E$, and $\text{card}(A) = \text{card}(R_A)$;
2. for every non-empty subset $A \subsetneq E$, $\text{card}(A) < \text{card}(R_A)$.

(b) Give an example in which I is true while II and III are false.

(c) By using the compactness theorem, prove III when E is infinite.

2 Boolean algebras

When we identify logically equivalent formulas of the propositional calculus, we obtain a set on which, in a natural way, we can define a unary operation and two binary operations which correspond respectively to negation, to conjunction, and to disjunction. The structure obtained in this way is an example of a Boolean algebra. Another example of a Boolean algebra is provided by the set of subsets of a given set together with the operations of complementation, intersection, and union (which, by the way, are often called Boolean operations).

There are diverse ways of approaching Boolean algebras. While we will begin with two purely algebraic presentations (as rings or as ordered sets), we will discover at the end of the chapter that we can just as well adopt a topological point of view: every Boolean algebra can be identified with the set of those subsets of some compact, zero-dimensional space that are both open and closed. The reader should not be concerned by these perhaps unfamiliar words; Section 2.1 will contain all the necessary reminders, both from algebra and topology (we nonetheless assume that the reader does know the definitions of ring, field, and topological space; if not, the reader should consult, for example, A Survey of Modern Algebra by Garrett Birkhoff and Saunders Maclane (A. K. Peters Ltd, 1997) and the text General Topology by John L. Kelley (Springer-Verlag, 1991).

Section 2.2 contains the algebraic definitions and corresponding basic properties. A Boolean algebra is a ring in which every element is equal to its square; but it is just as well a distributive, complemented lattice, i.e. an ordered set in which: (i) there is a least element and a greatest element, (ii) every pair of elements has a lower bound and an upper bound, each of these operations being distributive with respect to the other, and, finally, (iii) every element has a complement. We will establish the equivalence of these two points of view and study examples. Section 2.3 is devoted to atoms: non-zero elements that are minimal with respect to the order on the Boolean algebra. This important notion arises frequently in what follows, especially in several exercises.

In Section 2.4 we are interested in homomorphisms of Boolean algebras. As always in algebra, the kernels of these homomorphisms (which in this context are ideals) play an essential role. When we consider a Boolean algebra $\mathcal{A}$ as a lattice, we prefer to study not the ideals but rather the filters which are canonically associated with them (we obtain a filter by taking the complements of the elements of *an ideal). Study of these ideals and filters is the objective of Section 2.5. Particular attention is paid to maximal filters or ultrafilters, which obviously correspond to maximal ideals, but also correspond to homomorphisms of $\mathcal{A}$ into the Boolean algebra $\{0, 1\}$. The set of these homomorphisms is given a topology: this is then called the Stone space of $\mathcal{A}$, a space which is studied in the sixth and last section of this chapter. The compactness theorem for propositional calculus, which we will prove using topology in Section 2.1, is related in a natural way to the compactness of the Stone space of the algebra of equivalence classes of logically equivalent formulas (Exercise 13).*

2.1 Algebra and topology review

2.1.1 Algebra

Consider a commutative ring with identity $\mathcal{A} = \langle A, +, \times, 0, 1 \rangle$.

We will always assume that in such a ring, we have $0 \neq 1$. As is customary, either of the notations $a \times b$ or ab will denote the product of two elements a and b of A.

Definition 2.1 *An **ideal** of $\mathcal{A}$ is a subset I of A such that*

- *$\langle I, +, 0 \rangle$ is a subgroup of $\langle A, +, 0 \rangle$;*
- *For every element x of I and for every element y of A, $x \times y \in I$.*

The set A itself obviously satisfies these conditions. An ideal of $\mathcal{A}$ distinct from A is called a **proper ideal**. An ideal I of $\mathcal{A}$ is a proper ideal if and only if $1 \notin I$. (If $I = A$, then $1 \in I$; if $1 \in I$, then for every element y of A, $1 \times y = y \in I$, hence $A = I$).

We will only consider proper ideals here. So for us, an ideal of $\mathcal{A}$ will be a subset I of A which, in addition to the two conditions above, satisfies the following property:

- $1 \notin I$.

To adopt this point of view can sometimes be inconvenient: for example, given two ideals I and J of $\mathcal{A}$, there may not be a smallest ideal containing both I and J because the **sum** of two ideals I and J, i.e. the set

$$I + J = \{x \in A : (\exists y \in I)(\exists z \in J)(x = y + z)\}$$

which usually plays this role, may well not be a proper ideal. For instance, in the ring of integers $\mathbb{Z}$, the sum of the ideals $2\mathbb{Z}$ (the set of multiples of 2) and $3\mathbb{Z}$ is the entire ring $\mathbb{Z}$.

Nonetheless, these potential inconveniences are not bothersome in the current context. The reader who absolutely insists on preserving the usual definition of ideal should, in what follows, replace 'ideal' by 'proper ideal' everywhere.

Krull's theorem can be stated this way:

Theorem 2.2 *Every ideal in a commutative ring with identity is included in at least one **maximal** ideal.*

(An ideal is maximal if it is not strictly included in any other ideal.)

Proof The proof uses Zorn's lemma (see Chapter 7 in Part 2). Let I be an ideal in the ring $\mathcal{A}$. Let $\mathcal{E}$ denote the set of ideals in $\mathcal{A}$ that include I;

$$\mathcal{E} = \{J \in \wp(A) : J \text{ is an ideal and } I \subseteq J\}.$$

The theorem will be proved if we can establish the existence of at least one maximal element in the ordered set $\langle \mathcal{E}, \subseteq \rangle$. For this it suffices (Zorn's lemma) to show that this ordered set is non-empty (but this is clear since $I \in \mathcal{E}$) and that every non-empty totally ordered subset of $\mathcal{E}$ has an upper bound in $\mathcal{E}$. So let X be a subset of $\mathcal{E}$ that is totally ordered by the relation of inclusion (also known as a **chain** in $\langle \mathcal{E}, \subseteq \rangle$); we assume that X is not empty. Let I_0 be the union of the elements of X: $I_0 = \bigcup_{J \in X} J$. As X is not empty and as any element of X includes I, I is included in I_0, thus $\mathbf{0} \in I_0$. If x and y are elements of I_0, there are two ideals J and K in X such that $x \in J$ and $y \in K$. As X is totally ordered, we have either $J \subseteq K$ or $K \subseteq J$. If we are, for example, in the first situation, then $x \in K$ and $y \in K$, therefore $x - y \in K$ and $x - y \in I_0$. It follows that $\langle I_0, +, \mathbf{0} \rangle$ is a subgroup of $\langle A, +, \mathbf{0} \rangle$. As well, if $x \in I_0$ and $y \in A$, then for at least one ideal $J \in X$, we have $x \in J$, hence $xy \in J$ and $xy \in I_0$. Finally, we have $\mathbf{1} \notin I_0$, for in the opposite case, $\mathbf{1}$ would belong to one of the elements of X, which is forbidden. We have thus established that I_0 is an ideal of $\mathcal{A}$ which includes I, i.e. is an element of $\mathcal{E}$. For each J in X, $J \subseteq I_0$: it follows that I_0 is, in $\mathcal{E}$, an upper bound for the chain X. ∎

Let I be an ideal in the ring $\mathcal{A}$. We define an equivalence relation on A called **congruence modulo** I and denoted by $\equiv_I$:

for all elements x and y of A, $x \equiv_I y$ if and only if $x - y \in I$.

The fact that this is an equivalence relation is easily proved. Let $\bar{a}$ denote the equivalence class of the element $a \in A$. We have $\bar{\mathbf{0}} = I$. Congruence modulo I is compatible with the ring operations $+$ and $\times$: this means that if $a, b, c,$ and d are elements of A, if $a \equiv_I c$ and $b \equiv_I d$, then $a + b \equiv_I c + d$ and $a \times b \equiv_I c \times d$. This allows us to define two operations on the set $A/\equiv_I$ of equivalence classes, which we will continue to denote by $+$ and $\times$, defined by: for all elements x and y of A, $\bar{x} + \bar{y} = \overline{x + y}$ and $\bar{x} \times \bar{y} = \overline{x \times y}$. These two operations on the set $\mathcal{A}/\equiv_I$ give it the structure of a commutative ring with unit (the zero element is I, the unit element is $\bar{\mathbf{1}}$) called the quotient ring of $\mathcal{A}$ by the ideal I and denoted by $\mathcal{A}/I$ rather than by $\mathcal{A}/\equiv_I$. All required verifications are elementary. The most well-known example of what we have just described is given by the rings $\mathbb{Z}/n\mathbb{Z}$ (where n is a natural number greater than or equal to 2).

Theorem 2.3 *The quotient ring $\mathcal{A}/I$ is a field if and only if the ideal I is maximal.*

Proof If we suppose that I is not maximal, we can choose an ideal J of $\mathcal{A}$ such that $I \subsetneq J$ (strict inclusion). Let a be an element of J that does not belong to I. We have $\overline{a} \neq I$, hence $\overline{a}$ is a non-zero element in the quotient ring. If this element were invertible, there would be an element $b \in A$ such that $\overline{a} \times \overline{b} = \overline{1}$, which is to say $ab \equiv_I 1$, or equivalently $ab - 1 \in I$, so also $ab - 1 \in J$. Because $a \in J$ and J is an ideal, $ab \in J$. Thus the difference $ab - (ab - 1) = 1$ would belong to J, which is impossible. It follows that there is at least one non-zero, noninvertible element in the ring $\mathcal{A}/I$: it is therefore not a field.

Now suppose that I is maximal. Let a be an element of A such that $\overline{a} \neq \overline{0}$ (equivalently, $a \notin I$). Our goal is to show that $\overline{a}$ is an invertible element of the quotient ring $\mathcal{A}/I$. Consider the following set:

$$K = \{x \in A : (\exists y \in A)(\exists z \in I)(x = ay + z)\}.$$

It is easy to verify that $\langle K, +, 0 \rangle$ is a subgroup of $\langle A, +, 0 \rangle$: first of all, $0 \in K$ since $0 = (a \times 0) + 0$; also, if $x_1 \in K$ and $x_2 \in K$, then we can find elements y_1 and y_2 in A, and z_1 and z_2 in I, such that $x_1 = ay_1 + z_1$ and $x_2 = ay_2 + z_2$; we conclude from this that

$$x_1 - x_2 = a(y_1 - y_2) + z_1 - z_2,$$

that $y_1 - y_2 \in A$, and that $z_1 - z_2 \in I$, thus $x_1 - x_2 \in K$. On the other hand, if $x \in K$ and $t \in A$, then $xt \in K$: indeed, there are elements $y \in A$ and $z \in I$ such that $x = ay + z$, thus $xt = a(ty) + tz$; but $ty \in A$ and $tz \in I$, so $xt \in K$. This shows that the first two conditions in the definition of an ideal are satisfied by K. If the third of these conditions were also satisfied (thus if $1 \notin K$), K would be an ideal in $\mathcal{A}$. But the set K strictly includes the set I: indeed, every element x of I can be written $x = (a \times 0) + x$, and thus also belongs to K; and the element a, which can be written $(a \times 1) + 0$, belongs to K but not to I. As I is a maximal ideal, K could not then be an ideal in $\mathcal{A}$. We conclude from this that $1 \in K$. So we can then find two elements $y \in A$ and $z \in I$ such that

$$ay + z = 1.$$

So we have $1 - ay = z \in I$, or equivalently, passing to the equivalence classes for the relation $\equiv_I$, $\overline{1 - ay} = \overline{0}$, which translates as $\overline{a} \times \overline{y} = \overline{1}$. So the element $\overline{a}$ does have an inverse in the quotient ring $\mathcal{A}/I$.

We have thus shown that every non-zero element of this ring is invertible: $\mathcal{A}/I$ is therefore a field. ∎

Observe that the proof we have just given contains an illustration of what we said earlier concerning the sum of two ideals. Indeed, the set K which we considered is the sum of the ideal I and what is known as the principal ideal generated by a (i.e. the ideal consisting of multiples of a). We found ourselves precisely in the situation in which this sum of ideals is the entire ring.

2.1.2 Topology

Let X be a topological space and Y a subset of X. We give Y a topology, called **the topology induced on Y by that of** X, by taking as the open sets in this topology the intersections with Y of open subsets of X. In other words, for a subset $\Omega \subseteq Y$ to be open in the induced topology, it is necessary and sufficient that there exist an open set O in the topology on X such that $\Omega = O \cap Y$. We see immediately that the closed sets for the induced topology on Y are the intersections with Y of the closed subsets of X. When we speak of a **subspace** of a topological space X, we mean a subset with its induced topology.

A **basis** for the open sets of a topological space X is a family $(O_i)_{i \in I}$ of open sets in the topology such that every open set is a union of open sets from this family; in other words, for every open set G, there is at least one subset $J \subseteq I$ such that $G = \bigcup_{j \in J} O_j$. When a basis for the open sets of a topological space has been chosen, the elements of this basis are called **basic open sets**. The complements in X of the basic open sets are called **basic closed sets** and it is clear that every closed set is an intersection of basic closed sets. For the usual topology on the set $\mathbb{R}$ of real numbers, the bounded open intervals (i.e. sets of the form $]a, b[$ where $a \in \mathbb{R}$, $b \in \mathbb{R}$ and $a < b$) constitute a basis for the open sets. Moreover, it is obvious that in any topological space whatever, the family of all open sets is a basis for the open sets. The following property is immediate:

Lemma 2.4 *If $(O_i)_{i \in I}$ is a basis for the open sets of the topological space X and if Y is a subset of X, then the family $(O_i \cap Y)_{i \in I}$ is a basis for the topology on Y induced by that of X.*

This means that the intersections with Y of the basic open subsets of X are basic open subsets of Y.

Let X and Y be two topological spaces. A map f from X into Y is called **continuous** if and only if the inverse image under f of every open subset of Y is an open subset of X. In other words, f is continuous if and only if, for every open subset Ω of Y, the set $f^{-1}[\Omega] = \{x \in X : f(x) \in \Omega\}$ is an open subset of X.

Lemma 2.5 *Let $(\Omega_i)_{i \in I}$ be a basis for the open subsets of a topological space Y and let f be a map from X into Y. For f to be continuous, it is necessary and sufficient that for every index $i \in I$, $f^{-1}[\Omega_i]$ be an open subset of X.*

Proof This is necessary due to the definition of continuity (something that must hold for all the open subsets of Y must hold in particular for the basic open subsets). This is sufficient since, if Ω is any open subset of Y, then there is a subset $J \subseteq I$ such that $\Omega = \bigcup_{j \in J} O_j$, hence $f^{-1}[\Omega] = \bigcup_{j \in J} f^{-1}[O_j]$ (this last fact is a well-known property of inverse images); if all of the $f^{-1}[O_i]$ are open subsets of X, $f^{-1}[\Omega]$ will be a union of open subsets, and hence an open subset of X. ∎

Definition 2.6 *A **homeomorphism** of the topological space X onto the topological space Y is a bijective, continuous map from X into Y whose inverse is a*

*continuous map from Y into X. (We speak in this context of a bijective **bicontinu-**
ous map).*

Definition 2.7 *A topological space X is called **Hausdorff** (or **separated**) if and
only if for every pair of distinct elements x and y of X there exist disjoint open
sets G and H such that $x \in G$ and $y \in H$. It is immediate that every subspace of
a Hausdorff space is Hausdorff.*

Lemma 2.8 *Let X be a topological space that is Hausdorff and let Y be a subset
of X. Then the topology induced on Y by that of X makes Y a Hausdorff space.*

Proof If x and y are distinct points of Y, the intersections with Y of two disjoint
open subsets of X that contain x and y respectively will be two disjoint open
subsets of Y that contain x and y respectively. ∎

Definition 2.9 *A **covering** (or **cover**) of a topological space X is a family $(E_i)_{i \in I}$
of subsets of X such that $X = \bigcup_{i \in I} E_i$. If all of the E_i are open sets, we speak
of an **open covering** (or **open cover**). A **subcovering** of a covering $(E_i)_{i \in I}$ is
a subfamily $(E_j)_{j \in J}$ $(J \subseteq I)$ which is itself a covering of X. We will speak
of a **finite** covering (or subcovering) when the corresponding set of indices is
finite.*

Definition 2.10 *A topological space is called **compact** if and only if $1°$) it is
Hausdorff and $2°$) from every open covering of X we can extract a finite
subcovering.*

Lemma 2.11 *Let X be a Hausdorff space. For X to be compact, it is necessary and
sufficient that every family of closed subsets of X whose intersection is non-empty
have a finite subfamily whose intersection is non-empty.*

Proof It suffices to observe that if $(F_i)_{i \in I}$ is a family of closed subsets of X and
if, for each $i \in I$, we denote the complement of F_i in X by O_i (which is an open
set), then $\bigcap_{i \in I} F_i = \emptyset$ if and only if $\bigcup_{i \in I} O_i = X$. Thus, to each family of closed
subsets of X whose intersection is empty, there corresponds, by complementation,
an open covering of X, and vice versa. ∎

Lemma 2.12 *Let $(\Omega_i)_{i \in I}$ be a basis for the open sets of a Hausdorff space X.
For X to be compact, it is necessary and sufficient that from every covering of X
by basic open sets we can extract a finite subcovering.*

Proof The condition is obviously necessary. Assume it is satisfied and that
$(G_k)_{k \in K}$ is a covering of X by arbitrary open sets. We have $X = \bigcup_{k \in K} G_k$,
but since each G_k is a union of basic open sets, we have a covering of X by a
family of basic open sets $(\Omega_j)_{j \in J}$ $(J \subseteq I)$, with each Ω_j included in at least one
of the open sets G_k. So, given our assumption, we can extract from this cover-
ing a finite subcovering and we will have, for example, $X = \Omega_{j_1} \cup \Omega_{j_2} \cup \cdots \cup
\Omega_{j_n}$. It is now sufficient to choose open sets $G_{k_1}, G_{k_2}, \ldots, G_{k_n}$ from the family
$(G_k)_{k \in K}$ which include $\Omega_{j_1}, \Omega_{j_2}, \ldots, \Omega_{j_n}$ respectively; we will then have a finite

subcovering from $(G_k)_{k \in K}$ since $X = G_{k_1} \cup G_{k_2} \cup \cdots \cup G_{k_n}$. This proves that X is compact. ∎

Naturally, the previous property can be rephrased in terms of closed sets:

Lemma 2.13 *Let X be a Hausdorff space with a given basis for the open sets. For X to be compact, it is necessary and sufficient that from every family of basic closed sets whose intersection is empty, we can extract a finite subfamily whose intersection is already empty.*

Definition 2.14 *A subset of a topological space X that is simultaneously open and closed (i.e. an open subset whose complement in X is also open) will be called* **clopen**.

Definition 2.15 *A topological space which has a basis consisting of clopen sets is called* **zero-dimensional**. *For example, in the space $\mathbb{Q}$ of rational numbers, the bounded open intervals whose endpoints are irrational constitute a basis of clopen sets for the usual topology (as is easily verified): thus $\mathbb{Q}$ is a zero-dimensional topological space.*

Lemma 2.16 *For a topological space to be zero-dimensional, it is necessary and sufficient that the family of clopen subsets be a basis for the open sets.*

Proof It is obvious that any family of open sets that includes a basis for the open subsets of X is itself a basis for the open subsets of X. Thus, if X is zero-dimensional, then the family of all its clopen subsets is a basis for the open sets. The converse is immediate. ∎

Lemma 2.17 *Every subspace Y of a zero-dimensional topological space X is zero-dimensional.*

Proof Let $(O_i)_{i \in I}$ be a basis for the open subsets of X consisting of clopen sets. The family $(O_i \cap Y)_{i \in I}$ is a basis for the open subsets of Y (Lemma 2.4) but these open sets are also closed subsets of Y since they are the intersections with Y of closed subsets of X. ∎

Definition 2.18 *A compact zero-dimensional topological space is called a* **Boolean space**.

Definition 2.19 *Let $(X_i)_{i \in I}$ be a family of topological spaces. On the product $\prod_{i \in I} X_i$ of this family, we can define a topology by taking as the basic open sets all subsets of the form $\prod_{i \in I} O_i$ where, for each index $i \in I$, O_i is an open subset of X_i, but where for all but finitely many indices i, we have $O_i = X_i$. It is easy to verify that the collection consisting of unions of sets of this type is closed under intersections and arbitrary unions. It is this collection of sets that we take as the family of open sets for the topology on $\prod_{i \in I} X_i$. The topology defined in this way is called the* **product topology**.

Tychonoff's theorem asserts that:

Theorem 2.20 *The product of any family of compact topological spaces is a compact topological space.*

The proof makes use of Zorn's lemma and can be found, for example, in the text by J.L. Kelley (*General Topology*, Van Nostrand, 1955, republished by Springer-Verlag, Graduate Texts in Mathematics, 1991) (but it has the drawback of using the notion of filter which will be studied later in this chapter).

Let us now examine the special case which will interest us in this chapter (Section 2.6): the case in which each X_i in the family of spaces $(X_i)_{i \in I}$ is the set $\{0, 1\}$ with the **discrete** topology (the one in which all subsets are open).

In this case, the product $\prod_{i \in I} X_i$ is the set $\{0, 1\}^I$ of maps from I into $\{0, 1\}$.

To produce a basic open set Ω in the product topology, we must take a finite number of indices, $i_1, i_2, \ldots, i_k$ in I and open sets $O_{i_1}, O_{i_2}, \ldots, O_{i_k}$ from $\{0, 1\}$, which, in this case, are arbitrary subsets of $\{0, 1\}$. We then set

$$\Omega = \{0, 1\}^{I - \{i_1, i_2, \ldots, i_k\}} \times O_{i_1} \times O_{i_2} \times \cdots \times O_{i_k},$$

or alternatively

$$\Omega = \{f \in \{0, 1\}^I : f(i_1) \in O_{i_1} \text{ and } f(i_2) \in O_{i_2} \text{ and } \ldots \text{ and } f(i_k) \in O_{i_k}\}.$$

It is natural to suppose that we are really interested only in those indices i_j for which the corresponding open set is something other than the set $\{0, 1\}$ itself. It is also pointless to consider cases in which one of the O_{i_j} is the empty set, for we would then have $\Omega = \emptyset$. There remain only two possible choices for the O_{i_j}: $O_{i_j} = \{0\}$ or $O_{i_j} = \{1\}$.

Thus we see that to produce a basic open set Ω in the product topology on $\{0, 1\}$, we must take a finite number of indices $i_1, i_2, \ldots, i_k$ in I and the same number of elements $\varepsilon_1, \varepsilon_2, \ldots, \varepsilon_k$ in $\{0, 1\}$ and then set

$$\Omega = \{f \in \{0, 1\}^I : f(i_1) = \varepsilon_1 \text{ and } f(i_2) = \varepsilon_2 \text{ and } \ldots \text{ and } f(i_k) = \varepsilon_k\}.$$

A basic open set, therefore, is the set of all maps from I into $\{0, 1\}$ which assume given values at some finite number of given points.

Observe that the complement in $\{0, 1\}^I$ of the set Ω that we just considered is the following set:

$$\bigcup_{1 \leq j \leq k} \{f \in \{0, 1\}^I : f(i_j) = 1 - \varepsilon_j\}.$$

So it is the union of k basic open sets, which is obviously an open set. We conclude that Ω is a closed set.

The basic open sets in the topology on $\{0, 1\}$ are therefore clopen sets. Consequently, we have proved:

Lemma 2.21 *The topological space* $\{0, 1\}^I$ *is zero-dimensional.*

Since the discrete space $\{0, 1\}$ is clearly compact, we may conclude, using Tychonoff's theorem, that:

Theorem 2.22 *The space* $\{0, 1\}^I$ *is a Boolean topological space.*

2.1.3 An application to propositional calculus

Tychonoff's theorem allows us to give a very rapid proof of the compactness theorem for propositional calculus (Theorem 1.40).

Proof Let P be a set of propositional variables and let $\mathcal{F}$ denote the associated set of formulas. For each $F \in \mathcal{F}$, let $\Delta(F)$ denote the set of assignments of truth values that satisfy it:

$$\Delta(F) = \left\{\delta \in \{0, 1\}^P : \bar{\delta}(F) = 1\right\}.$$

If $A_1, A_2, \ldots, A_n$ are the variables that occur in the formula F, we see that the set $\Delta(F)$ is a union of sets of the form

$$\{\delta \in \{0, 1\}^I : \delta(A_1) = \varepsilon_1 \text{ and } \delta(A_2) = \varepsilon_2 \text{ and } \ldots \text{ and } \delta(A_k) = \varepsilon_k\},$$

where the ε_i are elements of $\{0, 1\}$.

Indeed, the satisfaction of a formula F by an assignment δ does not depend on the values that δ assumes outside the set $\{A_1, A_2, \ldots, A_n\}$ (Lemma 1.19).

So the set $\Delta(F)$ is a union of basic open sets in the topological space $\{0, 1\}^P$. This is a finite union: it involves at most 2^n sets. So we conclude that $\Delta(F)$ is itself a clopen set.

Now consider a set of formulas $T \subseteq \mathcal{F}$ that is not satisfiable. This means, precisely, that the intersection, $\bigcap_{F \in T} \Delta(F)$, is the empty set. Thus the family $(\Delta(F))_{F \in T}$ is, in the compact space $\{0, 1\}^P$, a family of closed sets whose intersection is empty. So it is possible to extract a finite subfamily whose intersection is already empty; so there is a finite subset $T_0 \subseteq T$ such that $\bigcap_{F \in T_0} \Delta(F) = \emptyset$. This means that there is some finite subset of T that is not satisfiable. This proves (version 2 of) the compactness theorem for propositional calculus. ∎

In Exercise 13, we will encounter another proof of this compactness theorem; this proof will make use of the results from Sections 2.5 and 2.6 and will avoid any appeal to Tychonoff's theorem, for which we have not given a proof.

2.2 Definition of Boolean algebra

Definition 2.23 *A **Boolean algebra** (sometimes called a **Boolean ring**) is a ring* $\langle A, +, \times, 0, 1 \rangle$ *in which each element is an idempotent for multiplication (i.e. is equal to its square).*

Example 2.24 The ring $\langle \mathbb{Z}/2\mathbb{Z}, +, \times, \mathbf{0}, \mathbf{1} \rangle$.

Example 2.25 The ring $\langle \wp(E), \Delta, \cap, \emptyset, E \rangle$, where E is an arbitrary non-empty set, Δ and $\cap$ are the operations of symmetric difference and intersection respectively on the set $\wp(E)$ of all subsets of E (see Exercise 2).

Example 2.26 Another interesting example is furnished by propositional calculus.

Consider a set of propositional variables and the corresponding set $\mathcal{F}$ of formulas. As we will make precise in Exercise 1, the set $\mathcal{F}/\sim$ of equivalence classes of logically equivalent formulas with the operations of $\not\Leftrightarrow$ and $\wedge$ has the structure of a Boolean algebra (these operations can be defined on this set because the relation $\sim$ is compatible with the propositional connectives). The class $\mathbf{0}$ of antilogies and the class $\mathbf{1}$ of tautologies are, respectively, the identity elements for the operations $\not\Leftrightarrow$ and $\wedge$. We will have occasion to return to this example, which is in fact our principal motivation for studying Boolean algebras.

2.2.1 Properties of Boolean algebras, order relations

Lemma 2.27

- *In any Boolean algebra, every element is its own additive inverse.*
- *Every Boolean algebra is commutative.*

Proof Let $\langle A, +, \times, \mathbf{0}, \mathbf{1} \rangle$ be a Boolean algebra and let x and y be elements of A. From the definition, we have $x^2 = x$, $y^2 = y$, and $(x + y)^2 = x + y$, while moreover, as in any ring, we have

$$(x + y)^2 = x^2 + xy + yx + y^2.$$

So we may conclude

$$x + y = x + xy + yx + y,$$

or after simplification, $xy + yx = \mathbf{0}$. Letting $y = \mathbf{1}$, we obtain in particular that $x + x = \mathbf{0}$ or $x = -x$, which proves the first point. For arbitrary x and y therefore, xy is the inverse of xy, but since $xy + yx = \mathbf{0}$, it is also the inverse of yx. From this we conclude that $xy = yx$ and that the algebra is commutative. ∎

Remark 2.28 *The Boolean ring $\langle \mathbb{Z}/2\mathbb{Z}, +, \times, \mathbf{0}, \mathbf{1} \rangle$ is the only Boolean ring that is a field, and even the only Boolean ring that is an integral domain: indeed, the relation $x^2 = x$, which is equivalent to $x(x - 1) = \mathbf{0}$, requires, in an integral domain, that $x = \mathbf{0}$ or $x = \mathbf{1}$.*

Let $\langle A, +, \times, \mathbf{0}, \mathbf{1} \rangle$ be a Boolean algebra. We define a binary relation $\leq$ on A as follows: for all elements x and y of A, $x \leq y$ if and only if $xy = x$.

We will now verify that this is indeed an order relation. For all elements x, y, and z of A, we have

- $x \leq x$, since $x^2 = x$ by definition;
- if $x \leq y$ and $y \leq z$, then $xy = x$ and $yz = z$; hence, $xz = (xy)z = x(yz) = xy = x$, so $x \leq z$;
- if $x \leq y$ and $y \leq x$, then $xy = x$ and $yx = y$, hence $x = y$ by commutativity.

So the relation $\leq$ is reflexive, transitive, and antisymmetric.

The following theorem lists the main properties of this order relation.

Theorem 2.29

(1) **0** *is* **a least element** *and* **1** *is* **a greatest element** *for the relation* $\leq$.

Proof Indeed, for every x, $\mathbf{0} \times x = \mathbf{0}$ and $\mathbf{1} \times x = x$, hence $\mathbf{0} \leq x$ and $x \leq \mathbf{1}$. ∎

(2) *Any two elements x and y of A have a* **greatest lower bound** *(i.e. a common lower bound that is greater than any other common lower bound), denoted by* $x \frown y$: *specifically, their product xy.*

Proof We have $(xy)x = x^2y = xy$ and $(xy)y = xy^2 = xy$, thus xy is a lower bound both for x and for y. Moreover, if z is a common lower bound for x and y, we have $zx = z$ and $zy = y$, hence $z(xy) = (zx)y = zy = z$, which means that $z \leq xy$; thus xy is the greatest of the common lower bounds for x and y. ∎

(3) *Any two elements x and y of A have a* **least upper bound** *(i.e. a common upper bound that is smaller than any other common upper bound), denoted by $x \smile y$: specifically, the element $x + y + xy$.*

Proof Indeed,

$$x(x + y + xy) = x^2 + xy + x^2y = x + xy + xy = x + \mathbf{0} = x,$$

and in analogous fashion, $y(x + y + xy) = y$. So we certainly have $x \leq x + y + xy$ and $y \leq x + y + xy$. On the other hand, if z is an element of A such that $x \leq z$ and $y \leq z$, which is to say $xz = x$ and $yz = y$, then

$$(x + y + xy)z = xz + yz + xyz = x + y + xy,$$

so $x + y + xy \leq z$; thus $x + y + xy$ is the least of the common upper bounds for x and y. ∎

(4) *The operations $\frown$ and $\smile$ thus defined on A are associative and commutative.*

Proof This is true (and very easy to prove!) in any ordered set that satisfies properties (2) and (3). ∎

(5) **0** *is an identity element for the operation* $\smile$ *and an absorbing element for the operation* $\frown$; *while* **1** *is an identity element for the operation* $\frown$ *and an absorbing element for the operation* $\smile$.

Proof To put this another way, for every element x of A, we have $x \smile \mathbf{0} = x$, $x \frown \mathbf{0} = \mathbf{0}$, $x \frown \mathbf{1} = x$ and $x \smile \mathbf{1} = \mathbf{1}$. This is true in any ordered set that satisfies properties (1), (2), and (3). It is trivial to verify this.   ∎

(6) *Every non-empty finite subset* $\{x_1, x_2, \ldots, x_k\}$ *of A* $(k \in \mathbb{N}^*)$ *has a greatest lower bound equal to* $x_1 \frown x_2 \frown \cdots \frown x_k$ *and a least upper bound equal to* $x_1 \smile x_2 \smile \cdots \smile x_k$.

Proof Except for the obvious case in which $k = 1$, this is a simple generalization of properties (2) and (3) which we naturally obtain by induction on k.

We wish to call your attention to the following fact: the expression $x_1 \frown x_2 \frown \cdots \frown x_k$ is not a new notation intended to denote some newly introduced object. It denotes an element of A that is legitimately defined (by induction) as soon as the operation $\frown$ has been defined (it is the element that we ought to denote by

$$((\ldots((x_1 \frown x_2) \frown x_3) \frown \cdots \frown x_{k-1}) \frown x_k),$$

an expression that contains $k - 1$ pairs of parentheses, which we have suppressed in view of associativity. Concerning the operation $\frown$, property (6) asserts two distinct facts: first, that the elements $x_1, x_2, \ldots, x_k$ have a greatest common lower bound; and second, that this greatest common lower bound is

$$x_1 \frown x_2 \frown \cdots \frown x_k.$$

The proof of these two facts is certainly extremely simple (indeed, we have avoided giving it!), but the difficulty perhaps lies in determining precisely what needs to be proved. (The same remark applies, of course, to the operation $\smile$.)   ∎

(7) *Each of the operations* $\smile$ *and* $\frown$ *is distributive over the other.*

Proof On the one hand,

$$x \frown (y \smile z) = x(y + z + yz) = xy + xz + xyz$$
$$= xy + xz + xy \cdot xz = (x \frown y) \smile (x \frown z)$$

for any elements x, y, and z of A which guarantees that $\frown$ distributes over $\smile$.

On the other hand, with x, y, and z still arbitrary elements of A,

$$(x \smile y) \frown (x \smile z) = (x + y + xy)(x + z + xz)$$
$$= x^2 + xz + x^2 z + yx + yz + yxz + x^2 y + xyz + x^2 yz$$
$$= x + yz + xyz$$

after obvious simplifications.

But $x + yz + xyz = x \smile (yz) = x \smile (y \frown z)$, hence the other distributive property.   ∎

(8) *For every element x in A, there is an element x' in A called the* **complement** *of x, such that $x \smile x' = 1$, and $x \frown x' = 0$.*

Proof If such an element x' exists, it satisfies $xx' = 0$ and $x + x' + xx' = 1$, hence also $x + x' = 1$, or again $x' = 1 + x$. It is easy to verify on the other hand that $x \smile (1 + x) = 1$ and $x \frown (1 + x) = 0$.

We have thus established not only the existence but also the uniqueness of the complement of x: it is $1 + x$.   ∎

(9) *The map $x \mapsto 1 + x$ from A into A is a bijection that reverses the order.*

Proof This map is even an involution (a bijection that is its own inverse) since, for every x, $1 + (1 + x) = x$. On the other hand, for any elements x and y, we have

$$(1 + x)(1 + y) = 1 + x + y + xy.$$

This element is equal to $1 + x$ if and only if $y + xy = 0$, or again $xy = y$. We see in this way that $1 + x \leq 1 + y$ if and only if $y \leq x$.   ∎

Remark 2.30 *The order relation in a Boolean algebra is compatible with multiplication: this means that if the elements a, b, c, and d satisfy $a \leq b$ and $c \leq d$, then $a \times c \leq b \times d$ (if $a \times b = a$ and $c \times d = c$, then $a \times c \times b \times d = a \times c$). The important fact to retain is that this order is not compatible with addition: for example, we have $0 \leq 1$, but we do not have $0 + 1 \leq 1 + 1$.*

Here is a property that we will frequently use:

Lemma 2.31 *For any elements x and y of A, we have $x \leq 1 + y$ if and only if $xy = 0$.*

Proof Indeed, $x \leq 1 + y$ means by definition that $x(1 + y) = x$, or again $x + xy = x$, which is certainly equivalent to $xy = 0$.   ∎

2.2.2 Boolean algebras as ordered sets

In fact, properties (1), (2), (3), (7), and (8) from Theorem 2.29 characterize Boolean algebras, as the following theorem shows; it also provides us with a second method for defining Boolean algebras.

Theorem 2.32 *Let $\langle A, \leq \rangle$ be an ordered set with the following properties:*

(a) *there is a least element (denoted by 0) and a greatest element (denoted by 1);*

(b) *any two elements x and y have a least upper bound (denoted by $x \smile y$) and a greatest lower bound (denoted by $x \frown y$);*

(c) *each of the operations $\smile$ and $\frown$ is distributive over the other;*

(d) *for every element x in A, there is at least one element x' in A such that $x \smile x' = 1$ and $x \frown x' = 0$.*

Then A can be given the structure of a Boolean algebra $\langle A, +, \times, 0, 1 \rangle$ in such a way that the given order $\leq$ on A will coincide with the order that is associated with its Boolean algebra structure (i.e. we will have $x \leq y$ if and only if $xy = y$).

The proof will be given in several stages.

- *Preliminary remarks*: An ordered set that has properties (a) and (b) of the theorem is called a **lattice**. If it also has property (c), it is called a **distributive** lattice. If it is properties (a), (b), and (d) that are satisfied, we speak of a **complemented** lattice, where the **complement** of an element x is the unique element x' for which $x \smile x' = 1$ and $x \frown x' = 0$. Uniqueness is easy to prove:

Proof Suppose that x' and x'' are each complements of x and consider the element $y = (x \frown x') \smile x''$. On one hand, y is equal to $0 \smile x''$, and hence to x''. On the other hand, distributivity leads us to

$$y = (x \smile x'') \frown (x' \smile x'') = 1 \frown (x' \smile x'') = x' \smile x''.$$

So we have $x'' = x' \smile x''$, which means $x' \leq x''$. By interchanging the roles of x' and x'' in this argument, we naturally obtain $x'' \leq x'$, and, in the end, $x' = x''$. ∎

The complement of the element x will be denoted by $x^{\complement}$. We obviously have $1^{\complement} = 0$ and $0^{\complement} = 1$. Observe that as a consequence of the uniqueness of the complement, the map $x \mapsto x^{\complement}$ from A into A is a bijection that is equal to its own inverse (for every x, $(x^{\complement})^{\complement} = x$).

Note in addition that, as we have already observed, when hypotheses (a), (b), (c), and (d) are satisfied, so too are properties (4), (5), and (6) from Theorem 2.29.

- We will now establish what are generally known as **de Morgan's laws**:

Lemma 2.33 *For any elements x and y of A,*

$$(x \frown y)^{\complement} = x^{\complement} \smile y^{\complement} \quad and$$
$$(x \smile y)^{\complement} = x^{\complement} \frown y^{\complement}.$$

Proof The second law follows from the first by replacing x with $x^{\complement}$ and y with $y^{\complement}$ and then using the properties of complementation.

To prove the first law, we show that $(x^C \smile y^C) \smile (x \frown y) = \mathbf{1}$ and that $(x^C \smile y^C) \frown (x \frown y) = \mathbf{0}$: to do this, we use the distributivity of the operations $\smile$ and $\frown$ as well as their associativity and commutativity:

$$\left(x^C \smile y^C\right) \smile (x \frown y) = \left(x^C \smile y^C \smile x\right) \frown \left(x^C \smile y^C \smile y\right)$$
$$= \left(\mathbf{1} \smile y^C\right) \frown \left(x^C \smile \mathbf{1}\right) = \mathbf{1} \frown \mathbf{1} = \mathbf{1};$$
$$\left(x^C \smile y^C\right) \frown (x \frown y) = \left(x^C \frown x \frown y\right) \smile \left(y^C \frown x \frown y\right)$$
$$= (\mathbf{0} \frown y) \smile (\mathbf{0} \frown x) = \mathbf{0} \smile \mathbf{0} = \mathbf{0}.$$

∎

- De Morgan's laws generalize immediately (by induction) as follows:

Lemma 2.34 *For any integer $k \geq 1$ and elements $x_1, x_2, \ldots, x_k$ from A,*

$$(x_1 \frown x_2 \frown \cdots \frown x_k)^C = x_1^C \smile x_2^C \smile \cdots \smile x_k^C \quad and$$
$$(x_1 \smile x_2 \smile \cdots \smile x_k)^C = x_1^C \frown x_2^C \frown \cdots \frown x_k^C.$$

- We will now define an addition $+$ and a multiplication $\times$ on the set A: for all x and y, we set

$$x \times y = x \frown y \quad and$$
$$x + y = \left(x \frown y^C\right) \smile \left(x^C \frown y\right).$$

We can obtain another expression for $x + y$ by using the fact that $\smile$ distributes over $\frown$:

$$x + y = \left(x \smile x^C\right) \frown (x \smile y) \frown \left(y^C \smile x^C\right) \frown \left(y^C \smile y\right)$$
$$= \mathbf{1} \frown (x \smile y) \frown \left(x^C \smile y^C\right) \frown \mathbf{1}, \quad \text{hence}$$
$$x + y = (x \smile y) \frown \left(x^C \smile y^C\right). \tag{$\star$}$$

We will now prove that $\langle A, +, \times, \mathbf{0}, \mathbf{1} \rangle$ is a Boolean algebra.

Proof

- The property 'for all x, $x^2 = x$' is immediate ($x \frown x = x$);
- $\langle A, +, \mathbf{0} \rangle$ is a commutative group:

  * commutativity follows immediately from that of $\frown$ and $\smile$.
  * $\mathbf{0}$ is an identity element for addition: for all x in A,

$$x + \mathbf{0} = \left(x \frown \mathbf{0}^C\right) \smile \left(x^C \frown \mathbf{0}\right) = (x \frown \mathbf{1}) \smile \mathbf{0} = x \smile \mathbf{0} = x.$$

* every element x of A has an inverse: namely, itself:

$$x + x = \left(x \wedge x^{C}\right) \vee \left(x^{C} \wedge x\right) = \mathbf{0} \vee \mathbf{0} = \mathbf{0}.$$

* Addition is associative: if x, y, and z are elements of A, we have

$$(x + y) + z = \left([x + y] \wedge z^{C}\right) \vee \left([x + y]^{C} \wedge z\right)$$

(using the definition of $+$)

$$= \left([(x \wedge y^{C}) \vee (x^{C} \wedge y)] \wedge z^{C}\right) \vee \left([x + y]^{C} \wedge z\right)$$

(using the definition of $+$)

$$= \left([(x \wedge y^{C}) \vee (x^{C} \wedge y)] \wedge z^{C}\right) \vee \left([(x \vee y) \wedge (x^{C} \vee y^{C})]^{C} \wedge z\right)$$

(using ($\star$))

$$= \left([(x \wedge y^{C}) \vee (x^{C} \wedge y)] \wedge z^{C}\right) \vee \left([(x \vee y)^{C} \vee (x^{C} \vee y^{C})^{C}] \wedge z\right)$$

(by de Morgan's laws)

$$= \left([(x \wedge y^{C}) \vee (x^{C} \wedge y)] \wedge z^{C}\right) \vee \left([(x^{C} \wedge y^{C}) \vee (x \wedge y)] \wedge z\right)$$

(by de Morgan's laws)

$$= \left[(x \wedge y^{C} \wedge z^{C}) \vee (x^{C} \wedge y \wedge z^{C})\right] \vee \left[(x^{C} \wedge y^{C} \wedge z)\right.$$
$$\left. \vee (x \wedge y \wedge z)\right]$$

($\wedge$ distributes over $\vee$).

At last, using the associativity of $\vee$, we obtain

$$(x + y) + z = \left(x \wedge y^{C} \wedge z^{C}\right) \vee \left(x^{C} \wedge y \wedge z^{C}\right) \vee \left(x^{C} \wedge y^{C} \wedge z\right) \vee (x \wedge y \wedge z).$$

The commutativity of $\vee$ and $\wedge$ implies that all permutations on x, y, and z produce this same result. In particular, $(x + y) + z = (y + z) + x$, but since the addition operation is commutative, we also have

$$(x + y) + z = x + (y + z).$$

- The multiplication $\times$ is associative and has $\mathbf{1}$ as an identity element: these are obvious properties of the operation $\wedge$.
- Multiplication distributes over addition: to prove this, we again invoke the associativity and commutativity of $\vee$ and $\wedge$, the distributivity of $\wedge$ over $\vee$, together with de Morgan's laws. This time we will omit the justifications of each step in the calculation; the reader will have no difficulty supplying them.

Let x, y, and z be elements of A. We have

$$xy + xz = (x \frown y) + (x \frown z)$$
$$= \left[(x \frown y) \frown (x \frown z)^{C}\right] \smile \left[(x \frown y)^{C} \frown (x \frown z)\right]$$
$$= \left[(x \frown y) \frown (x^{C} \smile z^{C})\right] \smile \left[(x^{C} \smile y^{C}) \frown (x \frown z)\right]$$
$$= \left[(x \frown y \frown x^{C}) \smile (x \frown y \frown z^{C})\right] \smile \left[(x^{C} \frown x \frown z) \smile (y^{C} \frown x \frown z)\right]$$
$$= (x \frown y \frown z^{C}) \smile (x \frown y^{C} \frown z)$$
$$= x \frown \left[(y \frown z^{C}) \smile (y^{C} \frown z)\right]$$
$$= x \frown [y + z]$$
$$= x(y + z).$$

This completes the proof that $\langle A, +, \times, \mathbf{0}, \mathbf{1} \rangle$ is a Boolean algebra. ∎

Let $\ll$ denote the order associated with this structure. For any elements x and y of A, $x \ll y$ if and only if $xy = x$, or again, $x \frown y = x$, but this last equality means precisely that x is less than or equal to y for the order $\leq$ given initially on A. It follows that these two orders coincide.

Therefore, a Boolean algebra is, optionally, a ring in which every element is equal to its square, or an ordered set which has the structure of a complemented, distributive lattice.

Without making this a strict rule, we will tend to adopt the second point of view in the rest of the course.

For the case given in Example 2.25 of the set of subsets of a given set, this point of view is more natural than the first. The order relation is the inclusion relation. The operations $\smile$ and $\frown$ are, respectively, union and intersection. The complement of an element is its set-theoretic complement (see Exercise 2).

In what follows, regardless of the point of view adopted, we will allow ourselves to use, simultaneously, the order relation $\leq$, multiplication and addition, and the operations $\smile$ and $\frown$.

2.3 Atoms in a Boolean algebra

Definition 2.35 *An element A in a Boolean algebra $\langle A, \leq, \smile, \frown, \mathbf{0}, \mathbf{1} \rangle$ is called an **atom** if and only if it is non-zero and has no non-zero strict lower bound.*

In other words, a is an atom if and only if $a \neq \mathbf{0}$ and, for every element b in A, if $b \leq a$, then either $b = a$ or $b = \mathbf{0}$.

Example 2.36 In the Boolean algebra $\wp(E)$ of subsets of the set E, the atoms are the **singletons** (i.e. subsets containing a single element).

Example 2.37 There are Boolean algebras without atoms: this is the situation for the Boolean algebra $\mathcal{F}/\sim$ of equivalence classes of formulas of propositional calculus when the set P of propositional variables is infinite (see Example 2.26).

Proof The order relation in this Boolean algebra is the following (Exercise 1):

if F and G are formulas,

then $\mathsf{cl}(F) \leq \mathsf{cl}(G)$ if and only if the formula $(F \Rightarrow G)$ is a tautology.

To prove that there are no atoms in $\mathcal{F}/\sim$, we will show that every non-zero element has a strict lower bound other than $\mathbf{0}$. So consider a formula F such that $\mathsf{cl}(F) \neq \mathbf{0}$, i.e. such that $\neg F$ is not a tautology, or equivalently that there is at least one assignment of truth values to P that satisfies F. Choose such a distribution and denote it by δ. Also choose a propositional variable X that does not occur in the formula F. This is possible because P is infinite. Let G denote the formula $(F \wedge X)$. We obviously have $\vdash^* ((F \wedge X) \Rightarrow F)$, hence $\mathsf{cl}(G) \leq \mathsf{cl}(F)$. The assignment of truth values λ defined by

for all $Y \in P$,

$$\lambda(Y) = \begin{cases} \delta(Y) & \text{if } Y \neq X \\ 1 & \text{if } Y = X \end{cases}$$

satisfies F (since X does not occur in F) and satisfies X, so it satisfies G. It follows that $\mathsf{cl}(G) \neq \mathbf{0}$. On the other hand, the assignment of truth values μ defined by

for all $Y \in P$,

$$\mu(Y) = \begin{cases} \delta(Y) & \text{if } Y \neq X \\ 0 & \text{if } Y = X \end{cases}$$

satisfies F (for the same reason that λ does) but does not satisfy G, so it does not satisfy the formula $(F \Rightarrow G)$. So we do not have $\mathsf{cl}(F) \leq \mathsf{cl}(G)$, which shows that $\mathsf{cl}(G)$ is a strict lower bound for $\mathsf{cl}(F)$; so it is a strict, non-zero lower bound. ∎

Definition 2.38 *A Boolean algebra is **atomic** if and only if every non-zero element has at least one atom below it.*

This is the situation, for example, with the Boolean algebra of all subsets of a given set (every non-empty set contains at least one singleton).

Theorem 2.39 *Every finite Boolean algebra is atomic.*

Proof Let $\langle A, \leq, \vee, \wedge, \mathbf{0}, \mathbf{1} \rangle$ be a finite Boolean algebra and let x be a non-zero element of A. Denote by $m(x)$ the set of non-zero strict lower bounds of x in A. If $m(x)$ is empty, then x is an atom. If $m(x)$ is not empty, then because it is finite, at least one of its elements is minimal in the ordering $\leq$, i.e. no element of $m(x)$

is strictly below it. It is easy to see that such a minimal element is an atom of A that is below x. ∎

Theorem 2.40 *Let* $\langle A, +, \times, \mathbf{0}, \mathbf{1} \rangle$ *be a Boolean algebra (finite or not). Then for every non-zero element a of A and for every integer* $k \geq 2$, *the following properties are equivalent:*

(1) *a is an atom;*

(2) *for every element x in A, we have* $a \leq x$ *or* $a \leq \mathbf{1} + x$;

(3) *for all elements* $x_1, x_2, \ldots, x_k$ *in A, if* $a \leq x_1 \vee x_2 \vee \cdots \vee x_k$, *then* $a \leq x_1$ *or* $a \leq x_2$ *or* $\ldots a \leq x_k$.

Proof First observe that by virtue of Lemma 2.31 and the definition of the order $\leq$, (2) is equivalent to

(2′) for every element x in A, we have $ax = a$ or $ax = \mathbf{0}$.

Now let us prove the theorem.

Let a be a non-zero element of A and k a natural number greater than or equal to 2.

- (1) $\Rightarrow$ (2′): for every element x in A we have $ax \leq a$, hence, since a is an atom, $ax = a$ or $ax = \mathbf{0}$.

- (2) $\Rightarrow$ (3): assume (2) and choose elements $x_1, x_2, \ldots, x_k$ in A such that $a \leq x_1 \vee x_2 \vee \cdots \vee x_k$. If none of $a \leq x_1, a \leq x_2, \ldots, a \leq x_k$ is true, then we conclude, from (2), that $a \leq \mathbf{1} + x_1$, and $a \leq \mathbf{1} + x_2$, and $\ldots$ and $a \leq \mathbf{1} + x_k$; so a would be a common lower bound for $\mathbf{1} + x_1, \mathbf{1} + x_2, \ldots, \mathbf{1} + x_k$, and hence would be less than or equal to their greatest lower bound $\mathbf{1} + (x_1 \vee x_2 \vee \cdots \vee x_k)$ (de Morgan). The element a would then be simultaneously less than or equal to both $x_1 \vee x_2 \vee \cdots \vee x_k$ and its complement, which is impossible since a is non-zero. So (3) is proved.

- (3) $\Rightarrow$ (1): assume (3) and let b be a lower bound for a. It is obvious that $a \leq b \vee (\mathbf{1} + b) = \mathbf{1}$. By taking $x_1 = b$ and $x_2 = x_3 = \cdots = x_k = \mathbf{1} + b$ in (3), we deduce that $a \leq b$ or $a \leq \mathbf{1} + b$. In the first case, we obtain $b = a$ and in the second case, $b = ab = \mathbf{0}$ (Lemma 2.31). We have thereby proved that a is an atom. ∎

2.4 Homomorphisms, isomorphisms, subalgebras

2.4.1 Homomorphisms and isomorphisms

We will generally refer to a homomorphism of rings with a unit (i.e. a map that preserves both addition and multiplication as well as the identity elements of these operations) as a homomorphism of Boolean algebras. We will give definitions, examples, counterexamples, and characterizations in terms of ordered sets.

Definition 2.41 *Let* $\mathcal{A} = \langle A, +, \times, \mathbf{0}, \mathbf{1} \rangle$ *and* $\mathcal{A}' = \langle A', +, \times, \mathbf{0}, \mathbf{1} \rangle$ *be Boolean algebras and* h *be a map from* A *into* A'. *We say that* h *is* **a homomorphism of Boolean algebras from** $\mathcal{A}$ **into** $\mathcal{A}'$ *if and only if for all elements* x *and* y *in* A, *we have*

$$h(x + y) = h(x) + h(y);$$
$$h(x \times y) = h(x) \times h(y);$$
$$h(\mathbf{1}) = \mathbf{1}.$$

Remark 2.42 *The condition* $h(\mathbf{0}) = \mathbf{0}$ *is not part of the definition since it can be deduced immediately from the first condition (just set* $x = y = \mathbf{0}$).

The situation is different for the multiplicative identity element: the third relation is not a consequence of the second, as the following example shows; take $A = \wp(\mathbb{N})$ and $A' = \wp(\mathbb{Z})$ with their natural Boolean algebra structures and take h to be the identity map from A into A (which we can consider as a map from A into A' since $A \subseteq A'$); it is very easy to verify that the first two relations above are satisfied, but the third is not since the identity element for multiplication in A is $\mathbb{N}$ while in A' it is $\mathbb{Z}$.

The reader will have observed that we committed an abuse of language by giving the same names to the operations (as well as to their identity elements) in $\mathcal{A}$ and $\mathcal{A}'$.

Remark 2.43 *The notion of homomorphism defined here is nothing more than the general notion of homomorphism for rings with unit, specialized to the case of Boolean rings. (We should note in passing that there can exist a homomorphism of rings with unit between a Boolean algebra and a ring with unit that is not a Boolean algebras.) Properties that hold for arbitrary homomorphisms of rings with unit continue to hold, obviously, for Boolean algebras: for example, the composition of two homomorphisms of Boolean algebras is a homomorphism of Boolean algebras. The same line of thought shows that we could actually dispense with Corollary 2.49 below.*

Lemma 2.44 *Let* $\mathcal{A} = \langle A, \leq, \mathbf{0}, \mathbf{1} \rangle$ *and* $\mathcal{A}' = \langle A', \leq, \mathbf{0}, \mathbf{1} \rangle$ *be two Boolean algebras and* h *be a homomorphism of Boolean algebras from* A *into* A'. *Then we have (continuing with the earlier notations and with the abuse of language mentioned in Remark 2.42):*

For all elements x *and* y *of* A,

$$h(x \frown y) = h(x) \frown h(y);$$
$$h\left(x^{\complement}\right) = (h(x))^{\complement};$$
$$h(x \smile y) = h(x) \smile h(y);$$
$$if \; x \leq y, \quad then \; h(x) \leq h(y).$$

Proof Because the operations $\times$ and $\frown$ are identical, the first relation is already a consequence of the definition of homomorphism. The second can be rewritten as $h(1 + x) = 1 + h(x)$, which is an immediate consequence of $h(1) = 1$ and the additivity of h. The third relation follows from the first two by de Morgan's laws. Finally, the last relation can be rewritten: if $xy = y$, then $h(x)h(y) = h(x)$; and the latter is true since $h(xy) = h(x)h(y)$. ∎

Theorem 2.45 *Let* $\mathcal{A} = \langle A, \leq, \mathbf{0}, \mathbf{1} \rangle$ *and* $\mathcal{A}' = \langle A' \leq, \mathbf{0}, \mathbf{1} \rangle$ *be two Boolean algebras and* h *be a map from* A *into* A'. *For* h *to be a homomorphism of Boolean algebras, it is necessary and sufficient that for all elements* x *and* y *in* A, *we have*

$$h(x \frown y) = h(x) \frown h(y),$$
$$h(x^{\complement}) = (h(x))^{\complement}.$$

Proof According to the lemma, the condition is necessary. Suppose it is satisfied and let x and y be elements of A. We then have:

$$h(xy) = h(x \frown y) = h(x) \frown h(y) = h(x)h(y),$$

$$
\begin{aligned}
h(x + y) &= h\left(\left(x \frown y^{\complement}\right) \smile \left(x^{\complement} \frown y\right)\right) = h\left(\left(\left(x \frown y^{\complement}\right)^{\complement} \frown \left(x^{\complement} \frown y\right)^{\complement}\right)^{\complement}\right) \\
&= \left(h\left(\left(x \frown y^{\complement}\right)^{\complement} \frown \left(x^{\complement} \frown y\right)^{\complement}\right)\right)^{\complement} = \left(h\left(\left(x \frown y\right)^{\complement}\right)^{\complement} \frown h\left(\left(x^{\complement} \frown y\right)^{\complement}\right)\right)^{\complement} \\
&= \left(\left(h\left(x \frown y^{\complement}\right)\right)^{\complement} \frown \left(h\left(x^{\complement} \frown y\right)\right)^{\complement}\right)^{\complement} = h\left(x \frown y^{\complement}\right) \smile h\left(x^{\complement} \frown y\right) \\
&= \left(h(x) \frown h\left(y^{\complement}\right)\right) \smile \left(h\left(x^{\complement}\right) \frown h(y)\right) \\
&= \left(h(x) \frown (h(y))^{\complement}\right) \smile \left((h(x))^{\complement} \frown h(y)\right) \\
&= h(x) + h(y).
\end{aligned}
$$

It follows that $h(\mathbf{0}) = \mathbf{0}$ (see Remark 2.42) and hence that

$$h(\mathbf{1}) = h\left(\mathbf{0}^{\complement}\right) = (h(\mathbf{0}))^{\complement} = \mathbf{0}^{\complement} = \mathbf{1}.$$

This shows that h is a homomorphism. ∎

Remark 2.46 *It is clear that in the statement of the preceding theorem, we could replace the operation* $\frown$ *by the operation* $\smile$ *everywhere.*

Definition 2.47 *An* **isomorphism of Boolean algebras** *is a homomorphism of Boolean algebras that is bijective.*

Theorem 2.48 *Let* $\mathcal{A} = \langle A, \leq, \mathbf{0}, \mathbf{1} \rangle$ *and* $\mathcal{A}' = \langle A' \leq, \mathbf{0}, \mathbf{1} \rangle$ *be two Boolean algebras and* h *be a surjective map from* A *onto* A'. *For* h *to be an isomorphism of Boolean algebras, it is necessary and sufficient that*

$(*)$ *for all elements* x *and* y *of* A, $\quad x \leq y \quad$ *if and only if* $h(x) \leq h(y)$.

Proof First let us suppose that h is an isomorphism and let x and y be elements of A. If $x \leq y$, then by Lemma 2.44, $h(x) \leq h(y)$. If $h(x) \leq h(y)$, then by the definition of $\leq$ and because h is a homomorphism, $h(x) = h(x)h(y) = h(xy)$. But since h is injective, this requires $x = xy$, which is to say, $x \leq y$. So $(*)$ is satisfied.

To prove the inverse, suppose $(*)$ holds and that u and v are two elements of A such that $h(u) = h(v)$. We have $h(u) \leq h(v)$ and $h(v) \leq h(u)$, thus, by $(*)$, $u \leq v$, and $v \leq u$, and so $u = v$. Thus, h is injective. Next, let x and y be arbitrary elements of A. Set $t = h(x) \frown h(y)$. Since h is a bijection, there is a unique element z in A such that $t = h(z)$. We have $h(z) \leq h(x)$ and $h(z) \leq h(y)$, hence, using $(*)$, $z \leq x$, and $z \leq y$, and so $z \leq x \frown y$. But since $x \frown y \leq x$ and $x \frown y \leq y$, we have, always using $(*)$, that $h(x \frown y) \leq h(x)$ and $h(x \frown y) \leq h(y)$, which implies

$$h(x \frown y) \leq h(x) \frown h(y) = h(z).$$

Using $(*)$ once more, we obtain $x \frown y \leq z$, and putting all this together, $z = x \frown y$, which proves

$$h(x \frown y) = h(x) \frown h(y).$$

When we replace $\frown$ by $\smile$ and $\leq$ by $\geq$ in the previous argument, we obtain

$$h(x \smile y) = h(x) \smile h(y).$$

Let u be an arbitrary element of A' and let t be its unique preimage in A under h. In A, we have $0 \leq t$ and $t \leq 1$. It follows, using $(*)$, that, in A', $h(0) \leq u$, and $u \leq h(1)$. This shows that $h(0)$ and $h(1)$ are, respectively, the least and greatest elements of A', or in other words, that $h(0) = 0$ and $h(1) = 1$.

So for every element x in A, we have

$$h\left(x^{\complement}\right) \frown h(x) = h\left(x^{\complement} \frown x\right) = h(0) = 0 \quad \text{and}$$
$$h\left(x^{\complement}\right) \smile h(x) = h\left(x^{\complement} \smile x\right) = h(1) = 1.$$

Therefore $h(x^{\complement})$ is the complement of $h(x)$, or in other words,

$$(h(x))^{\complement} = h\left(x^{\complement}\right).$$

We conclude, using Theorem 2.45, that h is a homomorphism of Boolean algebras. ∎

Notice that the relation $h(x \smile y) = h(x) \smile h(y)$ is not required to apply Theorem 2.45; rather, it was useful in proving that h commutes with the operation of taking complements.

Corollary 2.49 *The composition of two isomorphisms of Boolean algebras, as well as the inverse of an isomorphism of Boolean algebras, are isomorphisms of Boolean algebras.*

Proof Let $\mathcal{A} = \langle A, +, \times, \mathbf{0}, \mathbf{1} \rangle$, $\mathcal{B} = \langle B, +, \times, \mathbf{0}, \mathbf{1} \rangle$, and $\mathcal{C} = \langle C, +, \times, \mathbf{0}, \mathbf{1} \rangle$ be Boolean algebras, let ϕ be an isomorphism of Boolean algebras from $\mathcal{A}$ onto $\mathcal{B}$ and ψ an isomorphism of Boolean algebras from $\mathcal{B}$ onto $\mathcal{C}$. The mappings ϕ^{-1} and $\psi \circ \phi$ are obviously surjective. For all elements u and v of B, $\phi^{-1}(u) \leq \phi^{-1}(v)$ is equivalent to $\phi(\phi^{-1}(u)) \leq \phi(\phi^{-1}(v))$, i.e. to $u \leq v$. On the other hand, for all elements x and y of A, we have $x \leq y$ if and only if $\phi(x) \leq \phi(y)$, and $\phi(x) \leq \phi(y)$ if and only if $\psi(\phi(x)) \leq \psi(\phi(y))$. With the preceding theorem, we conclude that ϕ^{-1} and $\psi \circ \phi$ are isomorphisms of Boolean algebras, from $\mathcal{B}$ onto $\mathcal{A}$ and from $\mathcal{A}$ onto $\mathcal{C}$ respectively. ∎

Theorem 2.50 *Every finite Boolean algebra is isomorphic to the Boolean algebra of subsets of some set.*

Proof Let $\mathcal{A} = \langle A, +, \times, \mathbf{0}, \mathbf{1} \rangle$ be a finite Boolean algebra and let E be the set of its atoms. Note that E is not empty since there is at least one atom that is less than or equal to the non-zero element $\mathbf{1}$ (Theorem 2.39). We will show that $\mathcal{A}$ is isomorphic to the algebra of subsets of E.

To do this, consider the map h from A into $\wp(E)$ which, with each element x of A, associates the set of atoms that are below x:

$$\text{for each } x \in A, \ h(x) = \{a \in E : a \leq x\}.$$

- h is surjective: indeed, we first of all have $h(\mathbf{0}) = \emptyset$ (there is no atom below $\mathbf{0}$); as well, let $X = \{a_1, a_2, \ldots, a_k\}$ be a non-empty subset of E and set $M_X = a_1 \smile a_2 \smile \cdots \smile a_k$; we claim $h(M_X) = X$: the inclusion $X \subseteq h(M_X)$ follows immediately form the definition of h (every element of X is an atom which is below M_X); the reverse inclusion is shown using Theorem 2.40: if a is an element of $h(M_X)$, i.e. an atom which is below $M_X = a_1 \smile a_2 \smile \cdots \smile a_k$, then we have $a \leq a_i$ for at least one index i (this is clear if $k = 1$ and this is clause (3) of the theorem if $k \geq 2$), but since a and a_i are atoms, this entails $a = a_i$, and so $a \in X$.

- For all elements x and y of A, if $x \leq y$, then $h(x) \leq h(y)$: indeed, if $x \leq y$, every atom that is below x is an atom that is below y.

- For all elements x and y of A, if $h(x) \subseteq h(y)$, then $x \leq y$: indeed, if x is not less that or equal to y, then $x(\mathbf{1} + y) \neq \mathbf{0}$ (Lemma 2.31). As $\mathcal{A}$ is finite, it is atomic (Theorem 2.39) so we can find an atom $a \in E$ such that $a \leq x(\mathbf{1} + y)$. The atom a is thus below both x and $\mathbf{1} + y$; it cannot be below y as well since it is non-zero. So we have $a \in h(x)$ and $a \notin h(y)$, which shows that $h(x)$ is not included in $h(y)$.

We may now conclude, thanks to Theorem 2.48, that h is an isomorphism of Boolean algebras from $\mathcal{A}$ onto $\wp(E)$. ∎

Corollary 2.51 *The cardinality of any finite Boolean algebra is a power of 2.*

Proof If the finite set E has cardinality n, then the set of its subsets, $\wp(E)$, has cardinality 2^n. ∎

2.4.2 Boolean subalgebras

Definition 2.52 *Let* $\mathcal{A} = \langle A, +, \times, \mathbf{0}, \mathbf{1} \rangle$ *be a Boolean algebra. A subset B of A constitutes a **Boolean subalgebra** of* $\mathcal{A}$ *if and only if B contains the elements* $\mathbf{0}$ *and* $\mathbf{1}$ *and is closed under the operations* $+$ *and* $\times$ *(in other words,* $\mathbf{0} \in B$, $\mathbf{1} \in B$, *and if* $x \in B$ *and* $y \in B$, *then* $x + y \in B$ *and* $xy \in B$).

A Boolean subalgebra of $\mathcal{A}$ is thus a subring of $\mathcal{A}$ that contains the element $\mathbf{1}$. This distinction is essential: in a ring with unit, a subring can itself be a ring with unit without containing the unit element of the full ring: in this case, the role of the identity element for multiplication is played by some other element. Let us reexamine the ring $\langle \wp(\mathbb{Z}), \triangle, \cap, \emptyset, \mathbb{Z} \rangle$: $\wp(\mathbb{N})$ is a subset that is closed under the operations $\triangle$ and $\cap$ and it contains $\emptyset$; it is therefore a subring of $\wp(\mathbb{Z})$. Obviously, $\mathbb{Z} \notin \wp(\mathbb{N})$. Nonetheless, $\mathbb{N}$ is the identity element for the ring $\wp(\mathbb{N})$. Thus the Boolean ring $\wp(\mathbb{N})$ is a ring with unit and is a subring of $\wp(\mathbb{Z})$, but is not a substructure of $\wp(\mathbb{Z})$ as a ring with unit: it is therefore not a Boolean subalgebra of $\wp(\mathbb{Z})$.

Theorem 2.53 *Let* $\mathcal{A} = \langle A, +, \times, \mathbf{0}, \mathbf{1} \rangle$ *be a Boolean algebra and let B be a subset of A. For B to be a Boolean subalgebra of* $\mathcal{A}$, *it is necessary and sufficient that there exists a Boolean algebra* $\mathcal{A}' = \langle A', +', \times', \mathbf{0}', \mathbf{1}' \rangle$ *and a homomorphism of Boolean algebras, h, from* $\mathcal{A}'$ *into* $\mathcal{A}$ *such that the image of the map h is the subset B.*

Proof

- necessary: take $A' = B, +' = +, \times' = \times, \mathbf{0}' = \mathbf{0}, \mathbf{1}' = \mathbf{1}$, and $h =$ the identity map from B into A. It is immediate to verify that h is a homomorphism of Boolean algebras whose image is B.

- sufficient: choose $\mathcal{A}'$ and h as indicated. We have $h(\mathbf{0}') = \mathbf{0}$, hence $\mathbf{0} \in B$, and $h(\mathbf{1}') = \mathbf{1}$, so $\mathbf{1} \in B$. Moreover, if x and y are elements of B, then we can choose elements x' and y' in A' such that $x = h(x')$ and $y = h(y')$. We then have

$$x + y = h(x') + h(y') = h(x' + y'), \quad \text{hence } x + y \in \mathsf{Im}(h) = B; \quad \text{and}$$
$$xy = h(x')h(y') = h(x'y'), \quad \text{hence } xy \in \mathsf{Im}(h) = B.$$

So B is a Boolean subalgebra of $\mathcal{A}$.   ∎

Theorem 2.54 *In a Boolean algebra* $\mathcal{A} = \langle A, +, \times, \mathbf{0}, \mathbf{1} \rangle$, *for a subset B to be a Boolean subalgebra, it is necessary and sufficient that B contain* $\mathbf{0}$ *and be closed under the operations* $x \mapsto x^{\complement}$ *and* $(x, y) \mapsto x \frown y$.

Proof

- necessary: since $x^{\complement} = \mathbf{1} + x$ and $x \frown y = xy$, and since B must contain $\mathbf{0}$ and $\mathbf{1}$ and be closed under $+$ and $\times$, the result is immediate.

- sufficient: for every x and y in A we have $x \smile y = (x^{\complement} \frown y^{\complement})^{\complement}$. So the closure of B under complementation and $\frown$ guarantees its closure under $\smile$. Moreover, $\mathbf{1}^{\complement} = \mathbf{0}$ must belong to B. Since the operations $+$ and $\times$ can be defined exclusively in terms of $\frown$, $\smile$, and complementation, we conclude that B is closed under $+$ and $\times$ and that $\langle B, +, \times, \mathbf{0}, \mathbf{1} \rangle$ is a Boolean subalgebra of $\mathcal{A}$.   ∎

Example 2.55 Let E be an infinite set, let $A = \wp(E)$, the set of all its subsets, and let B be the subset of A consisting of those subsets of E that are finite or whose complement is finite. We will show, using the previous theorem, that B is a Boolean subalgebra of the Boolean algebra of subsets of E.

Proof A subset of E whose complement is finite will be called **cofinite**. The empty set, which is a finite subset of E, belongs to B. It is clear that B is closed under complementation: the complement (in this instance, the set-theoretic complement) of a finite subset of E is a cofinite subset and the complement of a cofinite set is finite. Also, B is closed under $\frown$ (in this instance, set-theoretic intersection): indeed, the intersection of a finite subset of E with any subset of E is a finite subset of E; as for the intersection of two cofinite subsets of E, this too is a cofinite subset of E: to see this, suppose that $U \subseteq E$ and $V \subseteq E$ are cofinite; this means that $E - U$ and $E - V$ are finite and hence, so is their union $(E - U) \cup (E - V)$ which is simply (de Morgan) $E - (U \cap V)$; it follows that $U \cap V$ is cofinite.   ∎

Example 2.56 Here is an example that will be of use in Section 2.6. Let X be a topological space and let $\mathcal{B}(X)$ be the subset of $\wp(X)$ consisting of subsets of X that are both open and closed in the topology on X (recall that we speak of clopen sets in this context). This set $\mathcal{B}(X)$ is a Boolean subalgebra of the Boolean algebra of subsets of X.

Proof First of all, the empty set ($\mathbf{0}$) and the whole space X ($\mathbf{1}$) itself are naturally clopen sets. Next, the complement of a clopen set is clopen and the intersection of two clopen sets is clopen. So Theorem 2.54 allows us, here as well, to arrive at the expected conclusion.   ∎

It can happen that the Boolean algebra $\mathcal{B}(X)$ under consideration here reduces to $\{\mathbf{0}, \mathbf{1}\}$ (this is the case, for example, when X is the space $\mathbb{R}$ with its usual topology: $\emptyset$ and $\mathbb{R}$ are the only subsets that are both open and closed); $\mathcal{B}(X)$ can also coincide with $\wp(X)$ (when the topology on X is the discrete topology (the topology in which all subsets are open) and, obviously, only is this case).

Let us now give two examples of homomorphisms of Boolean algebras.

Example 2.57 Consider the Boolean algebra $\mathcal{F}/\sim$ of equivalence classes of logically equivalent formulas of the propositional calculus built from a set of variables P (see Examples 2.26). Choose an assignment δ of truth values on P and as usual let $\bar{\delta}$ denote the extension of δ to the set $\mathcal{F}$ of all formulas. We can then define a map h_δ from $\mathcal{F}/\sim$ into $\{\mathbf{0}, \mathbf{1}\}$ by setting, for every formula F,

$$h_\delta(\mathrm{cl}(F)) = \bar{\delta}(F).$$

This definition is legitimate since $\overline{\delta}(F)$ assumes the same value on all formulas in a given equivalence class.

This map, h_δ, is a homomorphism of Boolean algebras from $\mathcal{F}/\!\!\sim$ into $\{0, 1\}$. By virtue of Theorem 2.45 and the definition of the operations for the Boolean algebra $\mathcal{F}/\!\!\sim$, it suffices to establish that for all formulas F and G in $\mathcal{F}$, we have

$$h_\delta(\mathsf{cl}(F \wedge G)) = h_\delta(\mathsf{cl}(F))h_\delta(\mathsf{cl}(G)) \quad \text{and}$$
$$h_\delta(\mathsf{cl}(\neg F)) = 1 - h_\delta(\mathsf{cl}(F)).$$

Now these relations are equivalent to

$$\overline{\delta}(F \wedge G) = \overline{\delta}(F)\,\overline{\delta}(G) \quad \text{and}$$
$$\overline{\delta}(\neg F) = 1 - \overline{\delta}(F),$$

which are true by definition of $\overline{\delta}$.

We will prove (Exercise 13) that all Boolean algebra homomorphisms from $\mathcal{F}/\sim$ into $\{0, 1\}$ are obtained in this manner.

Example 2.58 Let $\mathcal{A} = \langle A, +, \times, \mathbf{0}, \mathbf{1} \rangle$ be a Boolean algebra and let a be an atom in this algebra (we are assuming that one exists). Let us define a map h_a from A into $\{0, 1\}$ by

$$h_a(x) = \begin{cases} \mathbf{1} & \text{if } x \in A \text{ and } a \leq x \\ \mathbf{0} & \text{if } x \in A \text{ and } a \leq \mathbf{1} + x \end{cases}$$

(these two cases are mutually exclusive since a is different from $\mathbf{0}$, and there are no other cases since a is an atom: see Theorem 2.40).

We claim that h_a is a homomorphism of Boolean algebras from $\mathcal{A}$ into $\{0, 1\}$.

Proof To prove this, let us use Theorem 2.45: let x and y be two elements of A. We have $h_a(x \frown y) = \mathbf{1}$ if and only if $a \leq x \frown y$, but this is equivalent, by the definition of greatest lower bound, to $a \leq x$ and $a \leq y$, hence to $h_a(x) = \mathbf{1}$ and $h_a(y) = \mathbf{1}$, which is necessary and sufficient to have $h_a(x) \frown h_a(y) = \mathbf{1}$. It follows that

$$h_a(x \frown y) = h_a(x) \frown h_a(y).$$

Also, $h_a(x^{\complement}) = \mathbf{1}$ if and only if $a \leq x^{\complement}$, i.e. $a \leq \mathbf{1} + x$, which is equivalent to $h_a(x) = \mathbf{0}$. Since h_a only assumes the values $\mathbf{0}$ or $\mathbf{1}$, this means that

$$h_a(x^{\complement}) = (h_a(x))^{\complement}. \qquad \blacksquare$$

2.5 Ideals and filters

2.5.1 Properties of ideals

As mentioned in the reminders, 'ideal' means 'proper ideal'.

Theorem 2.59   *Let $\mathcal{A} = \langle A, \leq, \mathbf{0}, \mathbf{1} \rangle$ be a Boolean algebra and I a subset of A. For I to be an ideal, it is necessary and sufficient that the following three conditions be satisfied:*

  (i) $\mathbf{0} \in I$ *and* $\mathbf{1} \notin I$;

 (ii) *for all elements x and y of I, $x \smile y \in I$;*

(iii) *for all $x \in I$ and for all $y \in A$, if $y \leq x$, then $y \in I$.*

Proof   Suppose first that I is an ideal. Then in particular, it is a subgroup of the group $\langle A, +, \mathbf{0} \rangle$, so $\mathbf{0} \in I$. If $\mathbf{1}$ were in I, then I would be the entire ring and we have excluded this case; so (i) is verified. If x and y are in I, then so is their product xy and, consequently, the sum $x + y + xy = x \smile y$, which proves (ii). Finally, let us verify (iii): if $x \in I$ and $y \in A$, then $xy \in I$ and if $y \leq x$ as well, then $xy = y$ and, consequently, $y \in I$.

Conversely, suppose that (i), (ii), and (iii) are satisfied. We must show that I is an ideal in $\mathcal{A}$. If $x \in I$ and $y \in I$, then $x \smile y \in I$ by (ii), but since $x + y \leq x \smile y$ (this is trivial to check) and since $x + y = x - y$ (we are in a Boolean algebra), we conclude using (iii) that $x - y \in I$. Since $\mathbf{0} \in I$, we have all we need for $\langle I, +, \mathbf{0} \rangle$ to be a subgroup of $\langle A, +, \mathbf{0} \rangle$. Moreover, if $x \in I$ and $y \in A$, then since $xy \leq x$, we may conclude from (iii) that $xy \in I$. The set I is therefore an ideal in $\mathcal{A}$ ($I \neq A$ since $\mathbf{1} \notin I$). ∎

Corollary 2.60   *If I is an ideal in a Boolean algebra $\langle A, +, \times, \mathbf{0}, \mathbf{1} \rangle$, there is no element x in A that can satisfy both $x \in I$ and $\mathbf{1} + x \in I$.*

Proof   If the ideal I contains both x and $\mathbf{1} + x$, it would also contain the element $x \smile (\mathbf{1} + x) = \mathbf{1}$ (invoke property (ii) of the theorem). But this is not possible since $\mathbf{1} \notin I$ (property (i)). ∎

Corollary 2.61   *Let $\mathcal{A} = \langle A, +, \times, \mathbf{0}, \mathbf{1} \rangle$ be a Boolean ring and let I be an ideal in $\mathcal{A}$. For any integer $k \geq 1$ and any elements $x_1, x_2, \ldots, x_k$ in I, the least upper bound $x_1 \smile x_2 \smile \cdots \smile x_k$ belongs to I.*

Proof   This is a generalization of property (ii) of the theorem; its proof is immediate by induction on the integer k (the case $k = 1$ needs no argument). ∎

Example 2.62

(1) If E is an infinite set, the set $\wp_f(E)$ of finite subsets of E is an ideal in the Boolean algebra $\wp(E)$. It is easy to verify conditions (i), (ii), and (iii) of the theorem: $\emptyset$ is a finite subset of E while E itself is not one, the union of two finite subsets of E is a finite subset of E, and any subset of a finite subset of E is a finite subset of E.

(2) Let $\mathcal{A} = \langle A, +, \times, \mathbf{0}, \mathbf{1} \rangle$ be a Boolean algebra and let a be an element of A that is not equal to $\mathbf{1}$. The set $I_a = \{x \in A : x \leq a\}$ is an ideal in $\mathcal{A}$. Here too, the verification of properties (i), (ii), and (iii) is immediate: we have $\mathbf{0} \leq a$

and, since a is distinct from $\mathbf{1}$, we do not have $\mathbf{1} \leq a$; if $x \leq a$ and $y \leq a$, then $x \smallsmile y \leq a$; finally, if $x \leq a$ and $y \leq x$, then $y \leq a$. I_a is called **the principal ideal generated by** a. This agrees with the usual definition of a principal ideal in an arbitrary commutative ring since, in a Boolean ring, the set of elements below a given element is also the set of its multiples.

(3) In any Boolean algebra, $\{\mathbf{0}\}$ is obviously an ideal.

Lemma 2.63 *For any Boolean ring* $\mathcal{A} = \langle A, +, \times, \mathbf{0}, \mathbf{1} \rangle$ *and for any ideal* I *in* $\mathcal{A}$, *the quotient ring* $\mathcal{A}/I$ *is a Boolean ring.*

Proof For each element x in A, let $\bar{x}$ denote the equivalence class of x modulo I. We already know that $\mathcal{A}/I$ is a ring with unit. So it is sufficient to show that every element is an idempotent for multiplication. But this is an immediate consequence of the definition of multiplication in $\mathcal{A}/I$ and of the fact that $\mathcal{A}$ is a Boolean algebra: if $x \in A$, $\bar{x}^2 = \overline{x^2} = \bar{x}$.

Recall that in an arbitrary commutative ring with unit, the ideals are precisely the kernels of homomorphisms of rings with unit defined on the ring. (See Remark 2.43.) The theorem that follows merely takes up this result for Boolean rings, with this added precision: it shows that the ideals in a Boolean ring are precisely the kernels of homomorphisms of Boolean algebras defined on the ring. ∎

Theorem 2.64 *Let* $\mathcal{A} = \langle A, +, \times, \mathbf{0}, \mathbf{1} \rangle$ *be a Boolean ring and let* I *be a subset of* A. *The following properties are equivalent:*

(1) I *is an ideal in* $\mathcal{A}$;

(2) *there exists a homomorphism of Boolean algebras, h, defined on A whose kernel is I (in other words),*

$$I = h^{-1}[\{\mathbf{0}\}] = \{x \in A : h(x) = \mathbf{0}\};$$

(3) *there exists a homomorphism of commutative rings with unit defined on A whose kernel is I.*

Proof The equivalence of (1) and (3) is the result in the preceding reminder; (2) $\Rightarrow$ (3) is obvious. We will nonetheless prove (3) $\Rightarrow$ (1) and then (1) $\Rightarrow$ (2), which is, as observed above, more precise than (1) $\Rightarrow$ (3).

- (3) $\Rightarrow$ (1): suppose there exists a homomorphism h from $\mathcal{A}$ into a ring with unit $\mathcal{B} = \langle B, +, \times, \mathbf{0}, \mathbf{1} \rangle$ such that $I = h^{-1}[\{\mathbf{0}\}] = \{x \in A : h(x) = \mathbf{0}\}$. Let us verify that conditions (i), (ii), and (iii) from Theorem 2.59 are satisfied.

We have $h(\mathbf{0}) = \mathbf{0}$ and $h(\mathbf{1}) = \mathbf{1}$, hence $\mathbf{0} \in I$ and $\mathbf{1} \notin I$. If $x \in I$ and $y \in I$, then $h(x) = \mathbf{0}$ and $h(y) = \mathbf{0}$, so

$$h(x \smallsmile y) = h(x + y + xy) = h(x) + h(y) + h(x)h(y) = \mathbf{0}$$

and so $x \smile y \in I$. Finally, if $x \in I$, $y \in A$, and $y \leq x$, then $h(x) = \mathbf{0}$ and $xy = y$, hence $h(y) = h(x)h(y) = \mathbf{0}$, i.e. $y \in I$.

Thus, I is an ideal of $\mathcal{A}$.

(Of course, it would have made no sense to use an order relation or the $\smile$ and $\frown$ operations in $\mathcal{B}$).

- (1) $\Rightarrow$ (2): suppose that I is an ideal in $\mathcal{A}$ and consider the map h from A into A/I which, with each element x, associates its equivalence class, $\overline{x}$, modulo I (h is what is generally known as the canonical surjection of A onto A/I). h is a homomorphism of Boolean algebras (the **canonical homomorphism** of $\mathcal{A}$ onto A/I): to see this, we invoke Theorem 2.45; if $x \in A$ and $y \in A$, then

$$h(x \frown y) = h(xy) = \overline{xy} = \overline{x} \times \overline{y} = h(x) \times h(y) = h(x) \frown h(y) \quad \text{and}$$
$$h(x^{\complement}) = h(\mathbf{1} + x) = \overline{\mathbf{1} + x} = \overline{\mathbf{1}} + \overline{x} = \overline{\mathbf{1}} + h(x) = (h(x))^{\complement}.$$

Also, it is clear that $I = \overline{\mathbf{0}} = \{x \in A : h(x) = \overline{\mathbf{0}}\}$: I is the kernel of h.   ∎

2.5.2 Maximal ideals

Here is a collection of ways to characterize maximal ideals in a Boolean algebra:

Theorem 2.65 *For every Boolean ring $\mathcal{A} = \langle A, +, \times, \mathbf{0}, \mathbf{1} \rangle$, for every ideal I in $\mathcal{A}$, and for every integer $k \geq 2$, the following properties are equivalent:*

(1) *I is a maximal ideal;*

(2) *A/I is isomorphic to the Boolean algebra $\{\mathbf{0}, \mathbf{1}\}$;*

(3) *I is the kernel of a homomorphism of $\mathcal{A}$ into $\{\mathbf{0}, \mathbf{1}\}$;*

(4) *for every element x in A, $x \in I$ or $\mathbf{1} + x \in I$;*

(5) *for all elements x and y of A, if $xy \in I$, then $x \in I$ or $y \in I$;*

(6) *for all elements $x_1, x_2, \ldots, x_k$ in A, if $x_1 x_2 \cdots x_k \in I$, then $x_1 \in I$ or $x_2 \in I$ or $\cdots$ or $x_k \in I$.*

Proof

- (1) $\Rightarrow$ (2): In Section 1, we recalled that if the ideal I is maximal, then the quotient ring $\mathcal{A}/I$ is a field. But we also observed (Remark 2.28) that the only Boolean ring that is a field is $\{\mathbf{0}, \mathbf{1}\}$. So the result follows using Lemma 2.63.

- (2) $\Rightarrow$ (3): It suffices to note that I is always the kernel of the canonical homomorphism h from $\mathcal{A}$ into A/I. If there is an isomorphism ϕ from $\mathcal{A}/I$ onto $\{\mathbf{0}, \mathbf{1}\}$, then I will obviously be the kernel of the homomorphism $\phi \circ h$ from $\mathcal{A}$ into $\{\mathbf{0}, \mathbf{1}\}$.

- (3) $\Rightarrow$ (4): Consider a homomorphism h from $\mathcal{A}$ into $\{\mathbf{0}, \mathbf{1}\}$ whose kernel is I and let x be an arbitrary element of A. We have $h(x) = \mathbf{0}$ or $h(x) = \mathbf{1}$. In the first case, $x \in I$; in the second case we have $\mathbf{1} + h(x) = \mathbf{0}$, so $h(\mathbf{1} + x) = \mathbf{0}$ and $\mathbf{1} + x \in I$.

- $(4) \Rightarrow (5)$: Let x and y be elements of A such that $x \notin I$ and $y \notin I$. If (4) holds, then $\mathbf{1} + x \in I$ and $\mathbf{1} + y \in I$, so $(\mathbf{1} + x) \smile (\mathbf{1} + y) \in I$ (property (ii) of Theorem 2.59). But

$$(\mathbf{1} + x) \smile (\mathbf{1} + y) = \mathbf{1} + (x \frown y) = \mathbf{1} + xy,$$

  so, by Corollary 2.60 $xy \notin I$, and so (5) is proved.

- $(5) \Rightarrow (1)$: Suppose that I is not maximal. Let J be an ideal of $\mathcal{A}$ that strictly includes I and a be an element of J that does not belong to I. Using Corollary 2.60, $\mathbf{1} + a \notin J$, and so $\mathbf{1} + a \notin I$ since $I \subseteq J$. The ideal I contains neither a nor $\mathbf{1} + a$, but it does obviously contain the product $a(\mathbf{1} + a) = \mathbf{0}$. We conclude that (5) is not satisfied.

- $(5) \Rightarrow (6)$: We assume that (5) is satisfied and we argue by induction on the integer k. For $k = 2$, (6) coincides with (5). Assuming (6) is verified for k, we will prove that it is satisfied for $k + 1$. Let $x_1, x_2, \ldots, x_k, x_{k+1}$ be elements of A such that $x_1 x_2 \cdots x_k x_{k+1} \in I$. By (5) we must have either $x_1 x_2 \cdots x_k \in I$ or $x_{k+1} \in I$. In the first instance, by the induction hypothesis, we have $x_1 \in I$ or $x_2 \in I$ or $\ldots$ or $x_k \in I$. We then conclude that we must have $x_i \in I$ for at least one index i such that $1 \leq i \leq k + 1$; this proves (6) for $k + 1$.

- $(6) \Rightarrow (5)$: Let x and y be two elements of A such that $xy \in I$. Set $x_1 = x$ and $x_2 = x_3 = \ldots = x_k = y$. So we have $x_1 x_2 \cdots x_k = xy \in I$. So if (6) is true, we must have $x_i \in I$ for at least one index i between 1 and k; so either $x \in I$ or $y \in I$ and (5) is verified.   ∎

Remark 2.66 *In an arbitrary commutative ring, an ideal with property (5) from the preceding theorem is called a **prime ideal**. What we have just proved is that in a Boolean ring, the prime ideals are precisely the same as the maximal ideals. But there are rings for which this fails to be true. What is always true is that an ideal is prime if and only if the associated quotient ring is an integral domain (this is easy to prove); we conclude from this as well that a maximal ideal must necessarily be prime (it suffices to consider the corresponding quotient ring). Thus it is the converse of this that can fail (for example, in the ring $\mathbb{R}[X, Y]$ of polynomials in two variables with real coefficients: the ideal generated by the polynomial X, i.e. the set $\{XP : P \in \mathbb{R}[X, Y]\}$ is prime but is not maximal since it is strictly included in the ideal generated by the polynomials X and Y, i.e. the set $\{XP + YQ : P \in \mathbb{R}[X, Y], Q \in \mathbb{R}[X, Y]\}$).*

Remark 2.67 *In particular, we should take note of the equivalence between properties (1) and (3). Observe that if two homomorphisms h and g from a Boolean algebra $\mathcal{A} = \langle A, +, \times, \mathbf{0}, \mathbf{1} \rangle$ into $\{\mathbf{0}, \mathbf{1}\}$ have the same kernel, I, then they are equal: this is because for any element x in A, either $x \notin I$ and $g(x) = h(x) = \mathbf{0}$ or else $x \notin I$ and $g(x) = h(x) = \mathbf{1}$. We may conclude from this that there is a bijection between the set of maximal ideals in a Boolean algebra and the set of homomorphisms of Boolean algebras from this algebra into $\{\mathbf{0}, \mathbf{1}\}$.*

2.5.3 Filters

We will now introduce the notion that is dual to that of an ideal on a Boolean algebra: we will define filters.

Definition 2.68 *A filter in a Boolean algebra* $\mathcal{A} = \langle A, +, \times, \mathbf{0}, \mathbf{1} \rangle$ *is a subset F of A such that the set*

$$\left\{ x \in A : x^{\complement} \in F \right\}$$

is an ideal in $\mathcal{A}$.

Let F be a filter in a Boolean algebra $\mathcal{A} = \langle A, +, \times, \mathbf{0}, \mathbf{1} \rangle$. Let I be the ideal $\{x \in A : x^{\complement} \in F\}$. I is realized as the inverse image of F under the operation of complementation: $x \longmapsto x^{\complement}$. But, as this operation is an involution (see part (9) of Theorem 2.29), I is also the direct image of F under this operation: $I = \{x \in A : \exists y (y \in F \text{ and } x = y^{\complement})\}$. In other words, I is the set of complements of elements of F and F is the set of complements of elements of I. The ideal I is called the **dual ideal** of the filter F. It is easy to see that, given an arbitrary ideal J in $\mathcal{A}$, the set $G = \{x \in A : x^{\complement} \in J\}$ is a filter whose dual ideal is precisely J. To everyone's surprise, we will call G the **dual filter** of the ideal J. There is thus a bijection between the set of ideals and the set of filters in a Boolean algebra. For filters, we have the dual of Theorem 2.59.

Theorem 2.69 *Let* $\mathcal{A} = \langle A, \leq, \mathbf{0}, \mathbf{1} \rangle$ *be a Boolean algebra and F a subset of A. For F to be a filter, it is necessary and sufficient that the following three conditions be satisfied:*

(f) $\mathbf{0} \notin F$ *and* $\mathbf{1} \in F$;

(ff) *for all elements x and y of I,* $x \frown y \in F$;

(fff) *for all* $x \in F$ *and for all* $y \in A$, *if* $y \geq x$, *then* $y \in F$.

Proof Set $I = \{x \in A : x^{\complement} \in F\}$. If F is a filter, then I is its dual ideal and conditions (i), (ii), and (iii) from Theorem 2.59 are satisfied. So we have $\mathbf{0} \in I$, hence $\mathbf{0}^{\complement} = \mathbf{1} \in F$, and $\mathbf{1} \notin I$, hence $\mathbf{1}^{\complement} = \mathbf{0} \notin F$, which proves (f). If $x \in F$ and $y \in F$, then $x^{\complement} \in I$ and $y^{\complement} \in I$, so $x^{\complement} \smile y^{\complement} \in I$ (by (ii)), and, as $x^{\complement} \smile y^{\complement} = (x \frown y)^{\complement}$, we conclude that $x \frown y \in F$ and that (ff) is satisfied. Finally, if $x \in F$, $y \in A$, and $y \geq x$, then $x^{\complement} \in I$ and $y^{\complement} \leq x^{\complement}$, so, (by (iii)) $y^{\complement} \in I$ and $y \in F$, hence (fff).

Conversely, in a strictly analogous fashion, (i) follows from (f), (ii) from (ff) and (iii) from (fff). ∎

Corollary 2.70 *Let* $\mathcal{A} = \langle A, \leq, \mathbf{0}, \mathbf{1} \rangle$ *be a Boolean algebra and F a subset of A. For any integer* $k \geq 1$ *and any elements* $x_1, x_2, \ldots, x_k$ *in F, the greatest lower bound* $x_1 \frown x_2 \frown \cdots \frown x_k$ *belongs to F.*

Proof Analogue of Corollary 2.61. ∎

2.5.4 Ultrafilters

Definition 2.71 *In a Boolean algebra, an **ultrafilter** is a **maximal** filter, i.e. a filter which is not strictly included in any other filter.*

It is clear that, in view of the duality explained above, ultrafilters correspond to maximal ideals. In other words, the dual filter of a maximal ideal is an ultrafilter and the dual ideal of an ultrafilter is a maximal ideal.

There is an analogue for filters of Theorem 2.65:

Theorem 2.72 *For every Boolean ring $\mathcal{A} = \langle A, +, \times, \mathbf{0}, \mathbf{1} \rangle$, for every ideal F in $\mathcal{A}$, and for every integer $k \geq 2$, the following properties are equivalent:*

(1′) F is an ultrafilter;

(3′) there is a homomorphism h from $\mathcal{A}$ into $\{\mathbf{0}, \mathbf{1}\}$ such that

$$F = \{x \in A : h(x) = \mathbf{1}\};$$

(4′) for every element x in A, $x \in F$ or $\mathbf{1} + x \in F$;

(5′) for all elements x and y of A, if $x \smile y \in F$, then $x \in F$ or $y \in F$;

(6′) for all elements $x_1, x_2, \ldots, x_k$ in A, if $x_1 \smile x_2 \smile \cdots \smile x_k \in F$, then $x_1 \in F$ or $x_2 \in F$ or ... or $x_k \in F$.

Proof Given the algebra $\mathcal{A}$, the filter F and the integer k, let I denote the ideal dual to F. It is elementary to verify that properties (1′), (3′), (4′), (5′), and (6′) for the filter F are respectively equivalent to properties (1), (3), (4), (5), and (6) of Theorem 2.65 for the ideal I (we use the correspondence between I and F, de Morgan's laws, etc.) ∎

Remark 2.73 *The 'or' in property (4) of Theorem 2.65, as well as the one in property (4′) above, is in fact an 'exclusive or' (see Corollary 2.60). This means that if F is an ultrafilter in a Boolean algebra $\langle A, +, \times, \mathbf{0}, \mathbf{1} \rangle$ and if I is the maximal ideal dual to F, then I and F constitute a partition of A. Thus each of the sets I and F is simultaneously:*

- the complement of the other (viewed as subsets of A),
- the set of complements of the elements of the other (in the sense of complementation in the Boolean algebra under consideration).

The second of these properties holds whenever we have an ideal and a filter that are dual to one another, but the first holds only when the ideal and filter in question are maximal.

Remark 2.74 *Returning to Remark 2.67, we can now add to it and insist on the fact that for a Boolean algebra, $\mathcal{A}$, there are canonical one-to-one correspondences among (i) the maximal ideals in $\mathcal{A}$, (ii) the ultrafilters in $\mathcal{A}$ and (iii) homomorphisms of Boolean algebras from $\mathcal{A}$ into $\{\mathbf{0}, \mathbf{1}\}$.*

Examples: To have examples of filters, it obviously suffices to refer to the previously described examples of ideals (in Subsection 2.5.1) and to transform them by duality. The reader can, without any difficulty, provide the necessary verifications:

(1) If E is an infinite set, the set of all cofinite subsets of E (see Example 2.55) is a filter in the Boolean algebra $\langle \wp(E), \subseteq, \emptyset, E \rangle$. This filter is called the **Frechet filter** on E. It is not an ultrafilter since there are subsets of E which are infinite and whose complements are also infinite, so condition $(4')$ of Theorem 2.72 is not satisfied.

(2) If a is a non-zero element in a Boolean algebra $\langle A, \leq, \mathbf{0}, \mathbf{1} \rangle$, the set $F_a = \{x \in A : x \geq a\}$ is a filter called the **principal filter generated by** a. It is the dual filter of the ideal generated by $\mathbf{1} + a$.

(3) The set $\{\mathbf{1}\}$ is the dual filter of the ideal $\{\mathbf{0}\}$.

Theorem 2.75 *Let $\langle A, \leq, \mathbf{0}, \mathbf{1} \rangle$ be a Boolean algebra and let a be a non-zero element of A. For the principal filter generated by a to be an ultrafilter, it is necessary and sufficient that a be an atom.*

Proof By virtue of Theorem 2.40 and the definition of the filter F_a, a is an atom if and only if for every element x of A, $x \in F_a$ or $\mathbf{1} + x \in F_a$; but for this to happen, it is necessary and sufficient that F_a be an ultrafilter $((4') \Rightarrow (1')$ from Theorem 2.72). ∎

When the principal filter F_a generated by a non-zero element a of A is an ultrafilter (thus, when a is an atom), we say that this is a **trivial** ultrafilter. The homomorphism h_a with values in $\{\mathbf{0}, \mathbf{1}\}$ that is associated with it is also called a **trivial** homomorphism. As it is defined by : $h_a(x) = \mathbf{1}$ if $x \in F_a$ and $h_a(x) = \mathbf{0}$ if $x \notin F_a$, and as this is obviously equivalent to: $h_a(x) = \mathbf{1}$ if $a \leq x$ and $h_a(x) = \mathbf{0}$ if $a \leq \mathbf{1} + x$, we see that what is involved is precisely the homomorphism studied in Example 2.58.

Lemma 2.76 *Let $\mathcal{A}$ be a Boolean algebra and let $\mathcal{U}$ be an ultrafilter on $\mathcal{A}$. For $\mathcal{U}$ to be trivial, it is necessary and sufficient that it contain at least one atom.*

Proof If $\mathcal{U}$ is trivial, it is generated by an atom, a, and since $a \leq a, a \in \mathcal{U}$.

Conversely, if $\mathcal{U}$ contains an atom, b, it also contains all elements greater than or equal to b (condition (fff) of Theorem 2.69). It follows that the principal filter F_b generated by b is included in $\mathcal{U}$. But since b is an atom, F_b is maximal and cannot be strictly included in the filter $\mathcal{U}$. Hence $\mathcal{U} = F_b$ and $\mathcal{U}$ is a trivial ultrafilter. ∎

Theorem 2.77 *Let E be an infinite set and let $\mathcal{U}$ be an ultrafilter in the Boolean algebra $\wp(E)$. For $\mathcal{U}$ to be non-trivial, it is necessary and sufficient that it includes the Frechet filter on E.*

Proof The atoms of $\wp(E)$ are the singletons (subsets consisting of a single element); so they are finite sets. If $\mathcal{U}$ includes the Frechet filter, every cofinite subset of E belongs to $\mathcal{U}$, hence no finite subset of E can belong to $\mathcal{U}$ ($\mathcal{U}$ cannot

96 BOOLEAN ALGEBRAS

simultaneously contain a subset of E and its complement: see Remark 2.73). In particular, no atom can belong to $\mathcal{U}$. It follows, by the preceding lemma, that $\mathcal{U}$ is non-trivial.

If $\mathcal{U}$ does not include the Frechet filter, we can choose a cofinite subset X of E that does not belong to $\mathcal{U}$, and hence whose complement $E - X$ does belong to $\mathcal{U}$. As E is the identity element for the Boolean algebra $\wp(E)$, $E \in \mathcal{U}$; hence $X \neq E$. The complement of X in E is thus a non-empty finite subset of E: for example, $E - X = \{\alpha_1, \alpha_2, \ldots, \alpha_n\}$ $(n \geq 1)$. So we have $\{\alpha_1, \alpha_2, \ldots, \alpha_n\} \in \mathcal{U}$, which is to also say:

$$\{\alpha_1\} \cup \{\alpha_2\} \cup \cdots \cup \{\alpha_n\} = \{\alpha_1\} \vee \{\alpha_2\} \vee \cdots \vee \{\alpha_n\} \in \mathcal{U}.$$

If $n = 1$, $\{\alpha_1\} \in \mathcal{U}$. If $n \geq 2$, then by property (6') of Theorem 2.72, we have $\{\alpha_i\} \in \mathcal{U}$ for at least one index i between 1 and n. We see that in every case, $\mathcal{U}$ contains a singleton, i.e. an atom. The preceding lemma then shows that $\mathcal{U}$ is trivial. ∎

Remark: In Exercise 16, we will prove a property which includes this theorem as a special case.

2.5.5 Filterbases

Definition 2.78 *In a Boolean algebra $\langle A, \leq, 0, 1 \rangle$, a **basis for a filter** (**filterbase**) is a subset of A that has the following property, known as the **finite intersection property**: every non-empty finite subset of B has a non-zero greatest lower bound.*

In other words, $B \subseteq A$ is a filterbase if and only if: for any integer $k \geq 1$ and any elements $x_1, x_2, \ldots, x_k$ of B, $x_1 \cap x_2 \cap \cdots \cap x_k \neq 0$.

Lemma 2.79 *Let $\langle A, \leq, 0, 1 \rangle$ be a Boolean algebra and let X be a subset of A. For the existence of a filter on $\mathcal{A}$ that includes X, it is necessary and sufficient that X be a filterbase.*

Proof If X is included in a filter F, and if $x_1, x_2, \ldots, x_k$ are elements of X, then their greatest lower bound $x_1 \cap x_2 \cap \cdots \cap x_k$ belongs to F (Corollary 2.70), and as $0 \notin F$, this greatest lower bound is non-zero; thus X is a filterbase.

Now suppose that X is a filterbase.

- If $X = \emptyset$, $\{1\}$ is a filter on $\mathcal{A}$ that includes X.
- If X is not empty, we set

$$F_X = \{x \in A : (\exists k \in \mathbb{N}^*)(\exists x_1 \in X)(\exists x_2 \in X)\ldots(\exists x_k \in X)$$
$$(x \geq x_1 \cap x_2 \cap \cdots \cap x_k)\}.$$

So F_X consists of greatest lower bounds of non-empty finite subsets of X together with all elements greater than or equal to one of these greatest lower bounds. In particular, each element of X belongs to F_X, hence F_X includes X. It is easy to prove that F_X is a filter. We will restrict ourselves to the following remarks:

(f) $\mathbf{0} \notin F_X$ (otherwise the finite intersection property would not be true for X) and $\mathbf{1} \in F_X$ (because X is non-empty: at least one element of X is less than or equal to $\mathbf{1}$).

(ff) If $x \geq x_1 \cap x_2 \cap \cdots \cap x_h$ and $y \geq y_1 \cap y_2 \cap \cdots \cap y_k$, then we have

$$x \cap y \geq x_1 \cap x_2 \cap \cdots \cap x_k \cap y_1 \cap y_2 \cap \cdots \cap y_k.$$

(fff) If $x \geq x_1 \cap x_2 \cap \cdots \cap x_k$ and $y \geq x$, then $y \geq x_1 \cap x_2 \cap \cdots \cap x_k$. So we have found a filter that includes X. ∎

In the particular case of Boolean algebras, we can state Krull's theorem (see the beginning of this chapter) in terms of filters. It is then known as the **ultrafilter theorem**:

Theorem 2.80 *In a Boolean algebra, every filter is included in at least one ultrafilter.*

Proof Given a filter F, the ideal dual to F is included in at least one maximal ideal; its dual filter is then an ultrafilter that includes F.

Of course, for Boolean algebras, the formulation in terms of filters and the formulation in terms of ideals are equivalent.

The ultrafilter theorem allows us to give a slightly different restatement of Lemma 2.79: ∎

Lemma 2.81 *Let $\langle A, \leq, \mathbf{0}, \mathbf{1} \rangle$ be a Boolean algebra and let X be a subset of A. For the existence of an ultrafilter on $\mathcal{A}$ that includes X, it is necessary and sufficient that X be a filterbase.*

Proof The properties 'there exists an ultrafilter on $\mathcal{A}$ that includes X' and 'there exists a filter on $\mathcal{A}$ that includes X' are equivalent: the first clearly implies the second; the reverse implication follows from the ultrafilter theorem. The conclusion then follows from Lemma 2.79. ∎

2.6 Stone's theorem

The first example of a Boolean algebra that comes to mind is, without a doubt, that of the algebra of subsets of a given set. Is every Boolean algebra equal to (we really mean 'isomorphic to') the Boolean algebra of subsets of some set? We already have sufficient knowledge to answer 'no' to this question: we have encountered Boolean algebras without atoms (Example 2.37) and we know that the algebra of subsets of a set always contain atoms: the singletons; and any isomorphism transforms atoms into atoms (see Exercise 3); a Boolean algebra that does not contain atoms cannot therefore be isomorphic to an algebra that does.

None the less, Stone's theorem, to which this section is devoted, shows that there is always a link that connects a Boolean algebra to an algebra of subsets of a set.

More precisely, every Boolean algebra is isomorphic to a subalgebra of a Boolean algebra of subsets of a set.

2.6.1 The Stone space of a Boolean algebra

Consider a Boolean algebra $\mathcal{A} = \langle A, +, \times, \mathbf{0}, \mathbf{1} \rangle$.

Definition 2.82 *The set of homomorphisms of a Boolean algebra $\mathcal{A}$ into $\{\mathbf{0}, \mathbf{1}\}$ is denoted by $S(\mathcal{A})$ and is called the **Stone space** of $\mathcal{A}$.*

We could just have well chosen the set of maximal ideals or the set of ultrafilters on $\mathcal{A}$ (by virtue of Remark 2.74).

The set $S(\mathcal{A})$ is a subset of $\{\mathbf{0}, \mathbf{1}\}^A$, the set of maps from A into $\{\mathbf{0}, \mathbf{1}\}$, which we have considered earlier as a topological space by taking the discrete topology on $\{\mathbf{0}, \mathbf{1}\}$ and giving this space the product topology (see Definition 2.19). So we can give $S(\mathcal{A})$ the topology induced by that of $\{\mathbf{0}, \mathbf{1}\}^A$. The open subsets of $S(\mathcal{A})$ are then the intersections with $S(\mathcal{A})$ of the open subsets of $\{\mathbf{0}, \mathbf{1}\}^A$.

Lemma 2.83 *The topological space $S(\mathcal{A})$ is zero-dimensional.*

Proof We saw (Lemma 2.21) that $\{\mathbf{0}, \mathbf{1}\}^A$ is zero-dimensional. So it suffices to apply Lemma 2.17. ∎

We have exhibited, just before Lemma 2.21, a basis $(\Omega_i)_{i \in I}$ for the space $\{\mathbf{0}, \mathbf{1}\}^A$ consisting of clopen sets. Each of the Ω_i is the set of all maps from A into $\{\mathbf{0}, \mathbf{1}\}$ that take specified values at a specified finite number of points. If for each i, we set $\Gamma_i = \Omega_i \cap S(\mathcal{A})$, as in Lemma 2.17, then the family $(\Gamma_i)_{i \in I}$ is a basis for the open sets in $S(\mathcal{A})$ consisting of clopen sets. Each Γ_i is the set of homomorphisms of Boolean algebras from $\mathcal{A}$ into $\{\mathbf{0}, \mathbf{1}\}$ which assume given values at a finite number of given points.

From now on, it is exclusively this basis for the open sets that we will consider for the space $S(\mathcal{A})$. When we speak of a basic open set for the Stone space of $\mathcal{A}$, we mean one of the clopen sets in the family $(\Gamma_i)_{i \in I}$.

Lemma 2.84 *For a subset Δ of $S(\mathcal{A})$ to be a basic open set, it is necessary and sufficient that there exist an element a in A such that*

$$\Delta = \{h \in S(\mathcal{A}) : h(a) = \mathbf{1}\}.$$

Moreover, when this condition is realized, such an element is unique.

Proof

- *sufficient.* Suppose that $\Delta = \{h \in S(\mathcal{A}) : h(a) = \mathbf{1}\}$; Δ is the set of homomorphisms from A into $\{\mathbf{0}, \mathbf{1}\}$ that assume the value $\mathbf{1}$ at the point a: so it is one of the basic open sets in $S(\mathcal{A})$.

- *necessary.* Suppose Δ is a basic open subset of $S(\mathcal{A})$.

  * If $\Delta = \emptyset$, then $\Delta = \{h \in S(\mathcal{A}) : h(\mathbf{0}) = \mathbf{1}\}$.

∗ If $\Delta \neq \emptyset$, then there is an integer $n \geq 1$, elements $a_1, a_2, \ldots, a_n$ in A and elements $\varepsilon_1, \varepsilon_2, \ldots, \varepsilon_n$ in $\{\mathbf{0}, \mathbf{1}\}$ such that

$$\Delta = \{h \in S(\mathcal{A}) : h(a_1) = \varepsilon_1 \text{ and } h(a_2) = \varepsilon_2 \text{ and } \ldots \text{ and } h(a_n) = \varepsilon_n\}.$$

For every $k \in \{1, 2, \ldots, n\}$, set

$$b_k = \begin{cases} a_k & \text{if } \varepsilon_k = \mathbf{1}; \\ \mathbf{1} + a_k & \text{if } \varepsilon_k = \mathbf{0}. \end{cases}$$

For every homomorphism $h \in S(\mathcal{A})$ and for every $k \in \{1, 2, \ldots, n\}$, we have

$$h(b_k) = \begin{cases} h(a_k) & \text{if } \varepsilon_k = \mathbf{1}; \\ \mathbf{1} + h(a_k) & \text{if } \varepsilon_k = \mathbf{0}. \end{cases}$$

It follows that for $h \in S(\mathcal{A})$, $h \in \Delta$ if and only if $h(b_k) = \mathbf{1}$ for all $k \in \{1, 2, \ldots, n\}$. But this latter condition is equivalent to

$$h(b_1) \frown h(b_2) \frown \cdots \frown h(b_n) = \mathbf{1},$$

or again, since a homomorphism is involved, to

$$h(b_1 \frown b_2 \frown \cdots \frown b_n) = \mathbf{1}.$$

So we see that by setting $a = b_1 \frown b_2 \frown \cdots \frown b_n$, we have

$$\Delta = \{h \in S(\mathcal{A}) : h(a) = \mathbf{1}\}.$$

Now let us prove uniqueness: if a and b are distinct elements of A, then $a + b \neq \mathbf{0}$; so we may consider the principal filter generated by $a + b$ and, in view of the ultrafilter theorem, an ultrafilter that includes this filter. To such an ultrafilter, there is an associated homomorphism ϕ from $\mathcal{A}$ into $\{\mathbf{0}, \mathbf{1}\}$ that satisfies $\phi(a + b) = \mathbf{1}$, or again, $\phi(a) + \phi(b) = \mathbf{1}$, which means that one and only one of the two elements $\phi(a)$ and $\phi(b)$ is equal to $\mathbf{1}$. This proves that

$$\{h \in S(\mathcal{A}) : h(a) = \mathbf{1}\} \neq \{h \in S(\mathcal{A}) : h(b) = \mathbf{1}\}$$

since ϕ belongs to one of these two sets and not to the other. ∎

Corollary 2.85 *The set of basic closed subsets of $S(\mathcal{A})$ coincides with the set of its basic open subsets.*

Proof Let Γ be a basic closed subset of $S(\mathcal{A})$. Then $\Delta = S(\mathcal{A}) - \Gamma$ is a basic open subset; hence (by the preceding lemma) there is an element $a \in A$ such that

$$\Delta = \{h \in S(\mathcal{A}) : h(a) = \mathbf{1}\}.$$

Hence,

$$\Gamma = \{h \in S(\mathcal{A}) : h(a) \neq \mathbf{1}\}$$
$$= \{h \in S(\mathcal{A}) : h(a) = \mathbf{0}\}$$
$$= \{h \in S(\mathcal{A}) : h(\mathbf{1} + a) = \mathbf{1}\}.$$

So we see, thanks again to the preceding lemma, that Γ is a basic open subset. In the same way, we can show that every basic open set is a basic closed set. ∎

Lemma 2.86 *The topological space $S(\mathcal{A})$ is compact.*

Proof First of all, since the topology on $\{\mathbf{0}, \mathbf{1}\}^A$ is Hausdorff, so is the topology of $S(\mathcal{A})$.

Next, we must show that from any family of closed subsets of $S(\mathcal{A})$ whose intersection is empty, we can extract a finite subfamily whose intersection is already empty. But we have seen (Lemma 2.13) that it suffices to do this for families of basic closed sets. Now, as we have just seen, the basic closed sets coincide with the basic open sets. So consider an infinite family $(\Sigma_j)_{j \in J}$ of basic open subsets of $S(\mathcal{A})$ such that $\bigcap_{j \in J} \Sigma_j = \emptyset$. By the preceding lemma, there exists, for each $j \in J$, a unique element x_j in A such that

$$\Sigma_j = \{h \in S(\mathcal{A}) : h(x_j) = \mathbf{1}\}.$$

Set $X = \{x_j : j \in J\}$. To say that the intersection of the family $(\Sigma_j)_{j \in J}$ is empty is to say that there is no homomorphism of Boolean algebras from $\mathcal{A}$ into $\{\mathbf{0}, \mathbf{1}\}$ that assumes the value $\mathbf{1}$ for all elements of X, or again, that there is no ultrafilter on $\mathcal{A}$ that contains X. This means (Lemma 2.81) that X is not a filterbase. So there exists a finite subset $\{x_{j_1}, x_{j_2}, \ldots, x_{j_k}\} \subseteq X$ whose greatest lower bound is zero. So no ultrafilter on $\mathcal{A}$ could simultaneously contain x_{j_1}, x_{j_2}, and $\ldots$ and x_{j_k}. In other words, no homomorphism from $\mathcal{A}$ into $\{\mathbf{0}, \mathbf{1}\}$ can simultaneously assume the value $\mathbf{1}$ at the points x_{j_1}, x_{j_2}, and $\ldots$ and x_{j_k}. This amounts to saying that

$$\Sigma_{j_1} \cap \Sigma_{j_2} \cap \cdots \cap \Sigma_{j_k} = \emptyset.$$

We thus have a finite subfamily of the family $(\Sigma_j)_{j \in J}$ whose intersection is empty. ∎

Remark 2.87 *We can give a different proof that $S(\mathcal{A})$ is compact by invoking the fact that $\{\mathbf{0}, \mathbf{1}\}^A$ is itself compact (Theorem 2.20). It would then suffice to show that $S(\mathcal{A})$ is closed in $\{\mathbf{0}, \mathbf{1}\}^A$ (since every closed subset of a compact space is compact):*

For $a \in A$ and $b \in A$, set

$$\Omega(a, b) = \{f \in \{\mathbf{0}, \mathbf{1}\}^A : f(ab) = f(a)f(b) \text{ and } f(\mathbf{1} + a) = \mathbf{1} + f(a)\}.$$

From Theorem 2.45, we have $S(\mathcal{A}) = \bigcap_{\substack{a \in A \\ b \in A}} \Omega(a, b)$. But for all elements a and b of A, we can write:

$\Omega(a, b) =$

$\quad \{f \in \{0, 1\}^A : f(a) = 0 \text{ and } f(b) = 0 \text{ and } f(ab) = 0 \text{ and } f(1 + a) = 1\}$

$\quad \cup \{f \in \{0, 1\}^A : f(a) = 0 \text{ and } f(b) = 1 \text{ and } f(ab) = 0 \text{ and } f(1 + a) = 1\}$

$\quad \cup \{f \in \{0, 1\}^A : f(a) = 1 \text{ and } f(b) = 0 \text{ and } f(ab) = 0 \text{ and } f(1 + a) = 0\}$

$\quad \cup \{f \in \{0, 1\}^A : f(a) = 1 \text{ and } f(b) = 1 \text{ and } f(ab) = 1 \text{ and } f(1 + a) = 0\}.$

All four of the sets on the right side of this equality are basic open subsets of $\{0, 1\}^A$, hence are clopen. So in particular, their union is closed. So the intersection of all sets of the form $\Omega(a, b)$ as a and b range over A is a closed subset of $\{0, 1\}^A$. And as we have seen, this intersection is $S(\mathcal{A})$.

For us, this second proof suffers by depending on a theorem that we did not prove (Tychonoff's theorem, which was invoked to claim that $\{0, 1\}^A$ is compact). The first proof that we gave depends on Krull's theorem, which we did prove in Section 2.1.

Corollary 2.88 *The Stone space of $\mathcal{A}$ is a Boolean topological space.*

Proof Indeed, the space is compact (Lemma 2.86) and is zero-dimensional (Lemma 2.83). ∎

Lemma 2.89 *The set of clopen subsets of $S(\mathcal{A})$ coincides with the set of its basic open sets.*

Proof We already know that all the basic open sets are clopen (Lemma 2.83).

Conversely, let Γ be an arbitrary clopen subset of $S(\mathcal{A})$. As Γ is open, it is a union of basic open sets: for example, $\Gamma = \bigcup_{i \in J} \Gamma_i$ for some subset $J \subseteq I$. But since Γ is a closed subset of the compact space $S(\mathcal{A})$, it is itself compact. So from the open covering $(\Gamma_i)_{i \in J}$ of Γ, we can extract a finite subcovering, for example: $\Gamma = \Gamma_{j_1} \cup \Gamma_{j_2} \cup \cdots \cup \Gamma_{j_m}$. We know (Lemma 2.84) that we can find elements $x_1, x_2, \ldots, x_m$ in A such that

$$\text{for every } k \in \{1, 2, \ldots, m\}, \ \Gamma_{j_k} = \{h \in S(\mathcal{A}) : h(x_k) = 1\}.$$

Set $x = x_1 \vee x_2 \vee \cdots \vee x_k$ and set $\Delta = \{h \in S(\mathcal{A}) : h(x) = 1\}$; we will show that $\Gamma = \Delta$. Every element of Γ is a homomorphism that assumes the value **1** on at least one of the points $x_1, x_2, \ldots, x_m$; so it also assumes the value **1** at x which is their least upper bound. Thus, $\Gamma \subseteq \Delta$. On the other hand, any homomorphism that is not in Γ, and so does not assume the value **1** at any of the points $x_1, x_2, \ldots, x_m$, must assume the value **0** at each of these points, and so too at the point x which is their least upper bound; so it cannot belong to Δ. This proves that $\Delta \subseteq \Gamma$. Finally, $\Gamma = \Delta$, and as Δ is a basic open set (by Lemma 2.84), Γ must be one as well. ∎

2.6.2 Stone's theorem

Theorem 2.90 (*Stone's theorem*) *Every Boolean algebra is isomorphic to the Boolean algebra of clopen subsets of its Stone space.*

Proof The Boolean algebra of clopen subsets of $S(\mathcal{A})$ is denoted by $\mathcal{B}(S(\mathcal{A}))$ (see Example 2.56).

Let H denote the map from A into $\wp(S(\mathcal{A}))$ which, with each element a in A, associates

$$H(a) = \{h \in S(\mathcal{A}) : h(a) = \mathbf{1}\}.$$

Let us show that H is an isomorphism of Boolean algebras from $\mathcal{A}$ onto $\mathcal{B}(S(\mathcal{A}))$.

According to Lemmas 2.84 and 2.89, the map H assumes its values in $\mathcal{B}(S(\mathcal{A}))$ and its image is the whole of $\mathcal{B}(S(\mathcal{A}))$. So H is a surjection from A onto $\mathcal{B}(S(\mathcal{A}))$.

By virtue of Theorem 2.48, to show that H is an isomorphism of Boolean algebras, it suffices to guarantee that for all elements x and y in A, $x \le y$ if and only if $H(x) \subseteq H(y)$.

So let x and y be two elements of A. If x is less than or equal to y, then for any homomorphism h that satisfies $h(x) = \mathbf{1}$, we must also have $h(y) = \mathbf{1}$, which means that $H(x)$ is a subset of $H(y)$. If x is not less than or equal to y, then $x(\mathbf{1} + y) \neq \mathbf{0}$ (Lemma 2.31). So we may consider the principal filter generated by $x(\mathbf{1} + y)$, an ultrafilter that includes it (by the ultrafilter theorem) and the homomorphism $h \in S(\mathcal{A})$ associated with this ultrafilter. We have $h(x(\mathbf{1}+y)) = \mathbf{1}$, hence $h(x) = \mathbf{1}$ and $h(\mathbf{1} + y) = \mathbf{1}$, i.e. $h(y) = \mathbf{0}$. We conclude that $h \in H(x)$ and $h \notin H(y)$, and so $H(x)$ is not included in $H(y)$. ∎

Stone's theorem allows us to give a very simple proof of Theorem 2.50.

Corollary 2.91 *Every finite Boolean algebra is isomorphic to a Boolean algebra of subsets of some set.*

Proof If the set A is finite, then the topology on $\{\mathbf{0}, \mathbf{1}\}^A$ is the discrete topology. So this is also the case for the topology induced on the subset $S(\mathcal{A})$. All subsets of $S(\mathcal{A})$ are then both open and closed. So the Boolean algebra $\mathcal{B}(S(\mathcal{A}))$ coincides with $\wp(S(\mathcal{A}))$ and $\mathcal{A}$ is isomorphic to $\wp(S(\mathcal{A}))$. ∎

In the case of an arbitrary Boolean algebra, what Stone's theorem shows is that it is isomorphic to a Boolean subalgebra of the algebra of subsets of some set (Example 2.56).

2.6.3 Boolean spaces are Stone spaces

With each Boolean algebra, we have associated a Boolean topological space: its Stone space $S(\mathcal{A})$, and we have seen that $\mathcal{A}$ is isomorphic to the Boolean algebra of clopen subsets of this Boolean space. It is therefore natural to study the case in which $\mathcal{A}$ is given as the Boolean algebra of clopen subsets of some Boolean topological space X. The problem that then arises is to compare the space X to this other Boolean space that is the Stone space of $\mathcal{A}$, in other words, to compare X and $S(\mathcal{B}(X))$. The result of this comparison will reveal that these two objects very much resemble each other.

Theorem 2.92 *Every Boolean topological space X is homeomorphic to the Stone space $S(\mathcal{B}(X))$ of the Boolean algebra of clopen subsets of X.*

Proof Let X be a Boolean space. According to Lemma 2.16, we can take the Boolean algebra $\mathcal{B}(X)$ of clopen subsets of X as a basis for the open sets in the topology on X.

For each $x \in X$, let f_x denote the map from $\mathcal{B}(X)$ into $\{0, 1\}$ defined by

$$f_x(\Omega) = \begin{cases} 1 & \text{if } x \in \Omega; \\ 0 & \text{if } x \notin \Omega. \end{cases}$$

We will show that the map f which, with each $x \in X$, associates f_x, is a homeomorphism from the topological space X onto the topological space $S(\mathcal{B}(X))$. ∎

As f is, *a priori*, a map from X into $\{0, 1\}^{\mathcal{B}(X)}$, we must show, to begin with, that it really assumes values in $S(\mathcal{B}(X))$:

- For each $x \in X$, f_x is a homomorphism of Boolean algebras:

Proof For any clopen subsets Ω and Δ of X, we have $f_x(\Omega \cap \Delta) = 1$ if and only if $x \in \Omega \cap \Delta$, i.e. $x \in \Omega$ and $x \in \Delta$, which is equivalent to $f_x(\Omega) = 1$ and $f_x(\Delta) = 1$, and hence to $f_x(\Omega) f_x(\Delta) = 1$. We conclude from this that

$$f_x(\Omega \cap \Delta) = f_x(\Omega) \, f_x(\Delta).$$

On the other hand, $f_x(X - \Omega) = 1$ if and only if $x \in X - \Omega$, i.e. $x \notin \Omega$, or again, $f_x(\Omega) = 0$. Thus, $f_x(X - \Omega) = 1 + f_x(\Omega)$. We see that the conditions from Theorem 2.45 are satisfied: f_x is indeed a homomorphism. ∎

- The map f is injective:

Proof Let x and y be distinct elements of X. As X is Hausdorff, we can find an open set O such that $x \in O$ and $y \notin O$ (for example, we could take $O = X - \{y\}$). But O is a union of basic open sets from the basis $\mathcal{B}(X)$; so there is some clopen set $\Omega \in \mathcal{B}(X)$ such that $x \in \Omega$ and $y \notin \Omega$. We have $f_x(\Omega) = 1$ and $f_y(\Omega) = 0$, which proves that f_x is different from f_y. ∎

- The map f is surjective onto $S(\mathcal{B}(X))$:

Proof Let h be an element of $S(\mathcal{B}(X))$, i.e. a homomorphism from $\mathcal{B}(X)$ into $\{0, 1\}$. The ultrafilter on $\mathcal{B}(X)$ associated with h is

$$\mathcal{U} = \{\Omega \in \mathcal{B}(X) : h(\Omega) = 1\} = h^{-1}[\{1\}].$$

Since $\mathcal{U}$ has the finite intersection property (Lemma 2.81), since the elements of $\mathcal{U}$ are closed, and since the topological space X is compact, we may assert that the intersection of all the elements of $\mathcal{U}$ is non-empty. Let x be an element of this intersection.

For every clopen set $\Omega \in \mathcal{B}(X)$, we have: either $\Omega \in \mathcal{U}$, in which case $f_x(\Omega) = \mathbf{1}$ and $h(\Omega) = \mathbf{1}$, or else $\Omega \notin \mathcal{U}$, in which case $X - \Omega \in \mathcal{U}$ (Remark 2.73), so $f_x(\Omega) = \mathbf{0}$ and $h(\Omega) = \mathbf{0}$. Thus, for every $\Omega \in \mathcal{B}(X)$, $f_x(\Omega) = h(\Omega)$. It follows that $h = f_x = f(x)$. ∎

We may observe that the element x, which we have just shown is a preimage of h under the map f, is the unique element in the intersection of all the clopen sets belonging to $\mathcal{U}$. To see this, note that any element y in this intersection would, exactly as above, satisfy $h = f(y)$, and since f is injective, this would imply $x = y$. This observation will allow us to describe the inverse bijection f^{-1}: it is the map from $S(\mathcal{B}(X))$ into X that, to each homomorphism h of $\mathcal{B}(X)$ into $\{\mathbf{0}, \mathbf{1}\}$, associates the unique element in the intersection of all the clopen sets belonging to the ultrafilter $h^{-1}[\{\mathbf{1}\}]$.

- The map f is continuous.

Proof Let G be an open set belonging to the basis of clopen subsets of $S(\mathcal{B}(X))$. According to Lemma 2.84, there exists a unique element Ω in $\mathcal{B}(X)$ such that $G = \{h \in S(\mathcal{B}(X)) : h(\Omega) = \mathbf{1}\}$. The inverse image of G under the map f is

$$\{x \in X : f_x \in G\} = \{x \in X : f_x(\Omega) = \mathbf{1}\} = \{x \in X : x \in \Omega\} = \Omega.$$

So it is an open subset of X. ∎

- The inverse map f^{-1} is continuous.

Proof Let Ω be a basic open set in the space X (i.e. an element of $\mathcal{B}(X)$). Since f is a bijection, the inverse image of Ω under f^{-1} is its direct image under f. Thus it is the set $f[\Omega] = \{f_x : x \in \Omega\}$. We have to show that this is an open set in the space $S(\mathcal{B}(X))$.

Set $V = \{h \in S(\mathcal{B}(X)) : h(\Omega) = \mathbf{1}\}$.

The set V is open (it is even a basic open set) in $S(\mathcal{B}(X))$ (Lemma 2.84). If we show that $f[\Omega] = V$, this will complete the proof.

For every $x \in \Omega$, we have $f_x(\Omega) = \mathbf{1}$ by definition of f_x, so $f_x \in V$. Thus $f[\Omega] \subseteq V$.

Every $h \in V$ has a preimage $y \in X$ under the bijection f, say $h = f_y$. As $h \in V$, we have $h(\Omega) = f_y(\Omega) = \mathbf{1}$, so $y \in \Omega$ and $f_y = h \in f[\Omega]$. Hence, V is included in $f[\Omega]$. ∎

We should remark that the proof of this last point was superfluous: there is a famous theorem of topology asserting that any continuous bijection of a compact topological space into a Hausdorff space is a homeomorphism (the continuity of the inverse bijection is guaranteed).

We have definitely established a one-to-one correspondence between Boolean algebras and Boolean topological spaces (up to isomorphism on one side, up to homeomorphism on the other):

- every Boolean algebra is (isomorphic to) the Boolean algebra of clopen subsets of some Boolean topological space;
- every Boolean topological space is (homeomorphic to) the Stone space of some Boolean algebra.

In passing, let us observe that we had very good reasons for calling compact zero-dimensional spaces 'Boolean spaces'.

In a natural way, we have the following two properties (which are easily proved from all that has preceded):

- for any two Boolean algebras to be isomorphic, it is necessary and sufficient that their Stone spaces be homeomorphic;
- for any two Boolean topological spaces to be homeomorphic, it is necessary and sufficient that the Boolean algebras consisting of their respective clopen subsets be isomorphic.

E X E R C I S E S   F O R   C H A P T E R   2

1. (Refer to Example 2.26) Consider a set P of propositional variables and the associated set $\mathcal{F}$ of formulas. We are going to study the quotient set $\mathcal{F}/\sim$, i.e. the set of classes of logically equivalent formulas. The equivalence class of a formula F will be denoted by $\mathsf{cl}(F)$.

(a) Show that if, for all formulas F and G of $\mathcal{F}$, we set

$$\neg\mathsf{cl}(F) = \mathsf{cl}(\neg F) \qquad\qquad \mathsf{cl}(F) \Rightarrow \mathsf{cl}(G) = \mathsf{cl}(F \Rightarrow G)$$
$$\mathsf{cl}(F) \wedge \mathsf{cl}(G) = \mathsf{cl}(F \wedge G) \qquad \mathsf{cl}(F) \Leftrightarrow \mathsf{cl}(G) = \mathsf{cl}(F \Leftrightarrow G)$$
$$\mathsf{cl}(F) \vee \mathsf{cl}(G) = \mathsf{cl}(F \vee G) \qquad \mathsf{cl}(F) \nLeftrightarrow \mathsf{cl}(G) = \mathsf{cl}(F \nLeftrightarrow G),$$

this defines internal operations on $\mathcal{F}/\sim$ (denoted, abusively, by the same symbols for the corresponding connectives). Show that $\mathcal{F}/\sim$ is a Boolean algebra with respect to the operations $\nLeftrightarrow$ and $\wedge$. (Reminder: $(F \nLeftrightarrow G) = \neg(F \Leftrightarrow G)$).

(b) Show that the order relation in this Boolean algebra is the following: for all formulas F and G of $\mathcal{F}$,

$$\mathsf{cl}(F) \leq \mathsf{cl}(G) \text{ if and only if } \vdash^* (F \Rightarrow G).$$

(See Example 2.37).
   Explain how the operations of greatest lower bound, least upper bound and complementation are defined.

(c) Show that if the set P is finite, then the Boolean algebra $\mathcal{F}/\sim$ is atomic; describe the set of atoms.

2. Let E be an arbitrary set. On $\wp(E)$, we define a binary operation Δ (symmetric difference, see Exercise 16 from Chapter 1) as follows:

   for all elements X and Y of $\wp(E)$,
   $$X \Delta Y = (X \cup Y) - (X \cap Y) = (X \cap (E - Y)) \cup ((E - X) \cap Y).$$

(The symmetric difference of the subsets X and Y is the set of elements of E that belong to one and only one of these subsets).
   After observing that for all subsets X and Y of E we have

$$X \Delta Y = \{x \in E : x \in X \nLeftrightarrow x \in Y\},$$

show, using the properties of the usual connectives, for instance $\wedge$ and $\nLeftrightarrow$, that the set $\wp(E)$, with the two binary operations Δ and $\cap$, is a Boolean algebra. For this Boolean algebra, explain how the order relation and the operations of greatest lower bound, least upper bound and complementation are defined.

3. Let $\mathcal{A} = \langle A, +, \times, \mathbf{0}, \mathbf{1}\rangle$ and $\mathcal{B} = \langle B, +, \times, \mathbf{0}, \mathbf{1}\rangle$ be two Boolean algebras and let f be an isomorphism of Boolean algebras from $\mathcal{A}$ onto $\mathcal{B}$.

   (a) Show that an element $a \in A$ is an atom of $\mathcal{A}$ if and only if $f(a)$ is an atom of $\mathcal{B}$.

   (b) Show that $\mathcal{A}$ is atomless if and only if $\mathcal{B}$ is atomless.

   (c) Show that $\mathcal{A}$ is atomic if and only if $\mathcal{B}$ is atomic.

   (d) Show that a subset $I \subseteq A$ is an ideal of $\mathcal{A}$ if and only if $f[I]$, its direct image under f, is an ideal of $\mathcal{B}$.

   (e) Show that a subset $\mathcal{U} \subset A$ is an ultrafilter of $\mathcal{A}$ if and only if $f[\mathcal{U}]$ is an ultrafilter of $\mathcal{B}$.

4. We say that a Boolean algebra $\langle A, +, \times, \mathbf{0}, \mathbf{1}\rangle$ is **complete** if and only if every non-empty subset of A has a greatest lower bound.

   (a) Show that for a Boolean algebra to be complete, it is necessary and sufficient that every non-empty subset have a least upper bound.

   (b) Show that a Boolean algebra that is isomorphic to a complete Boolean algebra is complete.

   (c) Show that the Boolean algebra of subsets of a set is complete.

   (d) Show that if E is infinite, the Boolean algebra of subsets of E that are finite or cofinite (see Example 2.55) is not complete.

   (e) Is the Boolean algebra of classes of logically equivalent formulas of the propositional calculus (see Exercise 1 above) complete?

   (f) Show that for a Boolean algebra to be isomorphic to the Boolean algebra of subsets of some set, it is necessary and sufficient that it be atomic and complete.

5. Let $\mathcal{A} = \langle A, +, \times, \mathbf{0}, \mathbf{1}\rangle$ be a Boolean algebra and let $\mathcal{B} = \langle B, +, \times, \mathbf{0}, \mathbf{1}\rangle$ be a Boolean subalgebra of $\mathcal{A}$. Show that an element of B that is an atom of $\mathcal{A}$ must also be an atom of $\mathcal{B}$ but that there can be atoms of $\mathcal{B}$ that are not atoms of $\mathcal{A}$.

6. Let $\mathcal{A} = \langle A, +, \times, \mathbf{0}, \mathbf{1}\rangle$ be a Boolean algebra. Show that for every element $a \in A$, the set

$$B = \{x \in A : x \geq a \text{ or } x \leq 1 + a\}$$

   is a Boolean subalgebra of $\mathcal{A}$ and that it is a complete (see Exercise 4 above) Boolean algebra if $\mathcal{A}$ itself is complete.

7. Let $\mathcal{A} = \langle A, +, \times, \mathbf{0}, \mathbf{1}\rangle$ be a Boolean algebra. Consider a non-empty subset Z of $\wp(A)$ whose elements are filters on $\mathcal{A}$.

   (a) Show that the set $\bigcap_{F \in Z} F$, the intersection of the filters that belong to Z, is also a filter on $\mathcal{A}$, but that the union $\bigcup_{F \in Z} F$ of the elements of Z may not be a filter.

(b) Suppose, in addition, that Z is totally ordered by the inclusion relation. Show that in this case, $\bigcup_{F \in Z} F$ is a filter on $\mathcal{A}$.

8. Let E be a countable subset of $\wp(\mathbb{N})$ that has the finite intersection property (i.e. E is a filterbase in the Boolean algebra $\wp(\mathbb{N})$). The set of filters on $\wp(\mathbb{N})$ that include E is then non-empty and the intersection of the elements of this set (see Exercise 7 above) is called the **filter generated by** E.

Show that if the filter generated by E is an ultrafilter, then it is the trivial ultrafilter.

9. Consider a set P of propositional variables and its associated Boolean algebra $\mathcal{F}/\sim$ (see Exercise 1). We say that a class $x \in \mathcal{F}/\sim$ is **positive** if there is at least one formula F in x which does not contain any occurrences of the negation symbol.

(a) Show that for any class $x \in \mathcal{F}/\sim$, x is positive if and only if for every formula $F \in x$, $\delta_1(F) = \mathbf{1}$ (δ_1 is the assignment of truth values that assigns the value $\mathbf{1}$ to every variable: see Exercise 20 from Chapter 1.)

(b) Show that the set J of positive classes is an ultrafilter in the Boolean algebra $\mathcal{F}/\sim$.

(c) What is the homomorphism from $\mathcal{F}/\sim$ into $\{\mathbf{0}, \mathbf{1}\}$ associated with this ultrafilter?

10. Let X be a topological space. A subset $Y \subseteq X$ is said to be **dense in** X if and only if every non-empty open subset of X has a point in common with Y. (Some authors say **everywhere dense** instead of **dense**.) An element $x \in X$ is called an **isolated point** if and only if the singleton set $\{x\}$ is an open subset of X.

Let $\mathcal{A} = \langle A, +, \times, \mathbf{0}, \mathbf{1} \rangle$ be a Boolean algebra. Let S denote its Stone space and let H denote the isomorphism from $\mathcal{A}$ onto $\mathcal{B}(S)$ defined in the proof of Stone's theorem (Theorem 2.90).

(a) Show that for any element $x \in A$, x is an atom of $\mathcal{A}$ if and only if the set $H(x)$ is a singleton.

(b) Show that $\mathcal{A}$ has no atoms if and only if the topological space S has no isolated points.

(c) Show that $\mathcal{A}$ is atomic if and only if the set of isolated points of S is dense in S.

11. We say that a Boolean algebra $\langle A, +, \times, \mathbf{0}, \mathbf{1} \rangle$ is **dense** if and only if the order relation $\leq$ on A is dense, which means that for all elements a and b of A, if $a < b$, then there exists at least one element $c \in A$ such that $a < c < b$. Naturally, it is important not to confuse the notion of dense Boolean algebra with the notion of dense subset of a topological space introduced in Exercise 10. Show that a Boolean algebra is dense if and only if it has no atoms.

12. The purpose of this exercise is to show that, up to isomorphism, there is only one countable atomless Boolean algebra.

Consider a Boolean algebra $\mathcal{A} = \langle A, +, \times, \mathbf{0}, \mathbf{1} \rangle$ that has no atoms. Assume that the set A is countable and fix an enumeration $A = \{a_n : n \in \mathbb{N}\}$.

(a) Given a non-zero element x in A, a **splitting** of x is a pair $(y, z) \in A^2$ such that $y \neq \mathbf{0}$, $z \neq \mathbf{0}$, $y \frown z = \mathbf{0}$, and $y \smile z = x$. Show that (y, z) is a splitting of x if and only if $\mathbf{0} < y < x$ and $z = y + x$. Show that in A, every non-zero element has at least one splitting.

(b) Show that it is possible to define a family:

$$\{u_{\varepsilon_0 \varepsilon_1 \ldots \varepsilon_{n-1}} : (\varepsilon_0, \varepsilon_1, \ldots, \varepsilon_{n-1}) \in \{\mathbf{0}, \mathbf{1}\}^n, \ n \in \mathbb{N}\}$$

of non-zero elements of A such that $u_\emptyset = \mathbf{1}$ and, for every integer n,

$$(u_{\varepsilon_0 \varepsilon_1 \ldots \varepsilon_{n-1} 0}, \ u_{\varepsilon_0 \varepsilon_1 \ldots \varepsilon_{n-1} 1}) \text{ is a splitting of } u_{\varepsilon_0 \varepsilon_1 \ldots \varepsilon_{n-1}},$$

and if $u_{\varepsilon_0 \varepsilon_1 \ldots \varepsilon_{n-1}} \frown a_n \neq \mathbf{0}$ and $u_{\varepsilon_0 \varepsilon_1 \ldots \varepsilon_{n-1}} \frown (\mathbf{1} + a_n) \neq \mathbf{0}$, then

$$u_{\varepsilon_0 \varepsilon_1 \ldots \varepsilon_{n-1} 0} = u_{\varepsilon_0 \varepsilon_1 \ldots \varepsilon_{n-1}} \frown a_n$$

and

$$u_{\varepsilon_0 \varepsilon_1 \ldots \varepsilon_{n-1} 1} = u_{\varepsilon_0 \varepsilon_1 \ldots \varepsilon_{n-1}} \frown (\mathbf{1} + a_n).$$

(Thus, if $a_0 \notin \{\mathbf{0}, \mathbf{1}\}$, then $u_0 = a_0$ and $u_1 = \mathbf{1} + a_0$; otherwise, (u_0, u_1) is an arbitrary splitting of $\mathbf{1}$).

(c) Show that for every element $x \in A$ and every sequence $\varepsilon = (\varepsilon_n)_{n \in \mathbb{N}}$ of elements of $\{\mathbf{0}, \mathbf{1}\}$, one and only one of the following two conditions is satisfied:

  (i) for every $n \in \mathbb{N}$, $x \frown u_{\varepsilon_0 \varepsilon_1 \ldots \varepsilon_{n-1} \varepsilon_n} \neq \mathbf{0}$;
  (ii) for every $n \in \mathbb{N}$, $(\mathbf{1} + x) \frown u_{\varepsilon_0 \varepsilon_1 \ldots \varepsilon_{n-1} \varepsilon_n} \neq \mathbf{0}$.

(d) Consider two integers m and n satisfying $0 \leq n \leq m$ and $m + n + 2$ elements $\varepsilon_0, \varepsilon_1, \ldots, \varepsilon_n, \zeta_0, \zeta_1, \ldots, \zeta_m$ in $\{\mathbf{0}, \mathbf{1}\}$. Show that for $u_{\varepsilon_0 \varepsilon_1 \ldots \varepsilon_{n-1} \varepsilon_n} \frown u_{\zeta_0 \zeta_1 \ldots \zeta_{m-1} \zeta_m}$ to be non-zero, it is necessary and sufficient that $\varepsilon_0 = \zeta_0$ and $\varepsilon_1 = \zeta_1$ and ... and $\varepsilon_n = \zeta_n$.

(e) Let h be the map from A into $\wp(\{\mathbf{0}, \mathbf{1}\}^n)$ which, with each element $x \in A$, associates

$$h(x) = \{f \in \{\mathbf{0}, \mathbf{1}\}^n : (\forall n \in \mathbb{N})(x \frown u_{f(0) f(1) \ldots f(n)}) \neq \mathbf{0})\}.$$

Show that h is an isomorphism of Boolean algebras from $\mathcal{A}$ onto the Boolean algebra $\mathcal{B}(\{\mathbf{0}, \mathbf{1}\}^{\mathbb{N}})$ of clopen subsets of $\{\mathbf{0}, \mathbf{1}\}^{\mathbb{N}}$.

13. See Example 2.57 for the notation used here.

(a) Show that the map g from $\{0, 1\}^P$ into $\{0, 1\}^{\mathcal{F}/\sim}$ which, with each assignment of truth values δ, associates the homomorphism h_δ from $\mathcal{F}/\sim$ into $\{0, 1\}$, is a bijection from $\{0, 1\}^P$ onto the Stone space $S(\mathcal{F}/\sim)$.

(b) Show, without using the compactness theorem, that for any subset T of $\mathcal{F}$, T is satisfiable if and only if the set $T/\sim = \{\mathsf{cl}(G) : G \in T\}$ is a filterbase in the Boolean algebra $\mathcal{F}/\sim$.

(c) Use this result to provide a new proof of the compactness theorem for propositional calculus (Theorem 1.39).

14. Let $\mathcal{A} = \langle A, \leq, \mathbf{0}, \mathbf{1} \rangle$ be a Boolean algebra and let a be an element of A. Let B be the principal ideal generated by a (Example following Corollary 2.61) and I be the principal ideal generated by $a^{\complement}$:

$$B = \{x \in A : x \leq a\};$$
$$I = \left\{x \in A : x \leq a^{\complement}\right\}.$$

(a) Show that the order relation $\leq_B$, the restriction to B of the order relation $\leq$ on A, makes B into a Boolean algebra. Compare the operations in this Boolean algebra with the corresponding operations in the Boolean algebra $\mathcal{A}$.

(b) Show that the Boolean algebra $\langle B, \leq_B \rangle$ is isomorphic to the quotient Boolean algebra $\mathcal{A}/I$.

15. Let E be a non-empty finite set and let $\mathcal{A}$ be the Boolean algebra of subsets of E.

(a) Show that for a subset $I \subseteq \wp(E)$ to be an ideal of $\mathcal{A}$, it is necessary and sufficient that there exists a subset $X \subsetneq E$ (strict inclusion) such that $I = \wp(X)$.

(b) Let $\mathcal{C} = \langle C, \leq, \mathbf{0}, \mathbf{1} \rangle$ be an arbitrary Boolean algebra and let h be a homomorphism from $\mathcal{A}$ into $\mathcal{C}$. Show that there exists a unique subset $K \subseteq E$ such that for every element $Y \in \wp(E)$, $h(Y) = \mathbf{0}$ if and only if $Y \subseteq K$.

Let Z be the complement of K in E.

Show that the restriction of h to $\wp(Z)$ is an isomorphism of Boolean algebras from $\wp(Z)$ onto the image of $\mathcal{A}$ under h (which is a Boolean subalgebra of $\mathcal{C}$).

16. Let $\mathcal{A} = \langle A, \leq, \mathbf{0}, \mathbf{1} \rangle$ be a Boolean algebra.

(a) Show that if A is finite then every ideal of $\mathcal{A}$ is principal. Compare this result with that from Exercise 15(a).

(b) Show that if there exists an integer $k \geq 1$ and atoms $a_1, a_2, \ldots, a_k$ of $\mathcal{A}$ such that $a_1 \smile a_2 \smile \cdots \smile a_k = \mathbf{1}$, then A is finite.

(c) Assume that the Boolean algebra $\mathcal{A}$ is infinite. Show that the set

$$G = \{x \in A : \mathbf{1} + x \text{ is an atom}\}$$

is a filterbase on $\mathcal{A}$.

(d) Again suppose that $\mathcal{A}$ is infinite. Show that for an ultrafilter $\mathcal{U}$ on $\mathcal{A}$ to be non-trivial, it is necessary and sufficient that G be included in $\mathcal{U}$. Use this result to reprove Theorem 2.77.

(e) Show that for the existence of a non-trivial ultrafilter on $\mathcal{A}$, it is necessary and sufficient that $\mathcal{A}$ be an infinite Boolean algebra.

17. Let E be a set and let $\mathcal{A} = \langle \wp(E), \subseteq \rangle$ be the Boolean algebra of subsets of E.

(a) Consider a family $(E_i)_{i \in I}$ of subsets that constitutes a partition of E (which means each E_i is non-empty, that $\bigcup_{i \in I} E_i = E$, and that for $i \neq j$, $E_i \neq E_j$). Show that the set

$$ B = \left\{ X \in \wp(E) : (\exists J \subseteq I) \left(X = \bigcup_{j \in J} E_j \right) \right\} $$

is a Boolean subalgebra of $\mathcal{A}$ and that each of the E_i is an atom in this Boolean subalgebra.

(b) Suppose that E is finite and non-empty and that $\mathcal{C}$ is a Boolean subalgebra of $\wp(E)$. Show that the atoms of the Boolean algebra $\mathcal{C}$ constitute a family of subsets of E that partitions E.

(c) Show that if E is a non-empty finite set, there is a bijection between the set of partitions of E and the set of Boolean subalgebras of $\wp(E)$.

18. (a) Is every Boolean subalgebra of an atomic Boolean algebra an atomic Boolean algebra?

(b) Are there Boolean algebras all of whose Boolean subalgebras are atomic?

(c) Is every Boolean subalgebra of an atomless Boolean algebra an atomless Boolean algebra?

(d) Are there Boolean algebras all of whose Boolean subalgebras are atomless?

(e) Is every Boolean subalgebra of a complete Boolean algebra a complete Boolean algebra?

(f) Are there Boolean algebras all of whose Boolean subalgebras are complete?

19. Let $\mathcal{A}$ and $\mathcal{A}'$ be two Boolean algebras and let $S(\mathcal{A})$ and $S(\mathcal{A}')$ be their Stone spaces.

(a) Show that we can establish a bijection Φ between the set $\mathsf{Hom}(\mathcal{A}, \mathcal{A}')$ of homomorphisms of Boolean algebras from $\mathcal{A}$ into $\mathcal{A}'$ and the set $C^0(S(\mathcal{A}'), S(\mathcal{A}))$ of continuous maps from $S(\mathcal{A}')$ into $S(\mathcal{A})$.

(b) Show that for every homomorphism $\varphi \in \mathsf{Hom}(\mathcal{A}, \mathcal{A}')$, φ is injective (respectively, surjective) if and only if $\Phi(\varphi)$ is surjective (respectively, injective).

3 Predicate calculus

The fundamental work of the mathematician is to examine structures, to suggest properties that might pertain to these and to ask whether these properties are satisfied or not. Predicate calculus is, in some way, the first stage in the formalization of mathematical activity. Two aspects are involved: first, we provide ourselves with formal tools that are adequate to name these objects (these will be the terms) and to express (some of) their properties (these will be the formulas); second, we study the satisfaction of these properties in the structures under consideration.

As with the propositional calculus, the formulas are sequences of symbols that are taken from some alphabet and that obey precise syntactic rules.

There will not be a unique alphabet but rather an appropriate alphabet, called a language, for each type of structure under consideration. By structure, we mean a non-empty set M, provided with the following: a certain number of distinguished elements; for each positive integer p, a certain number of p-place relations on M (also called 'predicates', hence the expression 'predicate calculus'); and for each positive integer p, a certain number of functions from M^p into M. Obviously, we would not use the same language to speak, for example, of groups and of ordered sets.

Certain symbols are common to all the languages: these are the propositional connectives and parentheses, already used in propositional calculus, but also, and this is the essential innovation, the quantifiers $\forall$ ('for all') and $\exists$ ('there exists') and the variables $v_0, v_1, \ldots$.

The other symbols depend on the type of structure we have in mind; they will represent distinguished elements, predicates or functions. For example, for groups, we need a constant symbol (to represent the identity element) and a symbol for a binary function (to represent multiplication). For ordered sets, we only need a symbol for a binary relation.

The formulas that are involved here are called 'formulas of first order predicate calculus'. The reason for this name is that the quantifiers will range over elements of the structure. There are numerous mathematical properties for which this restriction is fatal. For example, to express the fact that a set A is well ordered, we need to say: for every subset B of A, if B is not empty, then B contains a least element. We note that the quantifier 'for all' in this definition ranges over subsets of A and not over elements of A. This is a second order quantifier. The concept of a well-ordered set is not expressible by a first order formula.

The problems of syntax, treated in Section 3.1, are substantially more complicated than for the propositional calculus. First of all, because we have to define terms before we can define formulas; second, because we have to introduce the notions of free and bound variable: we immediately sense that the status of the variable v_0 is not the same in the following two formulas: $v_0 + v_0 \simeq v_0$ and $\forall v_0(v_0 + v_0 \simeq v_0)$. We say that v_0 is free in the first and bound in the second. This distinction is fundamental in what follows.

In Section 3.2, we temporarily abandon logic to explain what we mean by structure. In Section 3.3, we define satisfaction of a formula in a structure (we also say that the structure is a model of the formula). The two facts mentioned above, particularly the fact that we will have to define satisfaction for formulas that contain free variables, will weigh us down considerably. But the reader must not be worried: despite its complication, the definition of satisfaction involves nothing more than what the reader would probably have guessed from the beginning.

In Section 3.4 we show that every formula is equivalent to (i.e. is satisfied in the same structures as) a formula that is written in a very particular form (with all the quantifiers at the beginning: this is called a prenex form). We will also see how to eliminate existential quantifiers by adding function symbols to the language (the Skolem form). In Section 3.5, we will present the ABCs of model theory, which is the study of the correspondence between a set of formulas and the class of models of this set. This will be studied in more depth in Chapter 8. Finally, in Section 3.6, we will examine the behaviour of equality, which is a binary predicate not quite like the others.

3.1 Syntax

3.1.1 First order languages

Definition 3.1 *A **first order language** (often we will simply say **language**) is a set L of symbols composed of two parts:*

- *the first part, common to all languages, consists, on the one hand, of a countably infinite set,*

$$\mathcal{V} = \{v_0, v_1, \ldots, v_n, \ldots\},$$

*of elements called **symbols for variables** or more simply **variables**, and, on the other hand, of the following nine symbols:*

- *the parentheses:), (and the symbols for the connectives: $\neg, \wedge, \vee, \Rightarrow, \Leftrightarrow$ used previously in propositional calculus,*
- *and two new symbols:*
  *$\forall$, which is called the **universal quantifier** and is read as 'for all', and*

> ∃, *which is called the* **existential quantifier** *and is read as 'there exists at least*
> *one' or 'for at least one' or simply 'there exists';*
> *(each of these quantifiers is called the* **dual** *of the other);*

- *the second part, which can vary from one language to another, is the union of*
  a set C *and of two sequences* $(\mathcal{F}_n)_{n \in \mathbb{N}^*}$ *and* $(\mathcal{R}_n)_{n \in \mathbb{N}^*}$ *of sets (pairwise disjoint*
  and all disjoint from C*);*

  * *the elements of* C *are called* **constant symbols;**

  * *for every integer* $n \geq 1$*, the elements of* $\mathcal{F}_n$ *are called n-***place** *(or n-***ary***
    or **of arity** *n, or* **with** *n* **arguments***) **function symbols** (or **functionals**) and*
    the elements of $\mathcal{R}_n$ *are called n-***place** *(or n-***ary** *or* **of arity** *n, or* **with** *n*
    arguments*) **relation symbols** (or **predicates**); (we say **unary**, **binary** and*
    ternary *respectively instead of 1-ary, 2-ary and 3-ary);*

  * *we consider one particular symbol with a special status that will be explained*
    later: the symbol $\simeq$*, called the* **equality symbol***, which, when a first order*
    language contains it, is an element of $\mathcal{R}_2$*, i.e. is a binary relation symbol.*
    Languages that contain this symbol are called **languages with equality***.*

Except in one important circumstance (see Chapter 4), the languages we
encounter in this text will be of this type and when we say 'language', with no
modifier, we always mean a language with equality.

Of course, all the symbols that we have just listed and which constitute the
language are assumed to be pairwise distinct.

Thus, to have a first order language, L, is to define the two sequences $(\mathcal{R}_n)_{n \in \mathbb{N}^*}$
and $(\mathcal{F}_n)_{n \in \mathbb{N}^*}$ and to consider the set:

$$L = \mathcal{V} \cup \{ \,), (, \neg, \wedge, \vee, \Rightarrow, \Leftrightarrow, \forall, \exists \} \cup C \cup (\mathcal{R}_n)_{n \in \mathbb{N}^*} \cup (\mathcal{F}_n)_{n \in \mathbb{N}^*}.$$

The symbols for the constants, the functions and the relations are sometimes
called the **non-logical symbols** of the language. Some other presentations differ
slightly from our own: it can happen that constant symbols are treated as 0-place
function symbols or that specific 0-place relation symbols are allowed, $\top$ ('true')
and $\bot$ ('false'). These slight variants do not modify anything that is essential in
what follows.

In most of the cases that we will examine, the languages will involve only a small
number of symbols for constants, functions and relations, so that in practice, most
of the sets $\mathcal{R}_n$ and $\mathcal{F}_n$ will be empty and those that are not empty will contain at
most two or three symbols. Under these circumstances, rather than give a fastidious
definition of the sequences $(\mathcal{R}_n)_{n \in \mathbb{N}^*}$ and $(\mathcal{F}_n)_{n \in \mathbb{N}^*}$, we will be content to provide
a list of the symbols that appear in these sequences, while indicating their status
(relation, function) and their arity, and to specify the elements of the set C.

As it is obviously pointless to repeat, for each language, the unchanged list of
symbols consisting of, for example, the variables, the connectives and quantifiers,
we will commit the abuse that consists in identifying a language with the list of its

symbols for constants, functions and relations. Thus, when we are led to say:

'Consider the language $L = \{R, c, f, g\}$ where R is a binary relation symbol, c is a constant symbol, and f and g are two unary function symbols', this will mean that we are interested in the set $\mathcal{C} = \{c\}$ and in the two sequences $(\mathcal{R}_n)_{n\in\mathbb{N}^*}$ and $(\mathcal{F}_n)_{n\in\mathbb{N}^*}$ defined by:

$$\mathcal{R}_2 = \{\simeq, R\}, \quad \mathcal{R}_n = \emptyset \quad \text{for } n \neq 2,$$
$$\mathcal{F}_1 = \{f, g\} \quad \text{and} \quad \mathcal{F}_n = \emptyset \quad \text{for } n \geq 2.$$

(In the absence of any mention to the contrary, the language will be one with equality.)

Starting from a first order language viewed as an alphabet, we will now construct, following the inductive method used previously for the propositional calculus, a family of words that we will call the first order formulas associated with our language. To do this will require an intermediate step in which we define, also inductively, another family of words called terms.

As our ultimate purpose is to use the language to formally describe certain properties of mathematical objects (or individuals), we may intuitively consider that terms will serve as names to denote these individuals whereas the formulas will serve as statements of facts concerning them.

Let us consider a first order language L.

3.1.2 Terms of the language

The symbols that serve as raw materials for the construction of terms are the variables and the function symbols. Note that the parentheses are not involved in writing terms.

Definition 3.2 *The set $\mathcal{T}(L)$ of **terms** of the language L is the smallest subset of $\mathcal{W}(L)$ that:*

- *contains the variables and the constant symbols (i.e. includes the set $\mathcal{V} \cup \mathcal{C}$);*
- *is closed under the operation*

$$(m_1, m_2, \ldots, m_n) \longmapsto f m_1 m_2 \ldots m_n$$

for every integer $n \geq 1$ and every element $f \in \mathcal{F}_n$.

In other words, terms are words that can be obtained by applying the following rules a finite number of times:

- variables and constant symbols are terms (recall that we do not distinguish between a symbol of the alphabet and the word of length 1 consisting of this symbol);
- if $n \in \mathbb{N}^*$, if f is an n-ary function symbol of L, and if $t_1, t_2, \ldots, t_n$ are terms, then the word $f t_1 t_2 \ldots t_n$ is a term.

After this definition 'from above', let us consider the equivalent definition 'from below':

Set $T_0(L) = V \cup C$, and for every integer k, set

$$T_{k+1}(L) = T_k(L) \cup \bigcup_{n \in \mathbb{N}^*} \{ f t_1 t_2 \ldots t_n : f \in \mathcal{F}_n, \ t_1 \in T_k(L),$$
$$t_2 \in T_k(L), \ldots, t_n \in T_k(L) \}.$$

Definition 3.3 *The **height** of a term $t \in T(L)$ is the least integer k such that $t \in T_k(L)$.*

Observe that in a language, there are always terms of height 0: the variables; but it is altogether possible that there are no terms whose height is non-zero (this will happen when there are no function symbols at all).

For terms, we can define the concept of a decomposition tree in a manner analogous to the one we used for formulas of the propositional calculus, with the difference that at each node, there can be any number of branches, this number being precisely the arity of the function symbol used at the given stage in the construction of the term. There will also be a unique readability theorem (3.7) which will guarantee the uniqueness of the decomposition tree.

Consider, for example, the language that has one constant symbol c, one unary function symbol f and a ternary function symbol g. Consider the word

$$W = gg f f v_0 g v_2 v_0 c f c f f g f c g v_2 f v_0 f f c f c f c.$$

Is this a term of the language?

After careful inspection, or a few probings, or with a bit of luck, by setting

$$r = g f f v_0 g v_2 v_0 c f c, \quad s = f f g f c g v_2 f v_0 f f c f c, \quad \text{and} \quad t = f c,$$

and, even better, by inserting spaces wisely,

$$r = g \ f f v_0 \ g v_2 v_0 c \ f c, \quad s = f \ f \ g \ f c \ g v_2 f v_0 f f c \ f c, \quad \text{and} \quad t = f c,$$

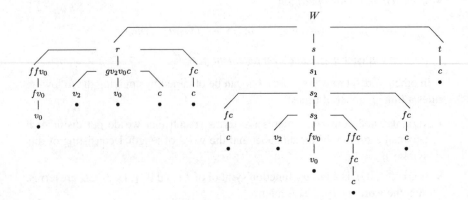

we discover that r, s and t really are terms and that the proposed word W, which is written $grst$ (a ternary function symbol followed by three terms), is itself a term (of height 7) whose decomposition tree is not difficult to sketch (to improve its legibility we have set $s_3 = gv_2fv_0ffc$, $s_2 = gfcs_3fc$ and $s_1 = fs_2$, so that $s = fs_1$).

The theorem that follows furnishes a simple test for determining whether a word in a given language is or is not a term. It also provides, when the test is positive, a method for finding the decomposition of the term. Finally, it provides us with a simple proof of the unique readability theorem. First we give a definition:

Definition 3.4 *The **weight** of a function symbol is equal to its arity minus 1. The **weight** of a variable or a constant symbol is -1. If W is a word written with variables, constant symbols and function symbols, the **weight** of W, denoted by $\rho(W)$, is the sum of the weights of the symbols that occur in the word W (the weight of the empty word is 0).*

*We say that a word W **satisfies the rule of weights** if and only if the weight of W is -1 and the weights of all its proper initial segments are greater than or equal to 0.*

Theorem 3.5 *For a word W, written with variables, constant symbols and function symbols, to be a term, it is necessary and sufficient that it satisfy the rule of weights.*

Proof First we will show, by induction, that all terms satisfy the rule of weights.

As far as variables and constant symbols are concerned, this is clear: their weight is equal to -1 and they do not admit any proper initial segments.

Consider an integer $n \geq 1$, an n-ary function symbol f and n terms $t_1, t_2, \ldots, t_n$ which are assumed (by the induction hypothesis) to satisfy the rule of weights. Set $t = ft_1t_2 \cdots t_n$.

We have $\rho(t) = \rho(f) + \rho(t_1) + \rho(t_2) + \cdots + \rho(t_n) = n - 1 + n \times (-1) = -1$.

Now let m be a proper initial segment of t. Then there exists an index $i \in \{1, 2, \ldots, n\}$ and an initial segment m_i of t_i such that

$$m = ft_1t_2 \cdots t_{i-1}m_i.$$

Note that if $i = n$, m_i is necessarily a proper initial segment of t_i but that if $i \neq n$, then m_i might also equal t_i or the empty word.

We have

$$\rho(m) = \rho(f) + \rho(t_1) + \rho(t_2) + \cdots + \rho(t_{i-1}) + \rho(m_i)$$
$$= n - 1 + (i - 1) \times (-1) + \rho(m_i) \quad \text{(by the induction hypothesis)}$$
$$= n - i + \rho(m_i).$$

Now, by the induction hypothesis, $\rho(m_i)$ is either -1 or is an integer greater than or equal to 0 (according as $m_i = t_i$ or not). It follows that $\rho(m) \geq n - i - 1$, which is a number greater than or equal to zero if i is strictly less than n. If $i = n$, then $m_i \neq t_i$ (otherwise m would equal t), and so $\rho(m) = \rho(m_i) \geq 0$.

We see then that, in all cases, the weight of m is greater than or equal to zero, which shows that t satisfies the rule of weights.

Now we have to show the converse: that every word that satisfies the rule of weights is a term. We will prove it by induction on the length of words.

The word of length 0 does not satisfy the rule of weights (its weight is 0).

If a word of length 1 satisfies the rule of weights, then its weight is -1. Thus we are dealing with either a variable or a constant symbol, so it is, in all cases, a term.

Consider an integer $k > 1$. Suppose (this is the induction hypothesis) that any word of length strictly less than k that satisfies the rule of weights is a term. Consider a word W of length k which satisfies the rule of weights. For example,

$$W = \alpha_1 \alpha_2 \ldots \alpha_k,$$

where the α_i are variables or constant symbols or function symbols (so that $\rho(\alpha_i) \geq -1$ for every i).

Thus we have $\rho(W) = -1$ and, for every $i \in \{1, 2, \ldots, k - 1\}$,

$$\rho(\alpha_1 \alpha_2 \ldots \alpha_i) \geq 0.$$

Since $k > 1$, α_1 is a strict initial segment of W; thus $\rho(\alpha_1) \geq 0$, which shows that α_1 must be a function symbol of arity at least equal to 1. Denote this arity by $n + 1$ ($n = \rho(\alpha_1) \geq 0$).

If $n = 0$, the word $\alpha_2 \alpha_3 \ldots \alpha_k$ satisfies the rule of weights (suppressing the initial symbol of weight 0 changes neither the total weight nor the weights of any proper initial segments); since the length of this word is $k - 1$, the induction hypothesis applies: therefore $\alpha_2 \alpha_3 \ldots \alpha_k$ is a term; and so is W since, in this case, α_1 is a unary function symbol.

Now let us examine the case $n > 0$.

Denote by ϕ the map from $\{1, 2, \ldots, k\}$ into $\mathbb{Z}$ which associates with each index, i, the value

$$\phi(i) = \rho(\alpha_1 \alpha_2 \ldots \alpha_i) = \rho(\alpha_1) + \rho(\alpha_2) + \cdots + \rho(\alpha_i).$$

We have $\phi(1) = n > 0$ and $\phi(k) = -1$. Since the passage from $\phi(i)$ to $\phi(i+1)$ adds an integer, $\rho(\alpha_{i+1})$, that is greater than or equal to -1, we see that the map ϕ, to get from its initial value n to its final value -1, must assume each of the intermediate values $n - 1, n - 2, \ldots, 1, 0$, at least once (this function cannot decrease by more than 1 at each step).

Denote by j_1 (and by $j_2, \ldots, j_n$ respectively) the first integer in $\{1, 2, \ldots, k\}$ for which the function ϕ takes the value $n - 1$ ($n - 2, \ldots, 0$ respectively). Extend this by setting $j_0 = 1$ and $j_{n+1} = k$ so that $\phi(j_0) = n$ and $\phi(j_{n+1}) = -1$.

We must have

$$j_0 = 1 < j_1 < j_2 < \cdots < j_n < j_{n+1} = k.$$

(The argument that we just gave can be repeated to prove that the function ϕ can not pass from the value $\phi(1) = n$ to the value $\phi(j_2) = n - 2$ without assuming at least once the value $n - 1$, hence $j_1 < j_2$, and so on.)

Set

$$t_1 = \alpha_2 \ldots \alpha_{j_1}, \quad t_2 = \alpha_{j_1+1} \ldots \alpha_{j_2}, \ldots,$$
$$t_n = \alpha_{j_{n-1}+1} \ldots \alpha_{j_n}, \quad t_{n+1} = \alpha_{j_n+1} \ldots \alpha_k.$$

We are going to show that each of these $n + 1$ words is a term; given that α_1 is a function symbol of arity $n + 1$ and that the word W is written $\alpha_1 t_1 t_2 \ldots t_{n+1}$, this will prove that W is also a term.

Let h be an integer such that $1 \leq h \leq n + 1$. We have

$$t_h = \alpha_{j_{h-1}+1} \ldots \alpha_{j_h};$$

hence

$$\rho(t_h) = \phi(j_h) - \phi(j_{h-1}) = n - h - (n - (h - 1)) = -1.$$

Moreover, if t_h had a proper initial segment whose weight is strictly negative, this would mean that there exists an index $i \in \{j_{h-1} + 1, \ldots, j_h - 1\}$ such that

$$\rho(\alpha_{j_{h-1}+1} \ldots \alpha_i) = \phi(i) - \phi(j_{h-1}) = \phi(i) - (n - (h - 1)) < 0;$$

or again,

$$\phi(i) \leq n - h.$$

But $\phi(j_{h-1}) = n - h + 1$. According to the argument already used twice previously, the value $n - h$ would then be assumed by the function ϕ for some index between j_{h-1} and i, i.e. strictly less than j_h, and this would contradict the definition of j_h.

We have thereby proved that t_h satisfies the rule of weights and, as its length is strictly less than k, we conclude from the induction hypothesis that it is a term. ∎

The unique readability theorem (for terms) will be proved in two stages.

Lemma 3.6 *For any term $t \in T(L)$, no proper initial segment of t is a term.*

Proof This is an immediate consequence of the rule of weights: if t is a term and u is a proper initial segment of t, then the weight of u is positive or zero, and so u cannot be a term. ∎

Theorem 3.7 (*unique readability theorem for terms*) *For any term* $t \in \mathcal{T}(L)$, *one and only one of the following three cases applies:*

- *t is a variable of L;*
- *t is a constant symbol of L;*
- *there is a unique integer $k \geq 1$, a unique k-ary function symbol f of the language L and a unique k-tuple $(u_1, u_2, \ldots, u_k) \in \mathcal{T}(L)^k$ such that $t = f u_1 u_2 \ldots u_k$.*

Proof Consider a term $t \in \mathcal{T}(L)$. By the definition of $\mathcal{T}(L)$, we are in one of the first two cases, or else in the third but without any guarantee in advance of uniqueness. It is clear, moreover, that these three cases are mutually exclusive (to see this, it suffices to examine the first symbol in the word t). So the only thing that we have to prove is the uniqueness for the third case.

For this, consider two natural numbers k and h, two function symbols f and g of the language L that are k-ary and h-ary respectively and $k + h$ terms $t_1, t_2, \ldots, t_k$, $u_1, u_2, \ldots, u_h$ in $\mathcal{T}(L)$ and suppose that:

$$t = f t_1 t_2 \ldots t_k = g u_1 u_2 \ldots u_h.$$

From this equality, we conclude that f and g are identical (they are the first symbols of the same word) hence their arities are equal. So we have

$$t = f t_1 t_2 \ldots t_k = f u_1 u_2 \ldots u_k.$$

Suppose now that there is some index $i \in \{1, 2, \ldots, k\}$ such that

$$t_1 = u_1, \quad t_2 = u_2, \ldots, t_{i-1} = u_{i-1}, \text{ and } t_i \neq u_i.$$

After simplifying, we obtain

$$t_i t_{i+1} \ldots t_k = u_i u_{i+1} \ldots u_k,$$

which proves that one of the two terms t_i and u_i is a proper initial segment of the other (this property was established at the beginning of the text in the section 'Notes to the Reader'). But this situation is precisely what is forbidden by the preceding lemma. ∎

Definition 3.8 *A term in which there is no occurrence of a variable is called a* **closed** *term.*

We immediately see that a closed term must contain at least one occurrence of a constant symbol. It follows that for a language that has no constant symbols, there are no closed terms.

Notation: Given a term $t \in \mathcal{T}(L)$ and pairwise distinct natural numbers $i_1, i_2, \ldots, i_n$, we will use the notation $t = t[v_{i_1}, v_{i_2}, \ldots, v_{i_n}]$ to indicate that the variables that have at least one occurrence in the term t are among $v_{i_1}, v_{i_2}, \ldots, v_{i_n}$.

We should note that for any term t, there is an integer m such that

$$t = t[v_0, v_1, \ldots, v_m];$$

(because t involves only a finite number of symbols, and hence a finite number of variables, it suffices to let m be the largest of the indices of all the variables that have at least one occurrence in t).

3.1.3 Substitutions in terms

Definition 3.9 *Let k be a natural number, let w_1, w_2, ..., w_k be pairwise distinct variables, and let t, u_1, u_2, ..., u_k be terms. The word $t_{u_1/w_1, u_2/w_2, \ldots, u_k/w_k}$ (read as 't sub u_1 replaces w_1, u_2 replaces w_2, ..., u_k replaces w_k') is the result of substituting the terms u_1, u_2, ..., u_k for the variables w_1, w_2, ..., w_k respectively for all occurrences of these variables in t and it is defined by induction (on t) as follows:*

- *if t is a constant symbol or a variable other than w_1, w_2, ..., w_k, then*

$$t_{u_1/w_1, u_2/w_2, \ldots, u_k/w_k} = t;$$

- *if $t = w_i$ ($1 \leq i \leq k$), then*

$$t_{u_1/w_1, u_2/w_2, \ldots, u_k/w_k} = u_i;$$

- *if $t = f t_1 t_2 \ldots t_n$ (where n is an integer greater than or equal to 1, f is an n-ary function symbol and t_1, t_2, ..., t_n are terms), then*

$$t_{u_1/w_1, u_2/w_2, \ldots, u_k/w_k}$$
$$= f {t_1}_{u_1/w_1, u_2/w_2, \ldots, u_k/w_k} {t_2}_{u_1/w_1, u_2/w_2, \ldots, u_k/w_k} \cdots {t_n}_{u_1/w_1, u_2/w_2, \ldots, u_k/w_k}.$$

Lemma 3.10 *For any natural number k, any pairwise distinct variables w_1, w_2, ..., w_k, and terms t, u_1, u_2, ..., u_k, the word $t_{u_1/w_1, u_2/w_2, \ldots, u_k/w_k}$ is a term.*

Proof The proof is obvious by induction on t. ∎

Notation: Given two natural numbers k and h, $k + h$ variables $z_1, z_2, \ldots, z_h$, $w_1, w_2, \ldots, w_k$, a term $t[z_1, z_2, \ldots, z_h, w_1, w_2, \ldots, w_k]$, and k terms $u_1, u_2, \ldots, u_k$, the term $t_{u_1/w_1, u_2/w_2, \ldots, u_k/w_k}$ will be denoted by

$$t[z_1, z_2, \ldots, z_h, u_1, u_2, \ldots, u_k].$$

Remark 3.11

- *Obviously, it is merely for convenience in writing that we have listed the variables in an order such that those involved in the substitution appear at the end of the list. It goes without saying that given, for example, a term*

$$t = t[w_1, w_2, w_3, w_4, w_5]$$

and two arbitrary terms u and u', we understand that the expression

$$t[w_1, u', w_3, u, w_5]$$

can be used to denote the term

$$t_{u'/w_2, u/w_4}.$$

- *The notation involving brackets presents some inconveniences analogous to those we discussed earlier concerning substitutions in formulas of the propositional calculus (see Chapter 1). We will use it, but with the usual precautions.*

- *As with the propositional calculus, it is appropriate to recall the fact that the substitutions defined above take place simultaneously; the same substitutions, implemented one after another, yield different results, in general, which depend, among other things, on the order in which they are performed.*

3.1.4 Formulas of the language

We will now undertake the definition, by induction, of the set of formulas of the language L. Here, first of all, are the formulas 'on the ground floor' (those that will have height 0); we call these the atomic formulas.

Definition 3.12 *A word $W \in \mathcal{W}(L)$ is an **atomic formula** if and only if there exist a natural number $n \in \mathbb{N}^*$, an n-ary relation symbol R, and n terms $t_1, t_2, \ldots, t_n$ of the language L such that*

$$W = Rt_1 t_2 \ldots t_n.$$

In the case where L is a language with equality and t and u are arbitrary terms in $\mathcal{T}(L)$, we agree to write

$$t \simeq u$$

for the atomic formula

$$\simeq tu.$$

We will denote the set of atomic formulas of the language L by $At(L)$.

Observe that we have unique readability for atomic formulas: we can easily convince ourselves by noting that an atomic formula becomes a term if we replace its first symbol (which is a relation symbol) by a function symbol of the same arity; it then suffices to apply the unique readability theorem for terms to obtain unique readability for atomic formulas (the convention we introduced for the equality symbol presents no difficulty in this matter).

We may now proceed with the definition of the set of formulas of L.

Definition 3.13 *The set $\mathcal{F}(L)$ of (first order) formulas of the language L is the smallest subset of $\mathcal{W}(L)$ that*

- *contains all the atomic formulas;*
- *whenever it contains two words V and W, it also contains the words*

$$\neg W, \quad (V \wedge W), \quad (V \vee W), \quad (V \Rightarrow W), \quad (V \Leftrightarrow W)$$

and, for every natural number n, the words

$$\forall v_n W \text{ and } \exists v_n W.$$

*Given two formulas F and G in $\mathcal{F}(L)$, the formulas $\neg F$, $(F \wedge G)$, and $(F \vee G)$, are called, respectively: the **negation** of the formula F, the **conjunction** of the formulas F and G and the **disjunction** of the formulas F and G.*

We are naturally led to compare the definition above, or at least the part of it that concerns the propositional connectives, to the definition of propositional formulas given in Chapter 1. We note in this context that the role played there by the propositional variables is played here by the atomic formulas. The major difference arises from the fact that there, the propositional variables were primary indecomposable ingredients whereas here, the atomic formulas are already the product of a fairly complicated construction. It is essential, in any case, not to imagine any analogy between the propositional variables of Chapter 1 and what, in this chapter, have been called variables; here they are certain symbols that are constituents of terms, which in turn are ingredients in the production of atomic formulas. The other obvious fundamental difference between the two situations is the appearance of quantifiers which provide new means for constructing formulas.

By analogy with Theorem 1.3, the following theorem, whose proof is left to the reader, provides a description 'from below' of the set of formulas.

Theorem 3.14 *Set*

$$\mathcal{F}_0(L) = At(L);$$

and, for every integer m,

$$\mathcal{F}_{m+1}(L) = \mathcal{F}_m(L) \cup \{\neg F : F \in \mathcal{F}_m(L)\}$$
$$\cup \{(F \alpha G) : F \in \mathcal{F}_m(L), G \in \mathcal{F}_m(L), \alpha \in \{\wedge, \vee, \Rightarrow, \Leftrightarrow\}\}$$
$$\cup \{\forall v_k F : F \in \mathcal{F}_m(L), k \in \mathbb{N}\} \cup \{\exists v_k F : F \in \mathcal{F}_m(L), k \in \mathbb{N}\}.$$

We then have

$$\mathcal{F}(L) = \bigcup_{n \in \mathbb{N}} \mathcal{F}_n(L).$$

As it should be, the **height** of a formula $F \in \mathcal{F}(L)$, denoted by $h[F]$, is the smallest integer k such that $F \in \mathcal{F}_k(L)$.

For first order formulas, there is a unique readability theorem:

Theorem 3.15 (*Unique readability theorem for formulas*) *For any formula $F \in \mathcal{F}(L)$, one and only one of the following five cases applies:*

- *F is an atomic formula (and there is then only one way in which it can be 'read');*

- *there is a unique formula $G \in \mathcal{F}(L)$ such that $F = \neg G$;*

- *there is a unique pair of formulas $(G, H) \in \mathcal{F}(L)^2$ and a unique symbol for a binary connective $\alpha \in \{\wedge, \vee, \Rightarrow, \Leftrightarrow\}$ such that $F = (G \, \alpha \, H)$;*

- *there exists a unique integer k and a unique formula $G \in \mathcal{F}(L)$ such that $F = \forall v_k G$;*

- *there exists a unique integer k and a unique formula $G \in \mathcal{F}(L)$ such that $F = \exists v_k G$.*

Proof   We can easily adapt the proof given for the case of formulas of the propositional calculus: here too, it is clear that the five cases are mutually exclusive and that at least one of the cases is applicable (ignoring the feature of uniqueness). In the last two cases, for which the propositional calculus has no analogue, uniqueness is obvious. In the first case, unique readability has already been noted (in the paragraph that follows Definition 3.13); in the second and third cases, the proof from Chapter 1 carries over with no problems (in particular, the relevant four lemmas concerning parentheses and proper initial segments remain valid).   ∎

Here too, we can speak of the decomposition tree of a formula: relative to the propositional calculus, the changes are, on the one hand, that the leaves are the atomic formulas and, on the other hand, that there are three kinds of unary branching instead of one:

$$
\begin{array}{ccc}
\neg F & \forall v_k F & \exists v_k F \\
| & | & | \\
F & F & F
\end{array}
$$

The sub-formulas of a first order formula are those that appear at the nodes of its decomposition tree. To be precise:

Definition 3.16   *The set* $\mathsf{sf}(F)$ *of* ***sub-formulas*** *of a formula $F \in \mathcal{F}(L)$ is defined by induction as follows:*

- *if F is atomic,*

$$\mathsf{sf}(F) = \{F\};$$

- *if $F = \neg G$,*

$$\mathsf{sf}(F) = \{F\} \cup \mathsf{sf}(G);$$

- *if $F = (G \, \alpha \, H)$ where α is a symbol for a binary connective,*

$$\mathsf{sf}(F) = \{F\} \cup \mathsf{sf}(G) \cup \mathsf{sf}(H);$$

- *if $F = \forall v_k G$ or if $F = \exists v_k(G)$,*

$$\mathsf{sf}(F) = \{F\} \cup \mathsf{sf}(G).$$

3.1.5 Free variables, bound variables, and closed formulas

Definition 3.17 *Given a natural number k and a formula $F \in \mathcal{F}(L)$, the occurrences, if any, of the variable v_k in the formula F can be of two kinds: **free** or **bound**. We proceed by induction:*

- *if F is atomic, then all occurrences of v_k in F are free;*
- *if $F = \neg G$, the free occurrences of v_k in F are the free occurrences of v_k in G;*
- *if $F = (G \, \alpha \, H)$ where α is a symbol for a binary connective, the free occurrences of v_k in F are the free occurrences of v_k in G and the free occurrences of v_k in H;*
- *if $F = \forall v_h G$ or if $F = \exists v_h(G)$, $(h \neq k)$, the free occurrences of v_k in F are the free occurrences of v_k in G;*
- *if $F = \forall v_k G$ or if $F = \exists v_k(G)$, none of the occurrences of v_k in F is free.*
  *The occurrences of v_k in F that are not free are called **bound** .*
  *Concerning the passage from the formula G to the formula $\forall v_k G$ ($\exists v_k G$, respectively) we say that the variable v_k has been **universally quantified** (**existentially quantified**, respectively) or that the formula G has been subjected to **universal quantification** (**existential quantification**, respectively) **with respect to** (or **over**) **the variable** v_k.*

Example 3.18 In the language $L = \{R, c, f\}$ where R is a binary relation symbol, c is a constant symbol and f is a unary function symbol, consider the formula

$F =$

$$\forall v_0 (\exists v_1 \forall v_0 (R v_1 v_0 \Rightarrow \neg v_0 \simeq v_3) \land \forall v_2 (\exists v_2 (R v_1 v_2 \lor f v_0 \simeq c) \land v_2 \simeq v_2)).$$

In F, all occurrences of v_0 and all occurrences of v_2 are bound; the first two occurrences of v_1 are bound while the third is free; finally, the unique occurrence of v_3 is free.

Definition 3.19 *The **free variables** in a formula $F \in \mathcal{F}(L)$ are those variables that have at least one free occurrence in F. A **closed formula** is one in which no variable is free.*

Thus, in the preceding example, the free variables in F are v_1 and v_3. Consequently, F is not a closed formula.

We should also note that a closed formula need not contain any quantifiers: the formula $Rfcc$ is a closed atomic formula in the language of Example 3.18.

However, in a language without constant symbols, there are no closed formulas without quantifiers.

Notation: Given a formula $F \in \mathcal{F}(L)$ and pairwise distinct natural numbers i_1, $i_2, \ldots, i_n$, we will use the notation $F = F[v_{i_1}, v_{i_2}, \ldots, v_{i_n}]$ to indicate that the free variables of the formula F are among $v_{i_1}, v_{i_2}, \ldots, v_{i_n}$.

As is the case for terms, we should note that for every formula $F \in \mathcal{F}(L)$, there exists an integer m such that

$$F = F[v_0, v_1, \ldots, v_m].$$

Definition 3.20 *Given a formula $F = F[v_{i_1}, v_{i_2}, \ldots, v_{i_n}]$ of the language L in which each of the variables $v_{i_1}, v_{i_2}, \ldots, v_{i_n}$ has at least one free occurrence, the formula*

$$\forall v_{i_1} \forall v_{i_2} \ldots \forall v_{i_n} F$$

*and all similar formulas obtained by permuting the order in which the variables $v_{i_1}, v_{i_2}, \ldots, v_{i_n}$ are quantified are called **universal closures** of the formula F.*

Observe that the universal closures of a formula are closed formulas.

Remark 3.21 *We hardly ever distinguish among the universal closures of a formula F and we speak of **the** universal closure of F, intending in this way to denote any one of them (the choice might be dictated by the order in which the free variables of F occur or by the order of their indices or by some other consideration). This abuse of language is unimportant: we will see that the various universal closures of a formula are all equivalent for our intended purposes, both from the semantic point of view (in the sections that follow) and for formal proofs (see Chapter 4).*

Let us return to the formula F in Example 3.18. We said that the third occurrence of the variable v_1, as opposed to the first two, is free. This is because, as the decomposition tree of F makes clear, the quantification $\exists v_1$ 'acts' on the first two occurrences but not on the third. We say that the first two occurrences of v_1 are within the **scope** of the quantifier $\exists v_1$. Anticipating a precise general definition, consider the occurrence in a formula F of a quantifier Qv (where Q denotes $\forall$ or $\exists$ and v the variable that necessarily follows Q in F). The word Qv is necessarily followed, in the word F, by a (unique) sub-formula G of F (the word QvG being, in turn, a sub-formula of F which could be characterized as the sub-formula of least height that contains the occurrence of Qv under consideration).

Definition 3.22 *Using the preceding notation, the occurrences of v that are within the **scope** of the quantifier Qv are the free occurrences of v in G as well as the occurrence of v that immediately follows the quantifier Q.*

For example, in the formula

$$\exists v(((v \simeq v) \vee \forall v \neg (v \simeq v)) \Rightarrow v \simeq v),$$

the occurrences of v that are in the scope of the first quantifier are the first three and the last two. The fourth, fifth and sixth occurrences are excluded, despite their being in the 'geographical range' of $\exists v$, because it is reasonable to agree that each occurrence of a variable is within the scope of at most one quantifier.

3.1.6 Substitutions in formulas

We will now define the notion of **substitution of terms for free variables in a formula**. The variables in a formula necessarily occur inside terms, and since substitution of terms for variables in a term has already been defined, our new definition would be automatic if it were not for an important restriction that will apply: substitution will only take place for the free occurrences of the variables under consideration. The definition is given, as we might expect, by induction.

Definition 3.23 *Consider a formula F, a natural number k, pairwise distinct variables w_1, w_2, ..., w_k and terms u_1, u_2, ..., u_k. The word $F_{u_1/w_1, u_2/w_2, ..., u_k/w_k}$ (read: 'F sub u_1 replaces w_1, u_2 replaces w_2, ..., u_k replaces w_k') is the result of substituting the terms u_1, u_2, ..., u_k respectively for all free occurrences of the variables w_1, w_2, ..., w_k in the formula F and it is defined as follows:*

• *if F is the atomic formula $Rt_1 t_2 \ldots t_n$ (where n is an integer greater than or equal to 1, R is an n-ary relation symbol and t_1, t_2, ..., t_n are terms), then*

$$F_{u_1/w_1, u_2/w_2, ..., u_k/w_k} =$$
$$Rt_{1_{u_1/w_1, u_2/w_2, ..., u_k/w_k}} t_{2_{u_1/w_1, u_2/w_2, ..., u_k/w_k}} \cdots t_{k_{u_1/w_1, u_2/w_2, ..., u_k/w_k}};$$

• *if $F = \neg G$, then*

$$F_{u_1/w_1, u_2/w_2, ..., u_k/w_k} = \neg G_{u_1/w_1, u_2/w_2, ..., u_k/w_k};$$

• *if $F = (G \alpha H)$ where α is a symbol for a binary connective, then*

$$F_{u_1/w_1, u_2/w_2, ..., u_k/w_k} = G_{u_1/w_1, u_2/w_2, ..., u_k/w_k} \alpha H_{u_1/w_1, u_2/w_2, ..., u_k/w_k};$$

• *if $F = \forall v G$ ($v \notin \{w_1, w_2, ..., w_k\}$), then*

$$F_{u_1/w_1, u_2/w_2, ..., u_k/w_k} = \forall v G_{u_1/w_1, u_2/w_2, ..., u_k/w_k};$$

• *if $F = \exists v G$ ($v \notin \{w_1, w_2, ..., w_k\}$), then*

$$F_{u_1/w_1, u_2/w_2, ..., u_k/w_k} = \exists v G_{u_1/w_1, u_2/w_2, ..., u_k/w_k};$$

• *if $F = \forall w_i G$ ($i \in \{1, 2, ..., k\}$), then*

$$F_{u_1/w_1, u_2/w_2, ..., u_k/w_k} = \forall w_i G_{u_1/w_1, u_2/w_2, ..., u_{i-1}/w_{i-1}, u_{i+1}/w_{i+1}, ..., u_k/w_k};$$

- *if $F = \exists w_i G$ ($i \in \{1, 2, \ldots, k\}$), then*

$$F_{u_1/w_1, u_2/w_2, \ldots, u_k/w_k} = \exists w_i \, G_{u_1/w_1, u_2/w_2, \ldots, u_{i-1}/w_{i-1}, u_{i+1}/w_{i+1}, \ldots, u_k/w_k}.$$

Notation: Given two natural numbers h and k, $h + k$ variables $z_1, z_2, \ldots, z_h$, $w_1, w_2, \ldots, w_k$, a formula $F[z_1, z_2, \ldots, z_h, w_1, w_2, \ldots, w_k]$, and k terms u_1, $u_2, \ldots, u_k$, the formula $F_{u_1/w_1, u_2/w_2, \ldots, u_k/w_k}$ will be denoted by

$$F[z_1, z_2, \ldots, z_h, u_1, u_2, \ldots, u_k].$$

This notation requires certain precautions similar to those already noted concerning terms. Remark 3.11 concerning the distinction between simultaneous and successive substitutions and the influence of the order of substitutions is relevant here too.

Example 3.24 Let us revisit the formula that we have used previously as an example:

$F =$

$\forall v_0 (\exists v_1 \forall v_0 (R v_1 v_0 \Rightarrow \neg v_0 \simeq v_3) \wedge \forall v_2 (\exists v_2 (R v_1 v_2 \vee f v_0 \simeq c) \wedge v_2 \simeq v_2)).$

Let t denote the term ffc. Then the word F_{t/v_1} is

$\forall v_0 (\exists v_1 \forall v_0 (R v_1 v_0 \Rightarrow \neg v_0 \simeq v_3) \wedge \forall v_2 (\exists v_2 (R ffc v_2 \vee f v_0 \simeq c) \wedge v_2 \simeq v_2)).$

The result of substituting terms for the free variables in a formula is always a formula:

Lemma 3.25 *Consider a formula F, a natural number k, pairwise distinct variables $w_1, w_2, \ldots, w_k$ and terms $u_1, u_2, \ldots, u_k$. The word $F_{u_1/w_1, u_2/w_2, \ldots, u_k/w_k}$ is a formula.*

Proof The proof is immediate by induction on F. ∎

Another kind of substitution, analogous to what we have already encountered with the propositional calculus, consists in replacing, in a given formula of the language, an occurrence of some sub-formula by some other formula. Without going into the details of a precise definition, we are content to point out what is essential: the result of these substitutions is always a first order formula.

More important, because they are more delicate, are substitutions that we will call **renaming a bound variable**. This involves, specifically, the substitution into a formula of a variable (and not an arbitrary term!) for some given variable at all the occurrences of this variable that are in the scope of some given quantifier. For example, in our formula

$F =$

$\forall v_0 (\exists v_1 \forall v_0 (R v_1 v_0 \Rightarrow \neg v_0 \simeq v_3) \wedge \forall v_2 (\exists v_2 (R v_1 v_2 \vee f v_0 \simeq c) \wedge v_2 \simeq v_2)),$

we can rename the bound variable v_2 by substituting v_5 for all occurrences of v_2 that are within the scope of the quantifier $\forall v_2$. This leads to the formula

$$F =$$

$$\forall v_0 (\exists v_1 \forall v_0 (R v_1 v_0 \Rightarrow \neg v_0 \simeq v_3) \wedge \forall v_5 (\exists v_2 (R v_1 v_2 \vee f v_0 \simeq c) \wedge v_5 \simeq v_5)).$$

In general, if, in a formula H, there is a sub-formula QvG, then changing the name of the variable v to w in the scope of the quantifier Qv consists in simply replacing the sub-formula QvG in H by the formula

$$QwG_{w/v}.$$

So we see that the result obtained is necessarily a formula.

Remark 3.26 *Renaming a bound variable is a procedure that deserves the greatest care. We may be tempted to believe that this is merely an anodine transformation that preserves the 'meaning' that we will later be giving to formulas (in other words, if we may anticipate, which transforms a formula into a logically equivalent formula). However, this may be false if we do not take certain precautions (we will see which ones at the appropriate time: Proposition 3.54 and Chapter 4). For example, in the formula $\exists w \forall v \, v \simeq w$, changing the name of the variable v to w leads to the formula $\exists w \forall w \, w \simeq w$ which, we may suspect, will not have the same meaning as the first.*

We have not reserved any special notation for these name changes of bound variables.

Before concluding this presentation of syntax, we will mention one more type of substitution that is of a slightly different nature from the ones we have treated so far but which is just as common. We will not dally over its definition nor on the fact, important but easy to verify, that the result of these substitutions is always a first order formula.

This involves starting from a formula J of the propositional calculus on an arbitrary set P of propositional variables and substituting for each propositional variable, at each of its occurrences in J, a first order formula of the language L. Here is an example involving the language $L = \{R, f, c\}$ used above. Suppose that A, B and C are propositional variables and consider the propositional formula

$$J = J[A, B, C] = ((A \wedge B) \Rightarrow (\neg A \vee C))$$

and the following three formulas of the language L:

$$F = \forall v_0 \neg R v_1 v_0;$$
$$G = (v_1 \simeq c \Rightarrow \exists v_2 R v_1 f v_2);$$
$$H = \neg f c \simeq c.$$

Then by substituting the formulas F, G, and H respectively for the variables A, B, and C in the propositional formula J, we obtain the first order formula

$$((\forall v_0 \neg R v_1 v_0 \wedge (v_1 \simeq c \Rightarrow \exists v_2 R v_1 f v_2)) \Rightarrow (\neg \forall v_0 \neg R v_1 v_0 \vee \neg f c \simeq c))$$

which we may choose to denote by $J[F, G, H]$ if this does not create an ambiguity.

Remark 3.27 *Every formula without quantifiers is obtained by a substitution of the type just described: it suffices to treat the set of atomic formulas of the language as the propositional variables.*

3.2 Structures

The word 'structure' is generally understood in mathematics to mean a set on which a certain number of functions and relations (or internal operations) are defined along with, upon occasion, what are habitually called 'distinguished elements'. For example, the ordered field of real numbers is the structure

$$\langle \mathbb{R}, \leq, +, \times, 0, 1 \rangle$$

(sometimes, it is considered superfluous to specify the two identity elements so these might be omitted); the additive group of integers is the structure $\langle \mathbb{Z}, + \rangle$ (or $\langle \mathbb{Z}, +, 0 \rangle$). The formulas that we have described in the previous section serve to express properties of such structures. For this purpose, the language must be adapted to the structure under consideration. Thus, we can easily guess that to speak of the ordered field of real numbers, the language will require a binary relation symbol R (intended to represent the order $\leq$), two binary function symbols f and g (for the two operations $+$ and $\times$), and, upon occasion, two constant symbols c and d (for 0 and 1). In this situation, we would express the fact that 1 is an identity element for multiplication by saying that the first order formula

$$\forall v_0 (g v_0 d \simeq v_0 \wedge g d v_0 \simeq v_0)$$

is satisfied in the given structure. As for the formula

$$\exists v_0 \forall v_1 (f v_1 v_0 \simeq v_1 \wedge f v_0 v_1 \simeq v_1),$$

it is satisfied because there exists an identity element for addition. But the formula:

$$\forall v_0 \forall v_1 (R v_0 v_1 \Rightarrow R v_1 v_0)$$

is not satisfied because the binary relation $\leq$ on $\mathbb{R}$ is not symmetric.

What about the formula $R c v_0$? We realize that the question of whether this formula is satisfied or not does not make sense in the absence of any specification concerning the individual v_0. However, it seems natural to say that $R c v_0$ is satisfied when the real represented by v_0 is π and that it is not satisfied when v_0 denotes -1.

So we see that the concept of satisfaction of a formula will need to be defined carefully and that the definition must take into account, in an essential way, the presence or absence of free variables in the formula under consideration. Another observation is forced upon us by these few examples: the syntax that we have defined requires us to break some entrenched habits; if the act of representing multiplication by some symbol other than $\times$ does not bother us much, the change from the usual way of writing $v_0 \, g \, v_1$ to writing $g v_0 v_1$ (so-called 'prefix' or 'Polish' notation) can be more disturbing. However, this Polish notation is necessary if we wish to have a uniform syntax, applicable in all situations and, in particular, to the representation of functions whose arity is greater than 2. The other considerable advantage of Polish notation is to free us from the use of parentheses which, with the standard way of writing binary operations, we could not do without. The same remark applies, to a lesser degree, to atomic formulas: we hardly ever write $\leq v_0 v_1$ instead of $v_0 \leq v_1$, but the prefix notation is none the less encountered occasionally.

The purpose of all these remarks is to prepare the reader for a series of definitions that are marked by the constraints of syntax. Following a path that is familiar to mathematicians, once these definitions have been given, we will immediately begin committing all sorts of abuses, writing $v_0 \times v_1$ and $1 \leq 0$ instead of $g v_0 v_1$ and Rdc and, more generally, taking any measures that render a formula more intelligible, at the risk of scandalizing those for whom rigour is sacrosanct. But we are not there yet!

We will, in an initial phase, give a series of purely algebraic definitions and properties that relate to structures, with syntax involved only incidentally (the language will serve to make precise the type of structure under consideration while the formulas will have no role to play in this first phase). After defining structures, we will examine certain tools that allow us to compare them: sub-structures, restrictions, homomorphisms, isomorphisms.

It is only in Section 3.3 that we will approach the purely logical aspect of things by presenting the notion of satisfaction of a formula in a structure.

3.2.1 Models of a language

Consider a first order language L that is not necessarily a language with equality.

Definition 3.28 *A **model of the language** L, or L-structure, is a structure* $\mathcal{M}$ *consisting of:*

- *a non-empty set M, called the **domain** or the **base set** or the **underlying set** of the structure* $\mathcal{M}$;

- *for each constant symbol c of L, an element* $\bar{c}^{\mathcal{M}}$ *of M called the **interpretation** of the symbol c in the model* $\mathcal{M}$;

- *for every natural number* $k \geq 1$ *and for every k-ary function symbol f of L, a mapping* $\bar{f}^{\mathcal{M}}$ *from* M^k *into M (i.e. a k-ary operation on the set M) called the **interpretation** of the symbol f in the model* $\mathcal{M}$;

- *for every natural number $k \geq 1$ and for every k-ary relation symbol R of L, a subset $\overline{R}^{\mathcal{M}}$ of M^k (i.e. a k-ary relation on the set M) called the **interpretation** of the symbol R in the model $\mathcal{M}$.*

- *In the case where L is a language with equality, we say that the model $\mathcal{M}$ **respects equality** if $\overline{\simeq}^{\mathcal{M}}$, the interpretation in $\mathcal{M}$ of the equality symbol of L, is the equality relation on M (i.e. is the set $\{(a,b) \in M^2 : a = b\}$, also called the **diagonal** of M^2).*

As we have already mentioned, only exceptionally (see Section 3.7) will we deal with languages without equality. Thus, in the absence of any indication to the contrary, 'language' and 'model' will always mean, respectively, 'language with equality' and 'model that respects equality'.

It is important to remember that the base set of a first order structure must be non-empty.

In practice, models will be described in the following way: we will denote models by calligraphic letters (usually $\mathcal{M}$ or $\mathcal{N}$) and will usually use the corresponding Latin letter to denote the underlying set; we will then provide the interpretations of the various symbols for constants, functions and relations (preferably in the same order that was used in the presentation of the language): this may range from a simple enumeration (in circumstances where there are standard symbols or names for these interpretations) to a more laboured definition. Thus, if the language $L = \{R, f, c\}$ involves a binary relation symbol R, a unary function symbol f and a constant symbol c, it will suffice to write

$$\mathcal{N} = \langle \mathbb{R}, \leq, \cos, \pi \rangle$$

to define the model of L in which the base set is the set of reals and in which the interpretations of the symbols R, f and c are respectively the usual order relation, the map $x \mapsto \cos x$, and the real number π. On the other hand, we would need a bit more space to define the L-structure

$$\mathcal{M} = \left\langle M, \overline{R}^{\mathcal{M}}, \overline{f}^{\mathcal{M}}, \overline{c}^{\mathcal{M}} \right\rangle$$

whose underlying set is the set of natural numbers that are not divisible by 5, and in which

- the relation $\overline{R}^{\mathcal{M}}$ is defined by the following: for every a and b in M, $(a,b) \in \overline{R}^{\mathcal{M}}$ if and only if $\gcd(a,b) = 3$;

- the map $\overline{f}^{\mathcal{M}}$ is the one which, to each element $a \in M$, associates the integer $a + 10^a$; and

- $\overline{c}^{\mathcal{M}}$ is the first prime number which, written in base 10, requires at least a million digits.

We emphasize that in this example, the language is one with equality and that the model $\mathcal{M}$ respects equality since there was no mention to the contrary.

It is obviously essential to make a clear distinction between a symbol of the language and its interpretations in various models; this explains the somewhat clumsy notation $\bar{s}^{\mathcal{M}}$ to denote the interpretation of the symbol s in the model $\mathcal{M}$. Despite this, we will omit mention of the model in contexts where there is no possible confusion. It sometimes happens that it is the symbols denoting the relations and operations of a particular structure that determine our choice of symbols of the language appropriate for that structure. Thus, for the structure

$$\mathcal{N} = \langle \mathbb{R}, \leq, \cos, \pi \rangle.$$

we might choose the language $\{\underline{\leq}, \underline{\cos}, \underline{\pi}\}$ where $\underline{\leq}$ is a binary relation symbol, $\underline{\cos}$ is a unary function symbol and $\underline{\pi}$ is a constant symbol. We will have understood that the underlining is, in a way, the inverse of overlining (underlining represents passing from the structure to the language, while overlining (or barring) represents passing from the language to the structure). For example, $\overline{\underline{\cos}}^{\mathcal{N}} = \cos$. This kind of notation will be used in particular for arithmetic (in Chapter 6).

In contexts where additional knowledge of the interpretations of the symbols for constants, functions and relations is not necessary, we will speak of 'an L-structure $\mathcal{M} = \langle M, \dots \rangle$'.

3.2.2 Substructures and restrictions

How can we pass from a structure to another structure that is 'larger'? There are two rather natural ways to imagine this passage: either we enlarge the underlying set and extend the given functions and relations in an appropriate way, keeping the language unchanged; (in this case, the result is called an extension of the original structure); or else, we keep the same underlying set and add new relations, new functions or new constants to this set; we are consequently obliged simultaneously to enrich the language by adding a matching collection of new symbols to it. (In this case, the resulting structure is called an expansion, or enrichment, of the original structure.) It is unfortunate that the use of the two very similar words, 'extension' and 'expansion', can be the source of some confusion; it is vital to avoid this confusion.

Definition 3.29 *Given two L-structures $\mathcal{M} = \langle M, \dots \rangle$ and $\mathcal{N} = \langle N, \dots \rangle$, $\mathcal{M}$ is an **extension** of $\mathcal{N}$ and $\mathcal{N}$ is a **substructure** (or **submodel**) of $\mathcal{M}$ if and only if the following conditions are satisfied:*

- *N is a subset of M;*
- *for every constant symbol c of L,*

$$\bar{c}^{\mathcal{N}} = \bar{c}^{\mathcal{M}};$$

- *for every natural number $k \geq 1$ and every k-ary function symbol f of L,*

$$\bar{f}^{\mathcal{N}} = \bar{f}^{\mathcal{M}} \restriction N^k;$$

- *for every natural number $k \geq 1$ and every k-ary relation symbol R of L,*

$$\overline{R}^{\mathcal{N}} = \overline{R}^{\mathcal{M}} \cap N^k.$$

Thus, for $\mathcal{N}$ to be a substructure of $\mathcal{M}$, the requirement is that the interpretations in $\mathcal{N}$ of the symbols of L be the restrictions to the subset N of their interpretations in $\mathcal{M}$. This has an important consequence for the constants and functions of the structure $\mathcal{M}$. On the one hand, if c is a constant symbol, the element $\overline{c}^{\mathcal{M}}$ of M must belong to the subset N (since $\overline{c}^{\mathcal{N}} = \overline{c}^{\mathcal{M}}$). On the other hand, if f is a k-ary function symbol of the language L, the restriction of the map $\overline{f}^{\mathcal{M}}$ to the subset N^k must be the map $\overline{f}^{\mathcal{N}}$, i.e. a map from N^k into N. We conclude from this that the subset N must be closed (or stable or invariant) under the k-ary operation $\overline{f}^{\mathcal{M}}$. In other words, given an L-structure $\mathcal{M} = \langle M, \dots \rangle$ and a subset $N \subseteq M$, the existence of an L-structure whose base set is N and which is a substructure of $\mathcal{M}$ requires, first of all, that the set N is non-empty, and second, that N contains all the interpretations in $\mathcal{M}$ of the constant symbols of L and is closed under all the functions of the structure $\mathcal{M}$. It is not difficult to verify that these conditions are also sufficient; when they are satisfied, the substructure is obviously unique.

Let us examine, once again, the structure

$$\mathcal{N} = \langle \mathbb{R}, \leq, \cos, \pi \rangle$$

that is a model of the language $L = \langle R, f, c \rangle$. This does not admit a substructure whose underlying set is $[-1, 1]$ because the interpretation of the constant symbol c is π, which does not belong to this subset of $\mathbb{R}$. Nor does it admit a substructure whose underlying set is $[0, \pi]$ since this subset of $\mathbb{R}$ is not closed under the cosine function, which is the interpretation in $\mathcal{N}$ of the function symbol f. There is, on the other hand, a substructure of $\mathcal{N}$ whose underlying set is $A = [-\pi, \pi]$; it is the substructure

$$\mathcal{A} = \langle A, \leq, \cos \upharpoonright A, \pi \rangle.$$

This constraint relating to functions has no counterpart for relations: if L is a language with no symbols for constants nor for functions and if $\mathcal{M} = \langle M, \dots \rangle$ is a model for this language, then for any non-empty subset N of M, there is one (and only one) L-structure whose base set is N and which is a substructure of $\mathcal{M}$: it is the structure in which the interpretation of every relation symbol is obtained by taking the trace on N of its interpretation in $\mathcal{M}$ (namely, for a symbol whose arity is k, its intersection with N^k).

In general, although there may not be a substructure of a given structure whose underlying set is some given subset N, there is, none the less, a substructure that is minimal, in a certain sense, whose underlying set includes the given subset N: this is called the **substructure generated by** N. This concept is described in detail in Exercise 3.12.

Let us now turn to expansions (or enrichments) of structures (and hence of languages).

Definition 3.30 *Let L and L' be two first order languages such that $L \subseteq L'$ (we say, in this case, that L' is an **expansion** or **enrichment** of L and that L is a restriction of L'). Let M be an L-structure and M' be an L'-structure. M' is an **enrichment** (or **expansion**) of M and M is a **restriction** of M' if and only if M and M' have the same underlying set and each symbol for a constant, a function or a relation of the language L has the same interpretation in the L-structure M as in the L'-structure M'.*

Very simply, this means that M' is an enrichment of M if and only if M' is obtained from the L-structure M by adding to it the interpretations of those symbols for constants, functions or relations of the language L' that were not already present in the language L.

We also say that M is the **reduction** (or **reduct**) of M' to the language L.

For example, where L is the language $\{R, f, c\}$ and where L_0 is the language $\{R\}$, the L-structure

$$\mathcal{N} = \langle \mathbb{R}, \leq, \cos, \pi \rangle$$

is an enrichment (or expansion) of the L_0-structure

$$\langle \mathbb{R}, \leq \rangle .$$

3.2.3 Homomorphisms and isomorphisms

A single language L is under consideration here. Let $\mathcal{M} = \langle M, \ldots \rangle$ and $\mathcal{N} = \langle N, \ldots \rangle$ be two L-structures and let ϕ be a map from M into N.

Definition 3.31 *The map ϕ is a **homomorphism of L-structures** from M into N if and only if the following conditions are satisfied:*

- *for every constant symbol c of L,*

$$\phi(\overline{c}^{\mathcal{M}}) = \overline{c}^{\mathcal{N}};$$

- *for every natural number $n \geq 1$, for every n-ary function symbol f of L and for all elements $a_1, a_2, \ldots, a_n$ belonging to M,*

$$\phi(\overline{f}^{\mathcal{M}}(a_1, a_2, \ldots, a_n)) = \overline{f}^{\mathcal{N}}(\phi(a_1), \phi(a_2), \ldots, \phi(a_n));$$

- *for every natural number $k \geq 1$, for every k-ary relation symbol R of L and for all elements $a_1, a_2, \ldots, a_k$ belonging to M,*

$$\text{if } (a_1, a_2, \ldots, a_k) \in \overline{R}^{\mathcal{M}}, \quad \text{then } (\phi(a_1), \phi(a_2), \ldots, \phi(a_k)) \in \overline{R}^{\mathcal{N}}.$$

Thus, a homomorphism from one L-structure into another is a map from the base set of the first into the base set of the second which 'respects' all the relations, functions and constants of these structures.

Definition 3.32 *A **monomorphism** of L-structures from $\mathcal{M}$ into $\mathcal{N}$ is a homomorphism from $\mathcal{M}$ into $\mathcal{N}$ which has the following property:*

$$(*) \quad \left| \begin{array}{l} \textit{for every natural number } k \geq 1, \textit{ for every } k\textit{-ary relation symbol } R \textit{ of } L \\ \textit{and for all elements } a_1, a_2, \ldots, a_k \textit{ belonging to } M, \\ (a_1, a_2, \ldots, a_k) \in \overline{R}^M \textit{ if and only if } (\phi(a_1), \phi(a_2), \ldots, \phi(a_k)) \in \overline{R}^{\mathcal{N}}. \end{array} \right.$$

As our definition of monomorphism, we could just as well taken Definition 3.31 and replaced its third clause by condition $(*)$ above.

Lemma 3.33 *Every monomorphism is injective.*

Proof We must remember here that we are only considering languages with equality and models that respect equality. If ϕ is a monomorphism from $\mathcal{M}$ into $\mathcal{N}$, property $(*)$ applied to the equality symbol $\simeq$ shows that for all elements a and b of M, we have

$$(a, b) \in \simeq^{\mathcal{M}} \quad \text{if and only if } (\phi(a), \phi(b)) \in \simeq^{\mathcal{N}},$$

which is to say that $a = b$ if and only if $\phi(a) = \phi(b)$. ∎

Lemma 3.34 *Let $\mathcal{N} = \langle N, \ldots \rangle$ be an L-structure and let N_1 be a subset of N; for the existence of a substructure of $\mathcal{N}$ whose underlying set is N_1, it is necessary and sufficient that there exists an L-structure $\mathcal{M} = \langle M, \ldots \rangle$ and a monomorphism ϕ from $\mathcal{M}$ into $\mathcal{N}$ such that the subset N_1 is the image of ϕ.*

Proof First, suppose that there is a substructure $\mathcal{N}_1$ of $\mathcal{N}$ whose base set is N_1. Then Definitions 3.29 and 3.32 clearly show that the identity map from N_1 into N_1 is a monomorphism from $\mathcal{N}_1$ into $\mathcal{N}$ whose image is N_1.

Conversely, suppose there exists an L-structure $\mathcal{M} = \langle M, \ldots \rangle$ and a monomorphism ϕ from $\mathcal{M}$ into $\mathcal{N}$ whose image is N_1. Then for every constant symbol c of L, we have $\overline{c}^{\mathcal{N}} = \phi(\overline{c}^{\mathcal{M}})$, hence $\overline{c}^{\mathcal{N}} \in N_1$; similarly, for every k-ary function symbol f of L ($k \geq 1$) and for all elements $a_1, a_2, \ldots, a_k$ belonging to N_1, we can find elements $b_1, b_2, \ldots, b_k$ of M such that $\phi(b_i) = a_i$ for $1 \leq i \leq k$, so we then have

$$\overline{f}^{\mathcal{N}}(a_1, a_2, \ldots, a_n) = \phi(\overline{f}^{\mathcal{M}}(b_1, b_2, \ldots, b_n)),$$

which proves that $\overline{f}^{\mathcal{N}}(a_1, a_2, \ldots, a_n)$ belongs to N_1. We may now conclude, based on the paragraphs that followed Definition 3.29, that there exists a substructure of $\mathcal{N}$ whose base set is N_1. ∎

Definition 3.35 *An **isomorphism** from an L-structure $\mathcal{M}$ onto another L-structure $\mathcal{N}$ is a monomorphism from $\mathcal{M}$ into $\mathcal{N}$ that is surjective.*

An **automorphism** *of an L-structure* $\mathcal{M}$ *is an isomorphism from* $\mathcal{M}$ *onto* $\mathcal{M}$.

It is clear that if a bijective map $\phi : M \to N$ is an isomorphism of the structure $\mathcal{M}$ onto the structure $\mathcal{N}$, then the inverse map $\phi^{-1} : N \to M$ is an isomorphism of $\mathcal{N}$ onto $\mathcal{M}$. If there is an isomorphism between two structures, they are said to be **isomorphic**.

Remark 3.36 *It follows from Lemma 3.34 that any monomorphism from a structure* $\mathcal{M} = \langle M, \ldots \rangle$ *into a structure* $\mathcal{N} = \langle N, \ldots \rangle$ *may be considered as an isomorphism of* $\mathcal{M}$ *onto a substructure of* $\mathcal{N}$.

Example 3.37 (in which the details are left to the reader)

- Where the language consists of one constant symbol c and one binary function symbol g, the structures $\langle \mathbb{R}_+^*, 1, \times \rangle$ and $\langle \mathbb{R}, 0, + \rangle$ are isomorphic; the map $x \longmapsto \ln x$ from $\mathbb{R}_+^*$ into $\mathbb{R}$ is witness to this fact.

- With this same language, the map $n \longmapsto (-1)^n$ is a homomorphism from the structure $\langle \mathbb{Z}, 0, + \rangle$ into the structure $\langle \{-1, 1\}, 1, \times \rangle$.

- In the language whose only symbol is the binary relation symbol R, the structures $\langle \mathbb{R}, \leq \rangle$ and $\langle (0, 1), \leq \rangle$ are isomorphic thanks to the map

$$x \longmapsto \frac{1}{2} + \frac{1}{\pi} \arctan x$$

from $\mathbb{R}$ into $(0, 1)$.

  However, the identity map from $(0, 1)$ into $\mathbb{R}$ is only a monomorphism. The reader will have noted the abuse of language that consists in using the same symbol to denote the order relation in $\mathbb{R}$ and the one in $(0, 1)$.

- With this same language, consider the structures $\mathcal{M} = \langle \{0, 1\}, = \rangle$ and $\mathcal{N} = \langle \{0, 1\}, \leq \rangle$. The identity map from $\{0, 1\}$ into $\{0, 1\}$ is obviously bijective and is a homomorphism from $\mathcal{M}$ into $\mathcal{N}$, but it is not an isomorphism. This shows that we cannot replace 'monomorphism' by 'homomorphism' in Definition 3.35.

3.3 Satisfaction of formulas in structures

3.3.1 Interpretation in a structure of the terms

We have already mentioned that the terms of a language will serve to denote objects. Suppose that the language involves a constant symbol c and two function symbols f and g that are unary and binary, respectively. We suspect that in a structure $\mathcal{M} = \langle M, \overline{c}, \overline{f}, \overline{g} \rangle$, the term ffc will be interpreted by $\overline{f}(\overline{f}(\overline{c}))$, which is an element of M, and the term $gfcgcc$ by the element $\overline{g}(\overline{f}(\overline{c}), \overline{g}(\overline{c}, \overline{c}))$. But to interpret the term fv_0, we first need to know what element is denoted by v_0. Now, we will not be giving fixed interpretations in a structure for the variables (otherwise, we wouldn't have called them variables). To be more precise, the interpretation in a structure of a symbol for a variable can *vary*. This leads us to the fact that the interpretation of the term $f v_0$ will depend on the interpretation given to v_0. Thus for any element a of M, we will say, when the interpretation of v_0 is a, that the interpretation of the term $f v_0$ in $\mathcal{M}$ is the element $\overline{f}(a)$. It is obvious that, when the interpretation of v_4 is a, the term $f v_4$ will have this same interpretation. As for the term $g f g v_2 c v_1$, it will be interpreted, when v_1 is interpreted by a and v_2 by b, by the element

$$\overline{g}(\overline{f}(\overline{g}(b, \overline{c})), a).$$

Definition 3.38 *Let n be a natural number, let $w_0, w_1, \ldots, w_{n-1}$ be n pairwise distinct variables, let $t = t[w_0, w_1, \ldots, w_{n-1}]$ be a term of the language L, let $\mathcal{M} = \langle M, \ldots \rangle$ be an L-structure and let $a_0, a_1, \ldots, a_{n-1}$ be n elements of M; the* **interpretation of the term t in the L-structure $\mathcal{M}$ when the variables $w_0, w_1, \ldots, w_{n-1}$ are interpreted respectively by the elements** *$a_0, a_1, \ldots, a_{n-1}$ is an element of M; it is denoted by*

$$\overline{t}^{\mathcal{M}}[w_0 \to a_0, w_1 \to a_1, \ldots, w_{n-1} \to a_{n-1}]$$

and is defined by induction on t as follows:

- *if $t = w_j$ $(0 \le j \le n - 1)$,*

$$\overline{t}^{\mathcal{M}}[w_0 \to a_0, w_1 \to a_1, \ldots, w_{n-1} \to a_{n-1}] = a_j;$$

- *if $t = c$ (a constant symbol of L),*

$$\overline{t}^{\mathcal{M}}[w_0 \to a_0, w_1 \to a_1, \ldots, w_{n-1} \to a_{n-1}] = \overline{c}^{\mathcal{M}};$$

- *if $t = f t_1 t_2 \ldots t_k$ (where $k \in \mathbb{N}^*$, f is a k-ary function symbol of L and $t_1, t_2, \ldots, t_k$ are terms of L),*

$$\overline{t}^{\mathcal{M}}[w_0 \to a_0, w_1 \to a_1, \ldots, w_{n-1} \to a_{n-1}] =$$
$$\overline{f}^{\mathcal{M}}(\overline{t_1}^{\mathcal{M}}[w_0 \to a_0, \ldots, w_{n-1} \to a_{n-1}],$$
$$\overline{t_2}^{\mathcal{M}}[w_0 \to a_0, \ldots, w_{n-1} \to a_{n-1}],$$
$$\ldots, \overline{t_k}^{\mathcal{M}}[w_0 \to a_0, w_1 \to a_1, \ldots, w_{n-1} \to a_{n-1}]).$$

In practice, we will denote the element $\overline{t}^{\mathcal{M}}[w_0 \to a_0, \ldots, w_{n-1} \to a_{n-1}]$ more simply by

$$\overline{t}^{\mathcal{M}}[a_0, a_1, \ldots a_{n-1}]$$

despite the fact that this notation is ambiguous: indeed, it contains no reference to a specific sequence of free variables (including those that have one or more occurrences in t); there are, in fact, an infinite number of such sequences which differ from one another either by containing extra variables (that do not occur in t) or by the order in which these variables are listed. Were it not for this ambiguity, it would have been practical to define, as is sometimes done, the interpretation of t

in the structure $\mathcal{M}$ as a map from M^n into M, i.e. the map that, to each n-tuple $(a_0, a_1, \ldots, a_{n-1})$, assigns the element

$$\bar{t}^{\mathcal{M}}[w_0 \to a_0, w_1 \to a_1, \ldots, w_{n-1} \to a_{n-1}].$$

None the less, this simplified notation will prevail in most concrete situations in which the context removes all ambiguity. For example, in the language under consideration at the beginning of this subsection, if t is the term $g v_0 f v_1$ and $\mathcal{M}$ is the structure $\langle \mathbb{R}, 0, \cos, + \rangle$, everyone will understand that for all reals a and b, $\bar{t}^{\mathcal{M}}[a, b]$ denotes the real $a + \cos b$ (however, the situation would already be less clear if the term had been $g v_1 f v_0$: it would be prudent in this instance to be precise, for example, by specifying $t = t[v_0, v_1]$ or by relying on the official notation).

Remark 3.39 *In the preceding definition, it is clear that the order in which the interpretations of the variables are specified is immaterial: precisely, for any permutation σ of the set $\{0, 1, \ldots, n-1\}$, we have*

$$\bar{t}^{\mathcal{M}}[w_0 \to a_0, \ldots, w_{n-1} \to a_{n-1}]$$
$$= \bar{t}^{\mathcal{M}}[w_{\sigma(0)} \to a_{\sigma(0)}, \ldots, w_{\sigma(n-1)} \to a_{\sigma(n-1)}].$$

Strictly speaking, this would require a proof by induction on t, but it is obvious.

The purpose of the lemma that follows is to show that the interpretation of a term in a structure does not depend on the values assigned to variables that do not occur in it.

Lemma 3.40 *Let m and n be two natural numbers, let $w_0, w_1, \ldots, w_{n-1}, z_0, z_1, \ldots, z_{m-1}$ be $m + n$ pairwise distinct variables and let t be a term of L whose variables are among $w_0, w_1, \ldots, w_{n-1}$ (while $z_0, z_1, \ldots, z_{m-1}$ do not occur in t) so that it makes sense to write*

$$t = t[w_0, w_1, \ldots, w_{n-1}] = t[w_0, w_1, \ldots, w_{n-1}, z_0, z_1, \ldots, z_{m-1}];$$

then for any L-structure $\mathcal{M} = \langle M, \ldots \rangle$ and for any elements $a_0, a_1, \ldots, a_{n-1}, b_0, b_1, \ldots, b_{m-1}$ of M, we have

$$\bar{t}^{\mathcal{M}}[w_0 \to a_0, w_1 \to a_1, \ldots, w_{n-1} \to a_{n-1}]$$
$$= \bar{t}^{\mathcal{M}}[w_0 \to a_0, \ldots, w_{n-1} \to a_{n-1}, z_0 \to b_0, \ldots, z_{m-1} \to b_{m-1}].$$

Proof The proof is immediate by induction on t.   ∎

We will now examine the effect of a substitution in a term on its interpretation.

Proposition 3.41 *Let n be a natural number and let $v, w_0, w_1, \ldots, w_{n-1}$ be $n+1$ pairwise distinct variables; consider two terms of L, $t = t[w_0, w_1, \ldots, w_{n-1}]$ and $u = u[v, w_0, w_1, \ldots, w_{n-1}]$ and let r denote the term $u[t, w_0, w_1, \ldots, w_{n-1}]$, i.e. r is the term $u_{t/v}$.*

Then for any L-structure $\mathcal{M} = \langle M, \ldots \rangle$ and for any elements $a_0, a_1, \ldots, a_{n-1}$ of M, we have

$$\bar{r}^{\mathcal{M}}[w_0 \to a_0, w_1 \to a_1, \ldots, w_{n-1} \to a_{n-1}]$$
$$= \bar{u}^{\mathcal{M}}\left[v \to \bar{t}^{\mathcal{M}}[w_0 \to a_0, \ldots, w_{n-1} \to a_{n-1}], w_0 \to a_0, \ldots, w_{n-1} \to a_{n-1}\right].$$

Proof Note first of all that the variables that occur in r are among $w_0, w_1, \ldots, w_{n-1}$, so that the term on the left side of the equality makes sense. The equality is proved by induction on the term u:

- if u is a constant symbol c of L, we have $r = u = c$ and each side of the equality denotes the element $\bar{c}^{\mathcal{M}}$;
- if u is the variable w_i ($0 \leq i \leq n-1$), $r = u = w_i$ and each side of the equality denotes the element a_i;
- if u is the variable v, we have $r = t$ and each side of the equality denotes the element $\bar{t}^{\mathcal{M}}[w_0 \to a_0, w_1 \to a_1, \ldots, w_{n-1} \to a_{n-1}]$;
- if $u = fu_1u_2\cdots u_k$ (where $k \in \mathbb{N}^*$, f is a k-ary function symbol of L and $u_1, u_2, \ldots, u_k$ are terms of L), then by setting $r_1 = u_{1_{t/v}}$, $r_2 = u_{2_{t/v}}$, $\ldots$, $r_k = u_{k_{t/v}}$, we have

$$r = fr_1r_2\cdots r_k;$$

the induction hypothesis is that for $i \in \{1, 2, \ldots, k\}$,

$$\bar{r_i}^{\mathcal{M}}[w_0 \to a_0, w_1 \to a_1, \ldots, w_{n-1} \to a_{n-1}] =$$
$$\bar{u_i}^{\mathcal{M}}\left[v \to \bar{t}^{\mathcal{M}}[w_0 \to a_0, \ldots, w_{n-1} \to a_{n-1}], w_0 \to a_0, \ldots, w_{n-1} \to a_{n-1}\right];$$

so we see, by referring to the definition above, that

$$\bar{r}^{\mathcal{M}}[w_0 \to a_0, w_1 \to a_1, \ldots, w_{n-1} \to a_{n-1}] =$$
$$\bar{u}^{\mathcal{M}}\left[v \to \bar{t}^{\mathcal{M}}[w_0 \to a_0, \ldots, w_{n-1} \to a_{n-1}], w_0 \to a_0, \ldots, w_{n-1} \to a_{n-1}\right].$$

∎

3.3.2 Satisfaction of the formulas in a structure

We are given a language L, an L-structure $\mathcal{M} = \langle M, \ldots \rangle$, a natural number n, n pairwise distinct variables $w_0, w_1, \ldots, w_{n-1}$, n elements $a_0, a_1, \ldots, a_{n-1}$ of M and a formula

$$F = F[w_0, w_1, \ldots, w_{n-1}] \in \mathcal{F}(L).$$

The definition that follows, which will be given by induction on the formula F, will give meaning to the following phrase:

'the formula F is satisfied in the structure $\mathcal{M}$ when the variables $w_0, w_1, \ldots, w_{n-1}$ are interpreted respectively by the elements $a_0, a_1, \ldots, a_{n-1}$.'

The notation for this property will be

$$\langle \mathcal{M} : w_0 \rightarrow a_0, w_1 \rightarrow a_1, \ldots, w_{n-1} \rightarrow a_{n-1} \rangle \models F$$

(the symbol $\models$ is read 'satisfies').

In practice, as with the interpretation of terms, we will most often resort to a way of speaking and a notation that are less cumbersome, but not without ambiguity. We will write

$$\mathcal{M} \models F[a_0, a_1, \ldots, a_{n-1}]$$

and will say

$$\mathcal{M} \text{ satisfies } F \text{ of } a_0, a_1, \ldots, a_{n-1};$$

or

the n-tuple (or the sequence) $(a_0, a_1, \ldots, a_{n-1})$ satisfies the formula F in $\mathcal{M}$;

or

the formula F is satisfied in $\mathcal{M}$ by the n-tuple $(a_0, a_1, \ldots, a_{n-1})$;

or

the n-tuple $(a_0, a_1, \ldots, a_{n-1})$ satisfies the formula $F[w_0, w_1, \ldots, w_{n-1}]$ in $\mathcal{M}$.

The last of these formulations is intended to recall our understanding that w_0, $w_1, \ldots, w_{n-1}$ are to be interpreted respectively by $a_0, a_1, \ldots, a_{n-1}$; the first three formulations ignore this. The ambiguity here is analogous to the one mentioned previously concerning terms: the choice of some fixed ordered list of variables that includes the free variables of F is left unspecified. But, just as for terms, the context will most often make things clear.

For the moment, the reader is invited to interpret the notation

$$\mathcal{M} \models F[a_0, a_1, \ldots, a_{n-1}]$$

as mere shorthand; one should not consider that $F[a_0, a_1, \ldots, a_{n-1}]$ denotes a formula. In fact, none of what has preceded would authorize this. None the less, such a point of view will be possible a bit later; we will at that point have the means to justify it (Theorem 3.86).

The negation of '$\langle \mathcal{M} : w_0 \to a_0, w_1 \to a_1, \ldots, w_{n-1} \to a_{n-1}\rangle \models F$' is written

$$\langle \mathcal{M} : w_0 \to a_0, w_1 \to a_1, \ldots, w_{n-1} \to a_{n-1}\rangle \nvDash F$$

or again, using the simplified notation

$$\mathcal{M} \nvDash F[a_0, a_1, \ldots, a_{n-1}].$$

Here is the promised definition.

Definition 3.42

- 1. *If F is the atomic formula $Rt_1 t_2 \cdots t_k$ where k is a natural number greater than or equal to 1, R is a k-ary relation symbol of L and $t_1, t_2, \ldots, t_k$ are terms of L (such that for each $i \in \{1, 2, \ldots, k\}$, $t_i = t_i[w_0, w_1, \ldots, w_{n-1}]$), we have $\langle \mathcal{M} : w_0 \to a_0, w_1 \to a_1, \ldots, w_{n-1} \to a_{n-1}\rangle \models F$ if and only if $(\overline{t_1}^{\mathcal{M}}[w_0 \to a_0, \ldots, w_{n-1} \to a_{n-1}], \ldots, \overline{t_k}^{\mathcal{M}}[w_0 \to a_0, \ldots, w_{n-1} \to a_{n-1}]) \in \overline{R}^{\mathcal{M}}$;*

  (in particular, if L is a language with equality and if $\mathcal{M}$ is a model that respects equality, we have
  $\langle \mathcal{M} : w_0 \to a_0, w_1 \to a_1, \ldots, w_{n-1} \to a_{n-1}\rangle \models t_1 \simeq t_2$ if and only if
  $\overline{t_1}^{\mathcal{M}}[w_0 \to a_0, \ldots, w_{n-1} \to a_{n-1}] = \overline{t_2}^{\mathcal{M}}[w_0 \to a_0, \ldots, w_{n-1} \to a_{n-1}]$.

- 2. *If $F = \neg G$:*
  $\langle \mathcal{M} : w_0 \to a_0, w_1 \to a_1, \ldots, w_{n-1} \to a_{n-1}\rangle \models F$ if and only if
  $\langle \mathcal{M} : w_0 \to a_0, w_1 \to a_1, \ldots, w_{n-1} \to a_{n-1}\rangle \nvDash G$.

- 3. *If $F = (G \wedge H)$:*
  $\langle \mathcal{M} : w_0 \to a_0, w_1 \to a_1, \ldots, w_{n-1} \to a_{n-1}\rangle \models F$ if and only if
  $\langle \mathcal{M} : w_0 \to a_0, w_1 \to a_1, \ldots, w_{n-1} \to a_{n-1}\rangle \models G$ and
  $\langle \mathcal{M} : w_0 \to a_0, w_1 \to a_1, \ldots, w_{n-1} \to a_{n-1}\rangle \models H$.

- 4. *If $F = (G \vee H)$:*
  $\langle \mathcal{M} : w_0 \to a_0, w_1 \to a_1, \ldots, w_{n-1} \to a_{n-1}\rangle \models F$ if and only if
  $\langle \mathcal{M} : w_0 \to a_0, w_1 \to a_1, \ldots, w_{n-1} \to a_{n-1}\rangle \models G$ or
  $\langle \mathcal{M} : w_0 \to a_0, w_1 \to a_1, \ldots, w_{n-1} \to a_{n-1}\rangle \models H$.

- 5. *If $F = (G \Rightarrow H)$:*
  $\langle \mathcal{M} : w_0 \to a_0, w_1 \to a_1, \ldots, w_{n-1} \to a_{n-1}\rangle \models F$ if and only if
  $\langle \mathcal{M} : w_0 \to a_0, w_1 \to a_1, \ldots, w_{n-1} \to a_{n-1}\rangle \nvDash G$ or
  $\langle \mathcal{M} : w_0 \to a_0, w_1 \to a_1, \ldots, w_{n-1} \to a_{n-1}\rangle \models H$.

- 6. *If $F = (G \Leftrightarrow H)$:*
  $\langle \mathcal{M} : w_0 \to a_0, w_1 \to a_1, \ldots, w_{n-1} \to a_{n-1}\rangle \models F$ if and only if
  $\langle \mathcal{M} : w_0 \to a_0, w_1 \to a_1, \ldots, w_{n-1} \to a_{n-1}\rangle \models G$ and
  $\langle \mathcal{M} : w_0 \to a_0, w_1 \to a_1, \ldots, w_{n-1} \to a_{n-1}\rangle \models H$;

  or else
  $\langle \mathcal{M} : w_0 \to a_0, w_1 \to a_1, \ldots, w_{n-1} \to a_{n-1}\rangle \nvDash G$ and
  $\langle \mathcal{M} : w_0 \to a_0, w_1 \to a_1, \ldots, w_{n-1} \to a_{n-1}\rangle \nvDash H$.

- 7. *If* $F = \forall v G$ *(where* $v \in V - \{w_0, w_1, \ldots, w_{n-1}\}$*):*
  $\langle \mathcal{M} : w_0 \to a_0, w_1 \to a_1, \ldots, w_{n-1} \to a_{n-1} \rangle \models F$ *if and only if for every element* $a \in M$,
  $$\langle \mathcal{M} : v \to a, w_0 \to a_0, w_1 \to a_1, \ldots, w_{n-1} \to a_{n-1} \rangle \models G.$$
- 8. *If* $F = \exists v G$ *(where* $v \in V - \{w_0, w_1, \ldots, w_{n-1}\}$*):*
  $\langle \mathcal{M} : w_0 \to a_0, w_1 \to a_1, \ldots, w_{n-1} \to a_{n-1} \rangle \models F$ *if and only if for at least one element* $a \in M$,
  $$\langle \mathcal{M} : v \to a, w_0 \to a_0, w_1 \to a_1, \ldots, w_{n-1} \to a_{n-1} \rangle \models G.$$
- 9. *If* $F = \forall w_i G$ *(where* $0 \le i \le n-1$*),*
  $\langle \mathcal{M} : w_0 \to a_0, w_1 \to a_1, \ldots, w_{n-1} \to a_{n-1} \rangle \models F$ *if and only if for every element* $a \in M$,
  $\langle \mathcal{M} : w_0 \to a_0, w_{i-1} \to a_{i-1}, w_i \to a, w_{i+1} \to a_{i+1}, w_{n-1} \to a_{n-1} \rangle \models G.$
- 10. *If* $F = \exists w_i G$ *(where* $0 \le i \le n-1$*),*
  $\langle \mathcal{M} : w_0 \to a_0, w_1 \to a_1, \ldots, w_{n-1} \to a_{n-1} \rangle \models F$ *if and only if for at least one element* $a \in M$,
  $\langle \mathcal{M} : w_0 \to a_0, w_{i-1} \to a_{i-1}, w_i \to a, w_{i+1} \to a_{i+1}, w_{n-1} \to a_{n-1} \rangle \models G.$

For a correct reading of this definition, it is appropriate to recall that in clauses 2, 3, 4, 5, and 6, the free variables of the formula G as well as those of H are among $w_0, w_1, \ldots, w_{n-1}$; in clauses 7 and 8, the free variables of G are among $v, w_0, w_1, \ldots, w_{n-1}$; and finally in clauses 9 and 10, the free variables of G are among $w_0, w_{i-1}, w_{i+1}, w_{n-1}$ (the variable w_i is no longer free in F, though this in no way prevents us from considering that $F = F[w_0, w_1, \ldots, w_{n-1}]$).

This definition notably applies to the case when the formula F is closed. In this context the property becomes

$$\mathcal{M} \models F$$

which is read: '$\mathcal{M}$ satisfies F'. When this property is satisfied, we also say that F is **true** in $\mathcal{M}$, or also that $\mathcal{M}$ **is a model of** F.

Remark 3.43 *The definition of satisfaction does not depend on the order in which the interpretations of the variables are specified. This means that for any permutation* σ *of the set* $\{0, 1, \ldots, n-1\}$, *we have*

$$\langle \mathcal{M} : w_0 \to a_0, w_1 \to a_1, \ldots, w_{n-1} \to a_{n-1} \rangle \models F \textit{ if and only if}$$

$$\left\langle \mathcal{M} : w_{\sigma(0)} \to a_{\sigma(0)}, w_{\sigma(1)} \to a_{\sigma(1)}, \ldots, w_{\sigma(n-1)} \to a_{\sigma(n-1)} \right\rangle \models F.$$

The proof is obvious: the argument is by induction on F; the case for atomic formulas is governed by Remark 3.39; the rest goes without saying.

We have observed that the list of variables that includes the free variables of a given formula can be artificially lengthened (by adding to this list variables that have no free occurrence in the formula). It is natural to ask whether this can have any effect on the notion of satisfaction that has just been defined. The answer, in the negative, is given by the next lemma.

Lemma 3.44 *Let m and n be two natural numbers, let w_0, w_1, ..., w_{n-1}, z_0, z_1, ..., z_{m-1} be $m+n$ pairwise distinct variables and let F be a formula of L whose free variables are among w_0, w_1, ..., w_{n-1} (while z_0, z_1, ..., z_{m-1} have no free occurrence in F) so that it makes sense to write*

$$F = F[w_0, w_1, \ldots, w_{n-1}] = F[w_0, w_1, \ldots, w_{n-1}, z_0, z_1, \ldots, z_{m-1}].$$

Then for any L-structure $\mathcal{M} = \langle M, \ldots \rangle$ and for any elements a_0, a_1, ..., a_{n-1}, b_0, b_1, ..., b_{m-1} of M, the following properties are equivalent:

(1) $\langle \mathcal{M} : w_0 \to a_0, w_1 \to a_1, \ldots, w_{n-1} \to a_{n-1} \rangle \models F$

(2) $\langle \mathcal{M} : w_0 \to a_0, \ldots, w_{n-1} \to a_{n-1}, z_0 \to b_0, \ldots, z_{m-1} \to b_{m-1} \rangle \models F$

Proof The proof is, of course, by induction on F.

• If F is the atomic formula $Rt_1 t_2 \ldots t_k$ where k is a natural number greater than or equal to 1, R is a k-ary relation symbol of L and t_1, t_2, ..., t_k are terms of L, then, by hypothesis, for each $i \in \{1, 2, \ldots, k\}$, we are free to write one or the other of

$$t_i = t_i[w_0, w_1, \ldots, w_{n-1}] \text{ or } t_i = t_i[w_0, w_1, \ldots, w_{n-1}, z_0, z_1, \ldots, z_{m-1}].$$

So we may conclude, using Lemma 3.40, that

$$\overline{t_i}^{\mathcal{M}}[w_0 \to a_0, w_1 \to a_1, \ldots, w_{n-1} \to a_{n-1}]$$
$$= \overline{t_i}^{\mathcal{M}}[w_0 \to a_0, \ldots, w_{n-1} \to a_{n-1}, z_0 \to b_0, \ldots, z_{m-1} \to b_{m-1}],$$

which, by the definition of satisfaction (Clause 1), yields the equivalence between (1) and (2).

• For those stages of the induction that refer to the symbols for connectives, the proof is obvious.

• It is also obvious for the cases where $F = \forall v G$ or $F = \exists v G$ when the variable v does not belong to $\{z_0, z_1, \ldots, z_{m-1}\}$.

• If $F = \forall z_h G$ where $h \in \{0, 1, \ldots, m-1\}$, then the free variables of G are among z_h, w_0, w_1, ..., w_{n-1}. Property (1) is verified if and only if for every element b of M,

$$\langle \mathcal{M} : z_h \to b, w_0 \to a_0, w_1 \to a_1, \ldots, w_{n-1} \to a_{n-1} \rangle \models G,$$

which is equivalent, by the induction hypothesis and Remark 3.43, to: for all $b \in M$,

$$\langle \mathcal{M} : w_0 \to a_0, \ldots, w_{n-1} \to a_{n-1}, z_0 \to b_0, \ldots,$$
$$\ldots, z_{h-1} \to b_{h-1}, z_h \to b, z_{h+1} \to b_{h+1}, \ldots, z_{m-1} \to b_{m-1} \rangle \models G,$$

but this means, by definition (Clause 9):

$$\langle \mathcal{M} : w_0 \to a_0, \ldots, w_{n-1} \to a_{n-1}, z_0 \to b_0, \ldots,$$
$$\ldots, z_{h-1} \to b_{h-1}, z_h \to b_h, z_{h+1} \to b_{h+1}, \ldots, z_{m-1} \to b_{m-1} \rangle \models \forall v_h G,$$

which is precisely property (2).

- The case $F = \exists v_h G$ where $h \in \{0, 1, \ldots, m - 1\}$ is treated analogously. ■

Here is a very useful consequence of the definition of satisfaction; it concerns substitutions into formulas.

Proposition 3.45 *Suppose n and p are natural numbers, v, w_0, w_1, ..., w_{n-1}, $u_0, u_1, \ldots, u_{p-1}$ are $n+p+1$ pairwise distinct variables, $t = t[w_0, w_1, \ldots, w_{n-1}]$ is a term of L and $F = F[v, w_0, w_1, \ldots, w_{n-1}, u_0, u_1, \ldots, u_{p-1}]$ is a formula of L. Suppose further that in the formula F, there is no free occurrence of v in the scope of any quantification $\forall w_i$ or $\exists w_i$ ($0 \le i \le n - 1$).*

Then for any L-structure $\mathcal{M} = \langle M, \ldots \rangle$ and for any elements $a_0, a_1, \ldots, a_{n-1}$, $b_0, b_1, \ldots, b_{p-1}$ of M, the following two properties are equivalent:

(1) $\langle \mathcal{M} : w_0 \to a_0, \ldots, w_{n-1} \to a_{n-1}, u_0 \to b_0, \ldots, u_{p-1} \to b_{p-1} \rangle \models F_{t/v}$;

(2) $\langle \mathcal{M} : v \to \bar{t}^{\mathcal{M}}[w_0 \to a_0, w_1 \to a_1, \ldots, w_{n-1} \to a_{n-1}], w_0 \to a_0, w_1 \to a_1, \ldots, w_{n-1} \to a_{n-1}, u_0 \to b_0, u_1 \to b_1, \ldots, u_{p-1} \to b_{p-1} \rangle \models F.$

Proof Note first of all that the free variables of the formula $F_{t/v}$ are among w_0, $w_1, \ldots, w_{n-1}, u_0, u_1, \ldots, u_{p-1}$, which shows that property (1) makes sense. The proof proceeds by induction on F.

- If F is the atomic formula $Rt_1 t_2 \ldots t_k$ where k is a natural number greater than or equal to 1, R is a k-ary relation symbol of L and $t_1, t_2, \ldots, t_k$ are terms of L, then, for each $i \in \{1, 2, \ldots, k\}$, we may write

$$t_i = t_i[v, w_0, w_1, \ldots, w_{n-1}, u_0, u_1, \ldots, u_{p-1}]$$

and, according to Clause 1 in the definition of satisfaction, if we set $r_i = t_{i_{t/v}}$, property (1) means

$$\left(\bar{r_1}^{\mathcal{M}}[w_0 \to a_0, \ldots, w_{n-1} \to a_{n-1}, u_0 \to b_0, \ldots, u_{p-1} \to b_{p-1}], \ldots \right.$$
$$\left. \ldots, \bar{r_k}^{\mathcal{M}}[w_0 \to a_0, \ldots, w_{n-1} \to a_{n-1}, u_0 \to b_0, \ldots, u_{p-1} \to b_{p-1}] \right)$$
$$\in \bar{R}^{\mathcal{M}};$$

now, by virtue of Proposition 3.41, if for each $i \in \{1, 2, \ldots, k\}$ we set

$$b_i = \bar{t_i}^{\mathcal{M}}\left[v \to \bar{t}^{\mathcal{M}}[w_0 \to a_0, \ldots, w_{n-1} \to a_{n-1}], w_0 \to a_0, \ldots\right.$$
$$\left. \ldots, w_{n-1} \to a_{n-1}, u_0 \to b_0, \ldots, u_{p-1} \to b_{p-1}]\right],$$

this becomes equivalent to

$$(b_1, b_2, \ldots, b_k) \in \overline{R}^{\mathcal{M}},$$

i.e. (by Clause 1 of the definition) to the following assertion, which is property (2):

$$\big\langle \mathcal{M} : v \to \overline{t}^{\mathcal{M}}[w_0 \to a_0, w_1 \to a_1, \ldots, w_{n-1} \to a_{m-1}], \ w_0 \to a_0,$$
$$w_1 \to a_1, \ldots, w_{n-1} \to a_{n-1}, u_0 \to b_0, u_1 \to b_1, \ldots, u_{p-1} \to b_{p-1} \big\rangle$$
$$\models F.$$

- The stages of the induction that concern the symbols for connectives are obvious.
- Suppose F is the formula $\exists z G$; the free variables of G are among z, v, w_0, $w_1, \ldots, w_{n-1}, u_0, u_1, \ldots, u_{p-1}$. If z is one of the w_i, then by hypothesis v has no free occurrence in G, nor in F; if $z = v$, v has no free occurrence in F; in both these cases, $F_{t/v} = F$; the equivalence of (1) and (2) is then a straightforward consequence of the previous lemma. If z is different from v and from all the w_i $(0 \leq i \leq n-1)$, then we have $F_{t/v} = \exists z G_{t/v}$; in the case where z differs as well from all the u_j $(0 \leq j \leq p-1)$ then z is not among the variables of t so property (1) is equivalent to the existence of an element $a \in M$ such that

$$\big\langle \mathcal{M} : z \to a, w_0 \to a_0, \ldots, w_{n-1} \to a_{n-1}, u_0 \to b_0, \ldots, u_{p-1} \to b_{p-1} \big\rangle$$
$$\models G_{t/v};$$

by the induction hypothesis, this is also equivalent to the existence of an element $a \in M$ such that

$$\big\langle \mathcal{M} : v \to \overline{t}^{\mathcal{M}}[w_0 \to a_0, \ldots, w_{n-1} \to a_{n-1}], z \to a, w_0 \to a_0, \ldots$$
$$\ldots, w_{n-1} \to a_{n-1}, u_0 \to b_0, \ldots, u_{p-1} \to b_{p-1} \big\rangle \models G,$$

or, in other words, to the following assertion which is precisely property (2):

$$\big\langle \mathcal{M} : v \to \overline{t}^{\mathcal{M}}[w_0 \to a_0, w_1 \to a_1, \ldots, w_{n-1} \to a_{n-1}], w_0 \to a_0, \ldots$$
$$\ldots, w_{n-1} \to a_{n-1}, u_0 \to b_0, \ldots, u_{p-1} \to b_{p-1} \big\rangle \models \exists z G.$$

In the case where $z = u_j$ $(0 \leq j \leq p-1)$, it suffices to repeat the five previous lines omitting '$u_j \to b_j$' from the assignments for the variables.

- The case involving universal quantification is similar.   ∎

In this proposition, the hypotheses that concern the variables are rather complicated. Most of the time, we will have stronger (and simpler) hypotheses at our disposition: this is the case, for example, when none of the variables $w_0, w_1, \ldots, w_{n-1}$ has a bound occurrence in the formula F.

Consider once again the language $L = \{R, f, c\}$ used earlier as well as the L-structure $\mathcal{N} = \langle \mathbb{R}, \leq, \pi, \cos \rangle$. Here is an assortment of examples of a formula

$F[v_0]$ of L along with the set of reals a such that $\mathcal{N} \models F[a]$ (which is, we should recall, another way to write $\langle \mathcal{N} : v_0 \to a \rangle \models F$).

Rcv_0	$[\pi, +\infty)$
$\exists v_1 f v_1 \simeq v_0$	$[-1, 1]$
$\exists v_1 f v_0 \simeq v_1$	$\mathbb{R}$
$f v_0 \simeq c$	$\emptyset$
$\exists v_1 (Rcv_0 \wedge f v_1 \simeq v_0)$	$\emptyset$
$\exists v_1 (Rcv_1 \wedge f v_1 \simeq v_0)$	$[-1, 1]$
$\forall v_1 Rv_0 f v_1$	$(-\infty, -1]$
$\forall v_1 Rf v_0 f v_1$	$\{(2k+1)\pi : k \in \mathbb{Z}\}$
$\forall v_1 \exists v_2 (Rv_1 v_2 \wedge f v_2 \simeq v_0)$	$[-1, 1]$
$\forall v_0 \exists v_1 f v_1 \simeq v_0$	$\emptyset$
$\exists v_1 \forall v_2 Rf v_2 v_1$	$\mathbb{R}$

The reader will have noticed that the last two illustrative formulas are closed. The last one is satisfied in $\mathcal{N}$ while the next-to-last is not. Listing these two formulas in the category of formulas with one free variable may appear preposterous, especially after encountering the fastidious verifications to which we are led by the definition we have adopted. It would seem much more natural to be satisfied by associating with each formula the list of variables that actually have at least one free occurrence in it. This is, by the way, what happens spontaneously in practice: when we are interested in the sequences of elements that satisfy the formula $(\forall v_0 Rv_0 c \Rightarrow (\neg Rv_1 v_2 \vee Rv_2 f v_3))$ in a given structure, we obviously tend to think of sequences of length 3. What justifies the more expansive definition that we have adopted in spite of this is, essentially, the fact that the subformulas of a formula do not necessarily involve all the free variables of the formula (and may even involve others which become bound in the whole formula); thus, to proceed otherwise would have led us to define the notion of satisfaction with a much more complicated induction. We realize that all these considerations are (merely) technical and that it is not necessary to pay undue attention to the purely formal subtleties that we might easily raise concerning this definition.

The essential notion to retain from all the preceding is that of the satisfaction of a closed formula in a structure.

3.4 Universal equivalence and semantic consequence

In this section, we will provide definitions and a basic vocabulary that are of constant use in model theory.

We are given a first order language (with or without equality).

Definition 3.46

- *A closed formula of L is called **universally valid** if and only if it is satisfied in every L-structure. Instead of '**universally valid formula**' we sometimes say '**valid formula**'.*

The notation for: 'F is universally valid' is

$$\vdash^* F,$$

while $\nvdash^* F$ stands for: 'F is not universally valid'.

- *A closed formula of L is called a **contradiction** (or is said to be **contradictory** or **inconsistent**) if and only if its negation is universally valid.*

- *A formula with free variables is **universally valid** if and only if its universal closure is universally valid (see Remark 3.47 below).*

- *Given two formulas F and G of L (whether closed or not), we say that F is **universally equivalent** (or **logically equivalent**, or simply **equivalent**) to G if and only if the formula $(F \Leftrightarrow G)$ is universally valid.*
  The notation for 'F is universally equivalent to G' is

$$F \sim G$$

- *A set of closed formulas of L is called a **theory of** L.*

- *Given a theory T and an L-structure $\mathcal{M}$, we say that $\mathcal{M}$ **is a model of** T (or that $\mathcal{M}$ **satisfies** T, or that T **is satisfied in** $\mathcal{M}$) if and only if $\mathcal{M}$ satisfies each formula that belongs to T.*
  The notation for '$\mathcal{M}$ is a model of T' is

$$\mathcal{M} \models T,$$

while $\mathcal{M} \nvDash T$ stands for :'$\mathcal{M}$ is not a model of T'.

- *A theory is **consistent** (or **non-contradictory**, or **satisfiable**) if it has at least one model.*
  *A theory that is not consistent is called **contradictory** (or **inconsistent**).*

- *A theory is **finitely consistent** (or **finitely satisfiable**) if all its finite subsets are consistent.*

- *Given a theory T and a closed formula F of L, F is **a semantic consequence of** T (or simply, a **consequence of** T) if and only if every L-structure that is a model of T is also a model of F.*
  The notation for 'F is a consequence of T' is

$$T \vdash^* F,$$

while $T \nvdash^* F$ stands for 'F is not a consequence of T'.

- *If T is a theory and F is a formula of L with free variables, F is a **consequence of** T if and only if the universal closure of F is a consequence of T. The notation is the same as for closed formulas.*

- *Two theories T_1 and T_2 are **equivalent** if and only if every formula in T_1 is a consequence of T_2 and every formula in T_2 is a consequence of T_1.*

Remark 3.47 *The definition that is given here for the universal validity of a formula with free variables is, a priori, incorrect. It only makes sense once it has been verified that the various universal closures of a formula are all universally equivalent. This fact, which is intuitively clear, is proved by referring to the definition of satisfaction (3.42). We will have to deal later with more or less this same issue to prove property (5) of Theorem 3.55.*

Remark 3.48 *We must be wary of the concept of universally equivalent formulas as it applies to formulas that are not closed: for two formulas to be equivalent, it is not sufficient that their universal closures be equivalent. Consider, for example, the following two formulas in the language whose only symbol is equality:*

$$F = \neg v_0 \simeq v_1 \quad and \quad G = \neg v_0 \simeq v_2.$$

Their universal closures are, respectively:

$$F_1 = \forall v_0 \forall v_1 \neg v_0 \simeq v_1 \quad and \quad G_1 = \forall v_0 \forall v_2 \neg v_0 \simeq v_2.$$

The formulas F_1 and G_1 are universally equivalent: indeed, they are both contradictory, for if we take an arbitrary structure $\mathcal{M}$ and an element a from its underlying set (necessarily non-empty), we have $\langle \mathcal{M} : v_0 \rightarrow a, v_1 \rightarrow a \rangle \nvDash F$ and $\langle \mathcal{M} : v_0 \rightarrow a, v_2 \rightarrow a \rangle \nvDash G$, which shows that $\mathcal{M}$ satisfies neither F_1 nor G_1, and hence does satisfy $(F_1 \Leftrightarrow G_1)$. However, the formula $(F \Leftrightarrow G)$ is not universally valid, for its universal closure,

$$\forall v_0 \forall v_1 \forall v_2 (\neg v_0 \simeq v_1 \Leftrightarrow \neg v_0 \simeq v_2),$$

is not satisfied in the structure whose base set is $\{0, 1\}$; to see this, note that

$$\langle \mathcal{M} : v_0 \rightarrow 0, v_1 \rightarrow 1, v_2 \rightarrow 0 \rangle \vDash F \text{ and} \langle \mathcal{M} : v_0 \rightarrow 0, v_1 \rightarrow 1, v_2 \rightarrow 0 \rangle \nvDash G.$$

It follows that

$$\langle \mathcal{M} : v_0 \rightarrow 0, v_1 \rightarrow 1, v_2 \rightarrow 0 \rangle \nvDash (\neg v_0 \simeq v_1 \Leftrightarrow \neg v_0 \simeq v_2).$$

So the universal closure of this last formula is false in the structure under consideration.

What is true, none the less, is that if two formulas are universally equivalent, then so are their universal closures (see Exercise 6).

It is time to indicate precisely which abuses of notation we will allow ourselves concerning first order formulas. In fact, we will be satisfied simply to recycle those that we have already decided to use for formulas of the propositional calculus:

- suppressing the outermost pair of parentheses;
- writing $(F \wedge G \wedge H)$ instead of $((F \wedge G) \wedge H)$;
- using 'large' conjunctions and disjunctions such as $\bigwedge_{i \in I} F_i$.

The justification for using these shorthand notations is essentially the same as it was for the propositional calculus. Their transfer to first order formulas is based on the following very simple and constantly used result:

Lemma 3.49 *Let $A_1, A_2, \ldots, A_k$ be propositional variables, let $J[A_1, A_2, \ldots, A_k]$ be a propositional formula and let $F_1, F_2, \ldots, F_k$ be first order formulas of a language L. If the formula J is a tautology, then the first order formula $J[F_1, F_2, \ldots, F_k]$ (the result of substituting the formulas $F_1, F_2, \ldots, F_k$ for the variables $A_1, A_2, \ldots, A_k$, respectively, in the formula J) is a universally valid formula.*

Proof Suppose that the free variables of the formulas $F_1, F_2, \ldots, F_k$ are among $v_0, v_1, \ldots, v_{n-1}$. Then this is also the case for the formula $F = J[F_1, F_2, \ldots, F_k]$. Consider an L-structure $\mathcal{M} = \langle M, \ldots \rangle$ and elements $a_0, a_1, \ldots, a_{n-1}$ of M. We define an assignment of truth values δ on $\{A_1, A_2, \ldots, A_n\}$ by setting, for $0 \leq i \leq k$:

$$\delta(A_i) = \begin{cases} 1 & \text{if } \mathcal{M} \models F_i[a_0, a_1, \ldots, a_{n-1}] \\ 0 & \text{if } \mathcal{M} \nvDash F_i[a_0, a_1, \ldots, a_{n-1}]. \end{cases}$$

Here we have used the simplified notation for satisfaction. The definition of satisfaction shows clearly (argue by induction on J) that the formula F is satisfied in $\mathcal{M}$ by the n-tuple $(a_0, a_1, \ldots, a_{n-1})$ if and only if the assignment of truth values δ assigns the value **1** to the formula J. We conclude immediately that when J is a tautology,

$$\mathcal{M} \models F[a_0, a_1, \ldots, a_{n-1}],$$

and that this holds for any n-tuple $(a_0, a_1, \ldots, a_{n-1})$, which proves that the universal closure of F is satisfied in any L-structure, i.e. that F is universally valid. ∎

Definition 3.50 *The universally valid formulas that are obtained from propositional tautologies by the method we have just described are called **tautologies of the predicate calculus**.*

So we see that everything that was said in Chapter 1 concerning tautologies (associativity of conjunction and disjunction, among other facts) transfers without difficulty to first order formulas; the concepts of tautology and of equivalent formulas become, respectively, those of universally valid formula and universally equivalent formulas.

Note finally that we adopt no particular abbreviations involving the quantifiers.

The properties asserted in the following theorem are immediate consequences of the definitions given above.

Theorem 3.51 *For any theories T and S of L, any integers m and $p \geq 1$, and any closed formulas G, H, $F_1, F_2, \ldots, F_m$, and $G_1, G_2, \ldots, G_p$ of L, the following properties are satisfied:*

- *The formula G is contradictory if and only if no L-structure satisfies it.*
- *The formula G is a consequence of T if and only if the theory $T \cup \{\neg G\}$ is contradictory.*
- *If T is consistent and if $S \subseteq T$, then S is consistent.*
- *If T is consistent, then T is finitely consistent.*
- *If T is contradictory and if $T \subseteq S$, then S is contradictory.*
- *If $T \vdash^* G$ and if $T \subseteq S$, then $S \vdash^* G$.*
- *$T \cup \{G\} \vdash^* H$ if and only if $T \vdash^* (G \Rightarrow H)$.*
- *$T \vdash^* (G \wedge H)$ if and only if $T \vdash^* G$ and $T \vdash^* H$.*
- *$\{F_1, F_2, \ldots, F_m\} \vdash^* G$ if and only if $\vdash^* ((F_1 \wedge F_2 \wedge \cdots \wedge F_m) \Rightarrow G)$.*
- *G is universally valid if and only if G is a consequence of the empty theory.*
- *G is universally valid if and only if G is a consequence of every theory of L.*
- *T is contradictory if and only if $T \vdash^* (G \wedge \neg G)$.*
- *T is contradictory if and only if every formula of L is a consequence of T.*
- *T is contradictory if and only if for every universally valid formula F, $\neg F$ is a consequence of T.*
- *T is contradictory if and only if there exists at least one universally valid formula F such that $\neg F$ is a consequence of T.*
- *The theory $\{F_1, F_2, \ldots, F_m\}$ is contradictory if and only if the formula $(\neg F_1 \vee \neg F_2 \vee \cdots \vee \neg F_m)$ is universally valid.*
- *The theories T and S are equivalent if and only if they have the same models (in other words, for T and S to be equivalent, it is necessary and sufficient that for every L-structure $\mathcal{M}$, $\mathcal{M}$ is a model of T if and only if $\mathcal{M}$ is a model of S).*
- *If every formula in T is replaced by a universally equivalent formula, the resulting theory is equivalent to T.*
- *If T is contradictory, then S is equivalent to T if and only if S is contradictory.*
- *The theory T is equivalent to the empty theory if and only if every formula belonging to T is universally valid.*
- *Every L-structure is a model of the empty theory.*
- *The empty theory is consistent.*
- *The set of all closed formulas of L is contradictory.*
- *The theories $\{G\}$ and $\{H\}$ are equivalent if and only if the formulas G and H are logically equivalent.*

- *The theories $\{F_1, F_2, \ldots, F_m\}$ and $\{G_1, G_2, \ldots, G_p\}$ are equivalent if and only if the formula*

$$((F_1 \wedge F_2 \wedge \cdots \wedge F_m) \Leftrightarrow (G_1 \wedge G_2 \wedge \cdots \wedge G_p))$$

  is universally valid.
- *Any finite theory is equivalent to a theory that consists of a single formula.*
- *The binary relation 'is universally equivalent to' is an equivalence relation on the set of formulas of L.*
- *The binary relation 'is equivalent to' is an equivalence relation on the set of theories of L, i.e. on the set of subsets of the set of closed formulas of L.*

Proof We invite the reader to supply proofs independently (this primarily involves arguments analogous to those used in the proof of Lemma 1.38). ∎

The next proposition expresses the fact that the equivalence relation $\sim$ for formulas is compatible with the operations that enter into the construction of formulas (the use of connectives and quantifiers).

Proposition 3.52 *For all formulas F, G, F' and G' and for every integer k, if F and G are equivalent to F' and G' respectively, then the formulas*

$$\neg F, \ (F \wedge G), \ (F \vee G), \ (F \Rightarrow G), \ (F \Leftrightarrow G), \ \forall v_k F \ \text{ and } \ \exists v_k F$$

are equivalent, respectively, to

$$\neg F', \ (F' \wedge G'), \ (F' \vee G'), \ (F' \Rightarrow G'), \ (F' \Leftrightarrow G'), \ \forall v_k F' \ \text{ and } \ \exists v_k F'.$$

Proof Let us treat, for example, the case of existential quantification. Suppose that the formulas F and F' are equivalent and that their free variables are among $v_0, v_1, \ldots, v_n$ $(n \geq k)$. We need to prove that the universal closure of the formula $(\exists v_k F \Leftrightarrow \exists v_k F')$ is satisfied in an arbitrary L-structure $\mathcal{M} = \langle M, \ldots \rangle$. So let us consider elements $a_0, a_1, \ldots, a_{k-1}, a_{k+1}, \ldots, a_n$ of M. If we suppose that

$$\langle \mathcal{M} : v_0 \to a_0, \ldots, v_{k-1} \to a_{k-1}, v_{k+1} \to a_{k+1}, \ldots, v_n \to a_n \rangle \models \exists v_k F,$$

then we can find an element $a \in M$ such that

$$\langle \mathcal{M} : v_0 \to a_0, \ldots, v_{k-1} \to a_{k-1}, v_k \to a, v_{k+1} \to a_{k+1}, \ldots, v_n \to a_n \rangle \models F;$$

but, since F and F' are equivalent, we also have

$$\langle \mathcal{M} : v_0 \to a_0, \ldots, v_{k-1} \to a_{k-1}, v_k \to a, v_{k+1} \to a_{k+1}, \ldots, v_n \to a_n \rangle \models F',$$

and hence

$$\langle \mathcal{M} : v_0 \to a_0, \ldots, v_{k-1} \to a_{k-1}, v_{k+1} \to a_{k+1}, \ldots, v_n \to a_n \rangle \models \exists v_k F'.$$

The converse is obtained in exactly the same manner. If

$$\langle \mathcal{M} : v_0 \to a_0, \ldots, v_{k-1} \to a_{k-1}, v_{k+1} \to a_{k+1}, \ldots, v_n \to a_n \rangle \models \exists v_k F',$$

then

$$\langle \mathcal{M} : v_0 \to a_0, \ldots, v_{k-1} \to a_{k-1}, v_{k+1} \to a_{k+1}, \ldots, v_n \to a_n \rangle \models \exists v_k F,$$

so we may conclude that

$$\langle \mathcal{M} : v_0 \to a_0, v_1 \to a_1, \ldots, v_{k-1} \to a_{k-1}, v_{k+1} \to a_{k+1}, \ldots, v_n \to a_n \rangle$$
$$\models (\exists v_k F \Leftrightarrow \exists v_k F').$$

The other cases are treated in a similar fashion.   ∎

Corollary 3.53 *Suppose that F is a formula, that G is a sub-formula of F and that G' is equivalent to G. Then the formula F', obtained from F by substituting G' for an arbitrary occurrence of the sub-formula G, is equivalent to F.*

Proof The argument is by induction on F. When F is atomic, G can only be equal to F, so $F' = G'$ and the result is immediate. For all other stages of the induction, it suffices to apply the preceding proposition.   ∎

We have thus justified an operation that we perform practically all the time when we manipulate first order formulas from a semantic point of view: we replace sub-formulas by equivalent formulas.

Changing the name of a bound variable, provided it is done subject to certain conditions, transforms a formula into an equivalent formula (see Remark 3.26):

Proposition 3.54 *For any integers k and h and any formula F, if the variable v_h does not occur in F, then the formulas*

$$\forall v_k F \quad and \quad \forall v_h F_{v_h/v_k} \quad (respectively, \ \exists v_k F \quad and \quad \exists v_h F_{v_h/v_k})$$

are equivalent.

Proof The result is trivial if $h = k$. So suppose h and k are distinct and that $F = F[v_{i_1}, v_{i_2}, \ldots, v_{i_n}, v_k]$, where the integers $i_1, i_2, \ldots, i_n$ are pairwise distinct and distinct from h and k (which the hypothesis allows). Given an L-structure $\mathcal{M} = \langle M, \ldots \rangle$ and arbitrary elements $a_1, a_2, \ldots, a_n$ of M, what needs to be proved is that

$$\langle \mathcal{M} : v_{i_1} \to a_1, v_{i_2} \to a_2, \ldots, v_{i_n} \to a_n \rangle \models \forall v_k F \qquad (*)$$

if and only if

$$\langle \mathcal{M} : v_{i_1} \to a_1, v_{i_2} \to a_2, \ldots, v_{i_n} \to a_n \rangle \models \forall v_h F_{v_h/v_k}, \qquad (**)$$

and the analogue for $\exists$.

Property (*) means that for any element $a \in M$, we have

$$\langle \mathcal{M} : v_k \to a, v_{i_1} \to a_1, v_{i_2} \to a_2, \ldots, v_{i_n} \to a_n \rangle \models F;$$

but according to Lemma 3.44, this is also equivalent to

$$\langle \mathcal{M} : v_k \to a, v_h \to a, v_{i_1} \to a_1, v_{i_2} \to a_2, \ldots, v_{i_n} \to a_n \rangle \models F.$$

We may then apply Proposition 3.45 (our hypotheses allow this), upon observing that

$$a = \overline{v_h}^{\mathcal{M}}[v_h \to a, v_{i_1} \to a_1, v_{i_2} \to a_2, \ldots, v_{i_n} \to a_n],$$

and we obtain the following property, which is equivalent to (*):

for every $a \in M$, $\langle \mathcal{M} : v_h \to a, v_{i_1} \to a_1, v_{i_2} \to a_2, \ldots, v_{i_n} \to a_n \rangle \models F v_h / v_k$,

which, by definition, means

$$\langle \mathcal{M} : v_{i_1} \to a_1, v_{i_2} \to a_2, \ldots, v_{i_n} \to a_n \rangle \models \forall v_h F v_h / v_k,$$

in other words, property (**). ∎

Using the tautologies and the equivalent propositional formulas from Chapter 1, we obtain, in a natural way, countless examples of universally valid formulas and others equivalent to them. The important properties that follow will supply us with examples that have no analogue in the propositional calculus since they involve quantifiers. These properties are extremely useful, in particular, for mastering the manipulation of quantifiers in statements of everyday mathematics: among the first exercises at the end of this chapter, several are aimed at precisely this.

Theorem 3.55 *For all integers h and k and for all formulas F and G, we have the following logical equivalences:*

$$\neg \forall v_k F \sim \exists v_k \neg F, \tag{1}$$
$$\forall v_k (F \wedge G) \sim (\forall v_k F \wedge \forall v_k G), \tag{2}$$
$$\exists v_k (F \vee G) \sim \exists v_k F \vee \exists v_k G, \tag{3}$$
$$\exists v_k (F \Rightarrow G) \sim (\forall v_k F \Rightarrow \exists v_k G), \tag{4}$$
$$\forall v_k \forall v_h F \sim \forall v_h \forall v_k F, \tag{5}$$
$$\exists v_k \exists v_h F \sim \exists v_h \exists v_k F. \tag{6}$$

Moreover, the following three formulas are universally valid:

$$\exists v_k (F \wedge G) \Rightarrow \exists v_k F \wedge \exists v_k G, \tag{7}$$
$$(\forall v_k F \vee \forall v_k G) \Rightarrow \forall v_k (F \vee G), \tag{8}$$
$$\exists v_k \forall v_h F \Rightarrow \forall v_h \exists v_k F. \tag{9}$$

In addition, if the variable v_k is not free in G, then:

$$\forall v_k G \sim \exists v_k G \sim G, \tag{10}$$

$$\forall v_k (F \wedge G) \sim (\forall v_k F \wedge G), \tag{11}$$

$$\exists v_k (F \vee G) \sim (\exists v_k F \vee G), \tag{12}$$

$$\forall v_k (F \vee G) \sim (\forall v_k F \vee G), \tag{13}$$

$$\exists v_k (F \wedge G) \sim (\exists v_k F \wedge G), \tag{14}$$

$$\exists v_k (G \Rightarrow F) \sim (G \Rightarrow \exists v_k F), \tag{15}$$

$$\forall v_k (G \Rightarrow F) \sim (G \Rightarrow \forall v_k F), \tag{16}$$

$$\exists v_k (F \Rightarrow G) \sim (\forall v_k F \Rightarrow G), \tag{17}$$

$$\forall v_k (F \Rightarrow G) \sim (\exists v_k F \Rightarrow G). \tag{18}$$

Proof   We obtain (1), (2), (5), (6), (7) and (9) without difficulty by referring to Definition 3.42; (3) and (8) are deduced from (2) and (7) respectively (applied to $\neg F$ and $\neg G$ together with (1) and familiar tautologies; (4) is an immediate consequence of (3) applied to $\neg F$ and G; when we take the additional hypothesis about v_k into account, (10) and (13) follow from the definition and Lemma 3.44; (11) and (15) can be deduced from (10) and from (2) and (4); it is again thanks to (1) and to some tautologies that we obtain (12) from (11), (14) from (13), (17) from (15) and (18) from (16); (16) is obtained from (13) applied to F and $\neg G$.   ■

It should be understood that these brief suggestions presuppose intensive use of Proposition 3.52 and Corollary 3.53.

We could say, loosely speaking, that the universal quantifier is 'distributive' over conjunction but not over disjunction, while the opposite is true for the existential quantifier. That is what is expressed by properties (2) and (3), together with the fact that the universal validity of formulas is not preserved, in general, when we replace the symbol $\Rightarrow$ by $\Leftrightarrow$ in (7) and (8): to see this, consider the language L that has two unary relation symbols A and B; let $F = Av_0$ and $G = Bv_0$ and take the model $\mathcal{M}$ whose underlying set is $\mathbb{N}$ and in which A and B are interpreted, respectively, by the relations 'is even' and 'is odd'; it is then clear that $\mathcal{M}$ satisfies the formulas $\exists v_0 F \wedge \exists v_0 G$ and $\forall v_0 (F \vee G)$ but does not satisfy the formula $\exists v_0 (F \wedge G)$ nor the formula $\forall v_0 F \vee \forall v_0 G$. The behaviour of the quantifiers with respect to implication is more complicated; what one can keep in mind is this: if we try to 'distribute' the quantifier in a formula of the form $Q v_k (F \Rightarrow G)$, where Q is $\forall$ or $\exists$, then, in those circumstances where this is possible,

- if the quantifier 'enters' on the right side of the symbol $\Rightarrow$, then it 'enters' as is; whereas,
- if the quantifier 'enters' on the left side of the symbol $\Rightarrow$, it must be replaced by its dual Q^* ($Q^* = \exists$ if $Q = \forall$ and $Q^* = \forall$ if $Q = \exists$).

We can permute two consecutive universal (existential, respectively) quantifiers (properties (5) and (6)), but not a $\forall$ with a $\exists$: (9) is no longer universally valid if we replace $\Rightarrow$ with $\Leftrightarrow$. To see this, consider once again the model $\mathcal{M}$ above and let F be the formula $(Av_0 \Leftrightarrow Bv_1)$; then $\mathcal{M} \models \forall v_0 \exists v_1 F$ (since for any integer a_0, we can find an integer a_1 whose parity is different from that of a_0) but $\mathcal{M} \not\models \exists v_1 \forall v_0 F$ (for we can hardly insist that an integer have a parity that is distinct from the parity of an arbitrary integer...). The formula $\forall v_h \exists v_k F$ expresses the existence of a v_k for each v_h (so it may vary with v_h) whereas the v_k whose existence is expressed by the formula $\exists v_k \forall v_h F$ must be 'the same for all v_h', which is what makes this formula 'stronger' than the preceding. There are classic illustrations of this remark in analysis involving the distinction between simple and uniform continuity and the distinction between simple and uniform convergence: we know well that the heart of the problem consists in determining whether 'the δ (or the N) depends on x or not'... and, ultimately, when we express these properties formally, they differ precisely by an inversion of the order of the quantifiers.

Using the results from Chapter 1 on complete sets of connectives, together with Lemma 3.49, Proposition 3.52 and Corollary 3.53, and property (1) of the preceding theorem, we immediately obtain

Theorem 3.56 *Every first order formula is universally equivalent to at least one formula whose only symbols for connectives and quantifiers are:* $\neg$, $\vee$ *and* $\exists$.

In this statement, we can obviously replace $\neg$ and $\vee$ by the elements of any complete set of connectives; we can also replace $\exists$ by $\forall$.

Remark 3.57 *An analogue of Remark 1.33 applies here as well: to prove that a certain property, compatible with the relation $\sim$, is true for every first order formula, it is sufficient, in a proof by induction, to 'restrict' oneself to the stages that relate to negation, disjunction and existential quantification.*

We conclude this section with a result that is very easy but indispensable. It concerns comparing the satisfaction of a formula in a structure for its own language with its satisfaction in a structure for a richer language.

Lemma 3.58 *Consider a first order language L and a language L^* that is an enrichment of L. Let $\mathcal{M} = \langle M, \ldots \rangle$ be an L-structure, let $\mathcal{M}^*$ be an enrichment of $\mathcal{M}$ for the language L^*, let $F = F[v_0, v_1, \ldots, v_{n-1}]$ be a formula of the language L and let $a_0, a_1, \ldots, a_{n-1}$ be elements of M.*
Under these conditions, we have

$$\mathcal{M} \models F[a_0, a_1, \ldots, a_{n-1}] \quad \text{if and only if } \mathcal{M}^* \models F[a_0, a_1, \ldots, a_{n-1}].$$

Proof The only thing that is perhaps not obvious is that these two properties make sense! We convince ourselves by noting that F is both a formula of L and a formula of L^*. For the rest, a quick glance at the definition of the satisfaction of F in $\mathcal{M}^*$ reveals that this depends only on the symbols of L and on the functions and relations of the structure $\mathcal{M}$. So the result is automatic (the reader who wishes to be completely rigourous should give a proof by induction on F). ∎

Observe, however, that this question would not have made sense if F contained symbols of L^* that did not belong to L.

3.5 Prenex forms and Skolem forms

What we do in this section will be amply used in the next chapter in which we will describe methods that allow us to answer questions of the form: 'is this closed formula universally valid?' or 'is it a consequence of such and such a theory?' The idea will be to reduce the question, at the cost of changing the language, to formulas whose syntactic construction is relatively simple: the Skolem forms. Beforehand, we will show that every formula F is equivalent to a formula (of the same language) structured as a sequence of quantifications followed by a quantifier-free formula (this will be called a prenex form of F). The interest and import of Theorem 3.60 about prenex forms extends beyond the context just described. It also presents a (slight) danger, in that it leads one to believe that a formula is 'easier to understand' when it is in prenex form: in fact, one quickly realizes that the contrary is true and that, to comprehend the property expressed by a closed formula, one is well advised to 'distribute' the quantifiers to the maximum extent possible, i.e. to do the exact opposite of putting it in prenex form.

3.5.1 Prenex forms

Definition 3.59 *A formula F is **prenex** if and only if there exist an integer k, variables w_1, w_2, ..., w_k, symbols for quantifiers Q_1, Q_2, ..., Q_k and a formula G without quantifiers such that*

$$F = Q_1 w_1 Q_2 w_2 \cdots Q_k w_k G.$$

*The word $Q_1 w_1 Q_2 w_2 \ldots Q_k w_k$ is then called the **prefix** of the prenex formula. A prenex formula is **polite** if and only if its prefix contains at most one occurrence of each variable.*
*If H is a formula, a prenex formula that is universally equivalent to H is called a **prenex form of** H.*
*A **universal** formula is a prenex formula with no existential quantifiers.*
*An **existential** formula is a prenex formula with no universal quantifiers.*

Naturally, the case $k = 0$ corresponds to the situation in which $F = G$, which is to say that formulas without quantifiers are special cases of prenex formulas (and they are, by the way, polite, universal, and existential).

We see immediately that any universal closure of a prenex formula is also a prenex formula.

Caution: a formula such as $\forall v_0 (\exists v_1 v_0 \simeq v_1 \Rightarrow v_0 \simeq v_1)$ is not prenex!

Theorem 3.60 *Every first order formula has at least one polite prenex form.*

Proof We will prove by induction that for any formula F, we can find a polite prenex formula F' that is universally equivalent to F. Because it is clear that this property is compatible with the relation $\sim$ we may, by Remark 3.57, restrict the number of cases to be considered.

- If F is atomic, it suffices to take $F' = F$.

- If $F = \neg G$, and if G is equivalent to $Q_1 v_1 Q_2 v_2 \ldots Q_k v_k G''$, where G'' is quantifier-free and the variables w_i are pairwise distinct, it suffices to take $F' = \overline{Q}_1 v_1 \overline{Q}_2 v_2 \ldots \overline{Q}_k v_k \neg G''$ where, for $1 \le h \le k$, $\overline{Q}_h$ is the dual of Q_h.

- If $F = (G \vee H)$, if G is equivalent to $G' = Q_1 v_1 Q_2 v_2 \ldots Q_k v_k G''$ and H to $H' = Q_1' z_1 Q_2' z_2 \ldots Q_m' z_m H''$ (where G'' and H'' are quantifier-free and where, in each prefix, the variables are pairwise distinct), then after first choosing $k+m$ pairwise distinct variables $x_1, x_2, \ldots, x_k, y_1, y_2, \ldots, y_m$ that have no occurrence in G' or in H', it suffices to set

$$G_1 = G''_{x_1/w_1, x_2/w_2, \ldots, x_k/w_k} \text{ and } H_1 = H''_{y_1/z_1, y_2/z_2, \ldots, y_m/z_m}$$

and to take

$$F' = Q_1 x_1 Q_2 x_2 \ldots Q_k x_k Q_1' y_1 Q_2' y_2 \ldots Q_m' y_m (G_1 \vee H_1).$$

To verify that this formula is indeed equivalent to F, we mainly apply ($k + m$ times) Proposition 3.54 and properties (12) and (13) of Theorem 3.55. It is appropriate to note that the order in which we repeatedly apply these two properties is of no importance: we might, for example, first bring all the Q_j' to the front, then all the Q_i, or else alternate them arbitrarily; on the other hand, what cannot change (except in special cases) is the order of the Q_i among themselves, or that of the Q_j' among themselves. We see, in any case, that we are far from having a unique prenex formula equivalent to F.

- If $F = \exists v G$ and if G is equivalent to $Q_1 w_1 Q_2 w_2 \ldots Q_k w_k G''$, where G'' is quantifier-free and the variables w_i are pairwise distinct, then either $v \notin \{w_1, w_2, \ldots, w_k\}$ and it suffices to take $F' = \exists v Q_1 w_1 Q_2 w_2 \ldots Q_k w_k G''$, or else $w = w_j$ for some index j between 1 and n: in this case, w is not free in the formula $Q_1 w_1 Q_2 w_2 \ldots Q_k w_k G''$, so we can take $F' = Q_1 w_1 Q_2 w_2 \ldots Q_k w_k G''$ since this polite prenex formula is equivalent to F by virtue of property (10) of Theorem 3.55. ∎

Remark 3.61 *The proof we have just given furnishes a procedure for constructing a prenex formula that is equivalent to a given formula F (we also speak of 'putting F in prenex form'). If we scrupulously follow this procedure, we must begin by eliminating the symbols $\wedge$, $\Rightarrow$, $\Leftrightarrow$ and $\forall$. In truth, the only indispensable step is to eliminate $\Leftrightarrow$. It is then possible, by judiciously changing the names of bound variables, to progressively 'bring the quantifiers to the front' by climbing the decomposition tree, using, step by step, the properties presented in Theorem 3.55 as we did above for the induction step involving disjunction. Of course, depending* *on the circumstances, we can be more or less efficient in changing the names of variables: it is rarely necessary to rename all the variables that occur, as we did above.*

The obvious purpose of these name changes of bound variables is to obtain a formula in which no variable occurs in the scope of more than one quantification, in order that Theorem 3.55 apply in all cases. A byproduct is a potential increase in the number of variables in use. However, if we wish to minimize the length (and height) of the prenex form, it can happen, on the contrary, that we are better served by substituting, for a given bound variable, a variable that already occurs in the scope of some other quantification. Let us take an example: the formula

$$F = \forall v_0 (\forall v_1 \neg v_1 \simeq v_1 \Rightarrow \exists v_2 v_0 \simeq v_2) \wedge \forall v_1 v_1 \simeq v_1$$

is equivalent to the prenex formula

$$G = \forall v_0 \exists v_1 \exists v_2 \forall v_3 ((\neg v_1 \simeq v_1 \Rightarrow v_0 \simeq v_2) \wedge v_3 \simeq v_3),$$

obtained as described above, by changing the name of the bound variable v_1 (into v_3) in the scope of the rightmost quantification in F; but we can find a simpler prenex form when we observe that properties (2) and (4) of Theorem 3.55 are applicable here: this leads us to substitute (in F) v_1 for the occurrences of v_2, then v_0 for the last three occurrences of v_1; the formula obtained in this way is $H = \forall v_0 (\forall v_1 (\neg v_1 \simeq v_1 \Rightarrow \exists v_1 v_0 \simeq v_1) \wedge \forall v_0 v_0 \simeq v_0$; it is indeed equivalent to F and the properties to which we just referred produce the following prenex form:

$$\forall v_0 \exists v_1 ((\neg v_1 \simeq v_1 \Rightarrow v_0 \simeq v_1) \wedge v_0 \simeq v_0),$$

whose height is 5 (that of G is 7) and which is obviously shorter than G.

Remark 3.62 *The last part of the proof of the preceding theorem clearly showed that to get from an arbitrary formula in prenex form to an equivalent prenex formula that is polite, it suffices, for each variable, to suppress in the prefix of F all but the rightmost quantification involving that variable. For example, if G is a formula without quantifiers, the formula*

$$\forall v_0 \exists v_1 \exists v_0 \exists v_2 \forall v_1 \exists v_0 \exists v_3 \exists v_0 G$$

is logically equivalent to the polite prenex formula

$$\exists v_2 \forall v_1 \exists v_3 \exists v_0 G.$$

Remark 3.63 *When we apply the method described above to obtain a prenex form of some given formula F, the result is a formula that has the same free variables as F (this is immediately verified). In particular, applying the method to a closed formula produces a prenex formula that is closed.*

Let F be a first order formula, let G be a prenex form of F, and let H be the subformula of G obtained by suppressing the prefix. As H is quantifier-free, it is obtained from some propositional formula J by a substitution of the kind described in Remark 3.27. As for any propositional formula, J has a conjunctive normal form J_1 and a disjunctive normal form J_2. The substitution that transforms J into H will transform J_1 and J_2, respectively, into first order formulas H_1 and H_2 which are clearly universally equivalent to H. The prenex formulas G_1 and G_2 obtained by placing the prefix of G at the front of H_1 and H_2, respectively, are equivalent to F. We say that G_1 (G_2, respectively) is a **conjunctive** (**disjunctive**, respectively) **prenex form** of F.

3.5.2 Skolem forms

Consider a polite prenex formula of a language L. Thus there exist pairwise distinct variables $w_1, w_2, \ldots, w_n$, quantifiers $Q_1, Q_2, \ldots, Q_n$ and a quantifier-free formula G of L such that $F = Q_1 w_1 Q_2 w_2 \cdots Q_n w_n G$.

Denote by $j_1, j_2, \ldots, j_p$ the indices of those Q that correspond to an existential quantification: that is,

$$\{j_1, j_2, \ldots, j_p\} = \{j \in \{1, 2, \ldots, n\} : Q_j = \exists\}$$

(we assume $1 \le j_1 < j_2 < \cdots < j_p \le n$).

With F, we will associate a language $L_{Sk}(F)$ that will be an enrichment of the language L obtained by adding p new symbols $f_1, f_2, \ldots, f_p$ for constants or functions (these are called the **symbols for the Skolem functions associated with** F and they correspond to the p occurrences of the symbol $\exists$ in the prefix of F). For $1 \le h \le p$, the arity of the symbol f_h must equal $j_h - h$, i.e. the number of times the quantifier $\forall$ occurs to the left of Q_{j_h} in the prefix of F (we adopt the convention of treating constant symbols as function symbols of arity 0; for f_h to be a constant symbol, it is necessary and sufficient that $j_h = h$, i.e. that the first h quantifications in the prefix of F be existential; naturally, in such a situation, f_1, $f_2, \ldots, f_{h-1}$ will also be constant symbols; moreover, the arity of f_h increases with h). For example, if the prefix of F is

$$\forall v_0 \forall v_1 \exists v_2 \forall v_3 \exists v_4 \forall v_5 \forall v_6 \exists v_7 \exists v_8 \forall v_9 \forall v_{10} \exists v_{11},$$

five new function symbols f_1, f_2, f_3, f_4 and f_5 whose respective arities are 2, 3, 5, 5, and 7 will be added to the language L.

Now that we have defined our new language, we will use it to define a formula F_{Sk}, which will be a polite universal formula of the language $L_{Sk}(F)$ and which we will call **the Skolem form of** F

First of all, for $1 \le h \le p$, we let u_h denote the term of $L_{Sk}(L)$ that consists of the symbol f_h followed by the $j_h - h$ universally quantified variables that occur to the left of the variable w_{j_h} in the prefix of F. In other words,

$$u_h = f_h w_1 w_2 \cdots w_{j_1-1} w_{j_1+1} \cdots w_{j_2-1} w_{j_2+1} \cdots w_{j_{h-1}-1} w_{j_{h-1}+1} \cdots w_{j_h-1}.$$

Then, for each index h between 1 and n, we replace each occurrence of the variable w_{j_h} in G by the term u_h.

Finally, in front of the formula thus formed, we place the prefix of G from which all existential quantifications have first been deleted.

We arrive in this way at the Skolem form of F, which is, therefore,

$$F_{Sk} = \forall w_1 \ldots \forall w_{j_1-1} \forall w_{j_1+1} \ldots \forall w_{j_h-1} \forall w_{j_h+1} \ldots \forall w_{j_p-1} \forall w_{j_p+1} \ldots \forall w_n$$

$$G_{u_1/w_{j_1}, u_2/w_{j_2}, \ldots, u_p/w_{j_p}}.$$

Example 3.64 Suppose the language L consists of a unary function symbol f and a binary relation symbol R. Consider the following formula F of L:

$$\exists v_0 \exists v_1 \forall v_2 \exists v_3 \forall v_4 \forall v_5 \exists v_6 ((R v_0 v_2 \wedge f v_5 \simeq v_3) \Rightarrow (R f v_6 v_2 \vee (R v_1 v_5 \wedge R v_4 v_3))).$$

The language $L_{Sk}(L)$ then contains, in addition to the symbols R and f, four new symbols: two constant symbols f_1 and f_2, a unary function symbol f_3 and a ternary function symbol f_4. The Skolem form of F is the formula

$$\forall v_2 \forall v_4 \forall v_5 ((R f_1 v_2 \wedge f v_5 \simeq f_3 v_2) \Rightarrow (R f f_4 v_2 v_4 v_5 v_2 \vee (R f_2 v_5 \wedge R v_4 f_3 v_2))).$$

Given an arbitrary formula F in a language L, we have seen how to find a polite prenex formula F' that is equivalent to F. The Skolem form of F' will also be called a Skolem form of F (so we do not have uniqueness of the Skolem form of an arbitrary formula).

It is very important not to lose sight of the fact that a Skolem form of a formula F of a language L is not (in general) a formula of L, but of a richer language. This will allow us to avoid a rather common error which consists in thinking that a formula is equivalent to its Skolem form. Such a statement makes sense only if we view F as a formula of the enriched language $L_{Sk}(F)$, which is, of course, possible but we realize right away that in order to support this, we would be led to examine arbitrary $L_{Sk}(F)$-structures and that the proposed equivalence would be rather pointless. An example would really be more convincing that an overly lengthy dissertation: the Skolem form of the formula $F = \forall v_0 \exists v_1 R v_0 v_1$ is the formula $\forall v_0 R v_0 g v_0$ (we have added the symbol for the unary Skolem function g to the initial language L consisting of the binary relation symbol R); the structure whose base set is $\mathbb{Z}$, in which R is interpreted by the usual order relation and g by the map $n \longmapsto n - 1$, obviously satisfies the first formula but not the second.

What is true, none the less, is that every formula F, considered as a formula of the language $L_{Sk}(F)$, is a semantic consequence of its Skolem form. As well, provided we allow the Axiom of Choice (see Chapter 7 in Volume 2), any L-structure that satisfies a (closed) formula F can be enriched to an $L_{Sk}(F)$-structure that satisfies the Skolem form of F. A notable consequence of this is that for a closed formula to be satisfiable, it is necessary and sufficient that its Skolem form be satisfiable (we sometimes say that F and F_{Sk} are **equisatisfiable**, in the absence of their being equivalent). It is this last property that will be mainly used in the next chapter.

Before proving everything that we have just claimed, let us verify it on a very simple example in which the essential ideas are clearly apparent.

Let us reconsider the formula $F = \forall v_0 \exists v_1 R v_0 v_1$ that we just used as an example. So we have $L = \{R\}$, $L_{Sk}(F) = \{R, g\}$ and $F_{Sk} = \forall v_0 R v_0 g v_0$. Let $\mathcal{M} = \langle M, \overline{R}, \overline{g} \rangle$ be an $L_{Sk}(F)$-structure that satisfies F_{Sk}. This means that for every element $a \in M$, we have $(a, \overline{g}(a)) \in \overline{R}$. Obviously then, for every element $a \in M$, we can find an element $b \in M$ (namely, $\overline{g}(a)$) such that $(a, b) \in \overline{R}$, which means that $\mathcal{M}$ satisfies the formula F. Thus, $F_{Sk} \Rightarrow F$ is a universally valid formula of $L_{Sk}(F)$.

Now consider an L-structure $\mathcal{N} = \langle N, \rho \rangle$ that is a model of the formula F. This means that for every element $a \in N$, the set of those elements $b \in N$ such that $(a, b) \in \rho$ is non-empty. It is here that the Axiom of Choice intervenes: it guarantees the existence of a map ϕ from the set of non-empty subsets of N into N (called **a choice function on** N) such that for every non-empty subset $X \subseteq N$, the value of ϕ at X is an element of X ($\phi(X) \in X$). With the help of such a choice function ϕ, we will enrich $\mathcal{N}$ into an $L_{Sk}(F)$-structure that will satisfy F_{Sk}. The problem is to provide an interpretation for the extra symbol g. For this, we will take the map γ defined on N in the following way: for all $a \in N$,

$$\gamma(a) = \phi(\{b \in N : (a, b) \in \rho\}).$$

It is then clear that the $L_{Sk}(F)$-structure $\langle N, \rho, \gamma \rangle$ is a model of F_{Sk}.

Let us now approach these properties in the general case.

Lemma 3.65 *Let $y_1, y_2, \ldots, y_n$ be pairwise distinct variables and let $F = F[y_1, y_2, \ldots, y_n]$ be a polite prenex formula of a language L. Then the formula $F_{Sk} \Rightarrow F$ of the language $L_{Sk}(F)$ is universally valid.*

Proof By induction on the number of occurrences of the existential quantifier in the prefix of F, we will prove, for every $L_{Sk}(F)$-structure $\mathcal{M} = \langle M, \ldots \rangle$ and for all elements $b_1, b_2, \ldots, b_n$ of M, that if the formula F_{Sk} is satisfied in $\mathcal{M}$ by the sequence $(b_1, b_2, \ldots, b_n)$, then the formula F is also satisfied in $\mathcal{M}$ by the same sequence (recall that the formulas F_{Sk} and F have the same free variables). The result is evident when the formula F has no existential quantifiers at all, for then $F_{Sk} = F$. Suppose (this is the induction hypothesis) that the result is true for all polite prenex formulas that have at most k existential quantifications and suppose that F has $k + 1$ of them. Thus

$$F = \forall x_1 \forall x_2 \ldots \forall x_m \exists x G[y_1, y_2, \ldots, y_n, x_1, x_2, \ldots, x_m, x],$$

where G itself is a polite prenex formula with at most k existential quantifications. There is then an m-ary function symbol f_1 in the language $L_{Sk}(F)$ such that the Skolem form of F is the formula obtained from the Skolem form of the formula

$$F' = \forall x_1 \forall x_2 \ldots \forall x_m G[y_1, y_2, \ldots, y_n, x_1, x_2, \ldots, x_m, x]$$

by substituting the term $f_1 x_1 x_2 \ldots x_m$ for the variable x.

So we have

$$F_{Sk} = F'_{Sk_{f_1 x_1 x_2 \ldots x_m / x}}.$$

(We should note that $L_{Sk}(F') = L_{Sk}(F) - \{f_1\}$. So we may consider F' as a formula of $L_{Sk}(F)$, whose satisfaction in an L_{Sk}-structure is equivalent to its satisfaction in the reduced language $L_{Sk}(F)$ (see Lemma 3.58)).

But we see immediately, by referring to the definition of the Skolem form, that starting from the formula F', the result of first taking its Skolem form and then substituting into it $f_1 x_1 x_2 \ldots x_m$ for x is exactly the same as the result of performing these two operations in the opposite order, i.e. of first doing the substitution in F' and then taking the Skolem form of the formula thereby obtained; this depends, essentially, on the fact that the variables involved in the substitution are not existentially quantified in F'. So we also have

$$F_{Sk} = F'_{f_1 x_1 x_2 \ldots x_m / x_{Sk}}.$$

The formula $F'_{f_1 x_1 x_2 \ldots x_m / x}$ is a polite prenex formula with at most k existential quantifiers, so we may apply the induction hypothesis to it:

if $\langle \mathcal{M} : y_1 \to b_1, y_2 \to b_2, \ldots, y_n \to b_n \rangle \models F_{Sk}$, then we also have

$$\langle \mathcal{M} : y_1 \to b_1, y_2 \to b_2, \ldots, y_n \to b_n \rangle \models F'_{f_1 x_1 x_2 \ldots x_m / x},$$

which means that for all elements $a_1, a_2, \ldots, a_m$ of M,

$$\langle \mathcal{M} : y_1 \to b_1, \ldots, y_n \to b_n, x_1 \to a_1, \ldots, x_m \to a_m \rangle \models G_{f_1 x_1 x_2 \ldots x_m / x}.$$

Again, this is equivalent (Proposition 3.45) to

$$\langle \mathcal{M} : y_1 \to b_1, \ldots, y_n \to b_n, x_1 \to a_1, \ldots, x_m \to a_m,$$
$$x \to \overline{f_1}^{\mathcal{M}} (a_1, a_2, \ldots, a_m) \rangle \models G.$$

Thus, for all $a_1, a_2, \ldots, a_m$, there exists an element $b \in M$, namely

$$\overline{f_1}^{\mathcal{M}} (a_1, a_2, \ldots, a_m),$$

such that

$$\langle \mathcal{M} : y_1 \to b_1, \ldots, y_n \to b_n, x_1 \to a_1, \ldots, x_m \to a_m, x \to b \rangle \models G,$$

which proves that

$$\langle \mathcal{M} : y_1 \to b_1, \ldots, y_n \to b_n, x_1 \to a_1, \ldots, x_m \to a_m, \rangle \models \exists x G,$$

and, finally,

$$\langle \mathcal{M} : y_1 \to b_1, y_2 \to b_2, \ldots, y_n \to b_n \rangle \models \forall x_1 \forall x_2 \ldots \forall x_m \exists x G,$$

which is the long awaited property.   ∎

Lemma 3.66 (*using the axiom of choice*) *Assume that L is any language, that $y_1, y_2, \ldots, y_n$ are pairwise distinct variables, that $F = F[y_1, y_2, \ldots, y_n]$ is in polite prenex form, that $\mathcal{N} = \langle N, \ldots \rangle$ is an L-structure, and that the n-tuple $(b_1, b_2, \ldots, b_n)$ of elements of N satisfies the formula F in $\mathcal{N}$; then it is possible to enrich the structure $\mathcal{N}$ to an $L_{Sk}(F)$-structure in which the n-tuple $(b_1, b_2, \ldots, b_n)$ satisfies the formula F_{Sk}, the Skolem form of F.*

Proof Once again, the proof is by induction on the number of existential quantifications in the formula F and, as before, we note that the case in which this number is 0 is trivial ($L_{Sk}(F) = L$, $F_{Sk} = F$ and $\mathcal{N}$ is sufficiently rich as is). So we assume that the result is true for all polite prenex formulas (in all languages) that have at most k existential quantifications and that F has $k + 1$ of these. Under the same conditions as in the proof of the previous lemma, we may set

$$F = \forall x_1 \forall x_2 \ldots \forall x_m \exists x\, G[y_1, y_2, \ldots, y_n, x_1, x_2, \ldots, x_m, x];$$
$$F' = \forall x_1 \forall x_2 \ldots \forall x_m\, G[y_1, y_2, \ldots, y_n, x_1, x_2, \ldots, x_m, x];$$

and, for the same reasons, we have

$$F_{Sk} = F'_{Sk_{f_1 x_1 x_2 \ldots x_m / x}} = F'_{f_1 x_1 x_2 \ldots x_m / x_{Sk}}.$$

Since our hypothesis is $\langle \mathcal{N} : y_1 \to b_1, y_2 \to b_2, \ldots, y_n \to b_n \rangle \models F$, it follows that for all elements $a_1, a_2, \ldots, a_m$ of N, the set

$$\{b \in N : \langle \mathcal{N} : y_1 \to b_1, \ldots, y_n \to b_n, x_1 \to a_1, \ldots, x_m \to a_m, x \to b \rangle \models G\}$$

is non-empty. So, using a choice function ϕ on N, we may define a map γ from N^m into N by setting

$$\gamma(a_1, a_2, \ldots, a_m) = \phi(\{b \in N : \mathcal{N} \models G[b_1, \ldots, b_n, a_1, \ldots, a_m, b]\})$$

for all $a_1, a_2, \ldots, a_m$ in N.

We may immediately assert that (*):

$$\text{for all } a_1, a_2, \ldots, a_m \text{ in } N$$

$$\langle \mathcal{N} : y_1 \to b_1, \ldots, y_n \to b_n, x_1 \to a_1, \ldots, x_m \to a_m, x \to \gamma(a_1, a_2, \ldots, a_m) \rangle$$
$$\models G.$$

Let L_1 denote the language obtained by adding to L the m-ary function symbol f_1 of the language $L_{Sk}(F)$ (the one corresponding to the first occurrence of $\exists$ in F). We can enrich $\mathcal{N}$ to an L_1-structure $\mathcal{N}_1$ if we interpret the symbol f_1 by the map γ.

The previously displayed assertion (*) then becomes (by applying Lemma 3.58):

$$\langle \mathcal{N}_1 : y_1 \to b_1, \ldots, y_n \to b_n, x_1 \to a_1, \ldots, x_m \to a_m,$$
$$x \to \overline{f}_1^{\mathcal{N}_1}(a_1, a_2, \ldots, a_m)\rangle \models G.$$

This is also equivalent (by Proposition 3.45) to

$$\langle \mathcal{N}_1 : y_1 \to b_1, \ldots, y_n \to b_n, x_1 \to a_1, \ldots, x_m \to a_m \rangle \models G_{f_1 x_1 x_2 \ldots x_m / x}.$$

Consequently,

$$\langle \mathcal{N}_1 : y_1 \to b_1, \ldots, y_n \to b_n \rangle \models \forall x_1 \forall x_2 \ldots \forall x_m [G_{f_1 x_1 x_2 \ldots x_m / x}].$$

Since x is distinct from the x_i, this last formula is none other than

$$[\forall x_1 \forall x_2 \ldots \forall x_m G]_{f_1 x_1 x_2 \ldots x_m / x} = F'_{f_1 x_1 x_2 \ldots x_m / x},$$

which is a polite prenex formula F_1 of the language L_1 with at most k existential quantifications. By the induction hypothesis, we can enrich the structure $\mathcal{N}_1$ to an $L_{1_{Sk}}(F_1)$-structure $\mathcal{N}^*$ such that

$$\langle \mathcal{N}^* : y_1 \to b_1, y_2 \to b_2, \ldots, y_n \to b_n \rangle \models F'_{f_1 x_1 x_2 \ldots x_m / x_{Sk}}.$$

We recognize this to be the formula F (in passing, we will have observed that the language $L_{1_{Sk}}(F_1)$ is exactly the language $L_{Sk}(F)$). ∎

Corollary 3.67 *For a closed formula to have a model, it is necessary and sufficient that any one of its Skolem forms have a model.*

Proof This is an immediate consequence of the two preceding lemmas and the theorem about prenex forms (Theorem 3.60). ∎

The exercises in Chapter 4 will provide other occasions to practise the art of putting a formula in prenex from or in Skolem form.

3.6 First steps in model theory

3.6.1 Satisfaction in a substructure

We will begin to take interest in what happens to the satisfaction of formulas when we pass from one structure to another. Though we should not expect to have much information when the two structures are completely arbitrary, we do have a few elementary results at our disposal in particular cases. In fact, we have already encountered one: the case where we compare the satisfaction of a formula in a structure with its satisfaction in an enrichment of the structure to a more extensive language (Lemma 3.58). We will now examine, successively, what can be said when we study the satisfaction of a formula in two structures, first, where one is an extension of the other, and second, where the two structures are isomorphic.

Knowing that a formula is satisfied in a certain structure, can we conclude (when this is meaningful) that it is satisfied in a substructure or in an extension? In general, the answer is no, but we do have, none the less, some useful information if the formula under consideration is sufficiently simple (see the next two theorems below). We need a lemma to begin with:

Lemma 3.68 *Assume that L is a language, $\mathcal{M} = \langle M, \ldots \rangle$ is an L-structure, $\mathcal{N} = \langle N, \ldots \rangle$ is an extension of $\mathcal{M}$, $t = t[v_0, v_1, \ldots, v_{m-1}]$ is a term of L, and $a_0, a_1, \ldots, a_{m-1}$ are elements of M. Then*

$$\bar{t}^{\mathcal{M}}[a_0, a_1, \ldots, a_{m-1}] = \bar{t}^{\mathcal{N}}[a_0, a_1, \ldots, a_{m-1}].$$

Proof This is proved by induction on t:

- if t is the variable v_j ($0 \leq j \leq m - 1$), then $\bar{t}^{\mathcal{M}}[a_0, a_1, \ldots, a_{m-1}] = \bar{t}^{\mathcal{N}}[a_0, a_1, \ldots, a_{m-1}] = a_j$;

- if t is a constant symbol c, then $\bar{t}^{\mathcal{M}}[a_0, a_1, \ldots, a_{m-1}] = \bar{c}^{\mathcal{M}} = \bar{c}^{\mathcal{N}} = \bar{t}^{\mathcal{N}}[a_0, a_1, \ldots, a_{m-1}]$ (since $\mathcal{N}$ is an extension of $\mathcal{M}$);

- if $t = ft_1 t_2 \ldots t_p$, where f is a p-ary function symbol and $t_1, t_2, \ldots, t_p$ are terms that satisfy $\bar{t_i}^{\mathcal{M}}[a_0, a_1, \ldots, a_{m-1}] = \bar{t_i}^{\mathcal{N}}[a_0, a_1, \ldots, a_{m-1}]$ for $1 \leq i \leq p$ (this is the induction hypothesis), then

$$\bar{t}^{\mathcal{M}}[a_0, a_1, \ldots, a_{m-1}]$$
$$= \bar{f}^{\mathcal{M}}(\bar{t_1}^{\mathcal{M}}[a_0, a_1, \ldots, a_{m-1}], \ldots, \bar{t_p}^{\mathcal{M}}[a_0, a_1, \ldots, a_{m-1}])$$
$$= \bar{f}^{\mathcal{N}}(\bar{t_1}^{\mathcal{M}}[a_0, a_1, \ldots, a_{m-1}], \ldots, \bar{t_p}^{\mathcal{M}}[a_0, a_1, \ldots, a_{m-1}]) \qquad (*)$$
$$= \bar{f}^{\mathcal{N}}(\bar{t_1}^{\mathcal{N}}[a_0, a_1, \ldots, a_{m-1}], \ldots, \bar{t_p}^{\mathcal{N}}[a_0, a_1, \ldots, a_{m-1}]) \qquad (**)$$
$$= \bar{t}^{\mathcal{M}}[a_0, a_1, \ldots, a_{m-1}],$$

where $(*)$ is justified by the fact that $\bar{f}^{\mathcal{M}}$ is the restriction of $\bar{f}^{\mathcal{N}}$ to M^k and $(**)$ is a consequence of the induction hypothesis. ∎

Theorem 3.69 *Assume that L is a language, $\mathcal{M} = \langle M, \ldots \rangle$ is an L-structure, $\mathcal{N} = \langle N, \ldots \rangle$ is an extension of $\mathcal{M}$, $F = F[v_0, v_1, \ldots, v_{m-1}]$ is a quantifier-free formula of L, and $a_0, a_1, \ldots, a_{m-1}$ are elements of M. Then F is satisfied in $\mathcal{M}$ by the sequence $(a_0, a_1, \ldots, a_{m-1})$ if and only if F is satisfied in $\mathcal{N}$ by this same sequence.*

Proof The proof is by induction on F.

- If F is atomic, there is a k-ary ($k \geq 1$) relation symbol R and terms $t_1, t_2, \ldots, t_k$ whose variables are among $v_0, v_1, \ldots, v_{m-1}$ such that $F = Rt_1 t_2 \cdots t_k$. So

$$\langle \mathcal{M} : v_0 \to a_0, v_1 \to a_1, \ldots, v_{m-1} \to a_{m-1} \rangle \models F$$

if and only if

$$(\overline{t_1}^{\mathcal{M}}[a_0, a_1, \ldots, a_{m-1}], \ldots, \overline{t_k}^{\mathcal{M}}[a_0, a_1, \ldots, a_{m-1}]) \in \overline{R}^{\mathcal{M}}. \qquad (\dagger)$$

Now, according to the lemma, we have

$$\overline{t_i}^{\mathcal{M}}[a_0, a_1, \ldots, a_{m-1}] = \overline{t_i}^{\mathcal{N}}[a_0, a_1, \ldots, a_{m-1}]$$

for $1 \leq i \leq k$ and, since $\overline{R}^{\mathcal{M}} = \overline{R}^{\mathcal{N}} \cap M^k$, $(\dagger)$ is equivalent to

$$(\overline{t_1}^{\mathcal{N}}[a_0, a_1, \ldots, a_{m-1}], \ldots, \overline{t_k}^{\mathcal{N}}[a_0, a_1, \ldots, a_{m-1}]) \in \overline{R}^{\mathcal{N}},$$

which means precisely that

$$\langle \mathcal{N} : v_0 \to a_0, v_1 \to a_1, \ldots, v_{m-1} \to a_{m-1} \rangle \models F.$$

• The induction steps that concern the connectives $\neg, \wedge, \vee, \Rightarrow$, and $\Leftrightarrow$ are obvious. There are no other cases to consider since F has no quantifiers. ∎

Theorem 3.70 *Assume that L is a language, $\mathcal{M} = \langle M, \ldots \rangle$ is an L-structure, $\mathcal{N} = \langle N, \ldots \rangle$ is an extension of $\mathcal{M}$, $F = F[v_0, v_1, \ldots, v_{m-1}]$ is a universal formula of L, $G = G[v_0, v_1, \ldots, v_{m-1}]$ is an existential formula of L, and $a_0, a_1, \ldots, a_{m-1}$ are elements of M. Under these conditions:*
if F is satisfied in $\mathcal{N}$ by the sequence $(a_0, a_1, \ldots, a_{m-1})$ then F is satisfied in $\mathcal{M}$ by this same sequence;
if G is satisfied in $\mathcal{M}$ by the sequence $(a_0, a_1, \ldots, a_{m-1})$ then G is satisfied in $\mathcal{N}$ by this same sequence.

Proof The second assertion follows immediately from the first: if G is existential, $\neg G$ is equivalent to a universal formula F' (property (1) of Theorem 3.55). If G is satisfied in $\mathcal{M}$ by $(a_0, a_1, \ldots, a_{m-1})$, then F' is not, hence (by the contrapositive of the first assertion) F' is not satisfied in $\mathcal{N}$ by $(a_0, a_1, \ldots, a_{m-1})$, which proves that G is.

We prove the first assertion by induction on the number of universal quantifiers in the prefix of F. If this number is 0, the result follows from Theorem 3.69. Otherwise, $F = \forall v_k H$ (where H is universal and has one fewer universal quantifier than F, so, by the induction hypothesis, it satisfies the assertion). Subject to replacing k by $h = \sup(k, m)$ and H by H_{v_h/v_k} (which produces a formula that is equivalent to F), we may assume $k \geq m$ (and even, if we wish, $k = m$). We then have $H = H[v_0, v_1, \ldots, v_{m-1}, v_k]$ and the fact that

$$\langle \mathcal{N} : v_0 \to a_0, v_1 \to a_1, \ldots, v_{m-1} \to a_{m-1} \rangle \models F$$

means that for every element $a \in N$,

$$\langle \mathcal{N} : v_0 \to a_0, v_1 \to a_1, \ldots, v_{m-1} \to a_{m-1}, v_k \to a \rangle \models H;$$

in particular, this must be true for every element $a \in M$. Thanks to the induction hypothesis, we may now conclude

$$\langle \mathcal{M} : v_0 \rightarrow a_0, v_1 \rightarrow a_1, \ldots, v_{m-1} \rightarrow a_{m-1} \rangle \models \forall v_k H.$$
■

The content of Theorem 3.70 can be summarized by saying that universal formulas are preserved by substructures while existential formulas are preserved by extensions. Some more refined preservation properties will be presented in Chapter 8 (we will also have a converse to Theorem 3.70: every formula that is preserved by substructures (extensions, respectively) is equivalent to a universal (existential, respectively) formula.)

There is one preservation property that we have every reason to expect should apply to arbitrary formulas: preservation by isomorphisms. This will be guaranteed by the next theorem.

Lemma 3.71 *Assume that L is a language, $\mathcal{M} = \langle M, \ldots \rangle$ and $\mathcal{N} = \langle N, \ldots \rangle$ are two L-structures, and that $h : M \longmapsto N$ is a homomorphism from $\mathcal{M}$ into $\mathcal{N}$. Then for every term $t = t[v_0, v_1, \ldots, v_{m-1}]$ and for all elements $a_0, a_1, \ldots, a_{m-1}$ of the set M, we have*

$$h\big(\bar{t}^{\mathcal{M}}[a_0, a_1, \ldots, a_{m-1}]\big) = \bar{t}^{\mathcal{N}}[h(a_0), h(a_1), \ldots, h(a_{m-1})].$$

Proof The proof is by induction on t:

- if t is the variable v_j ($0 \le j \le m - 1$), then each side of the proposed equality is equal to $h(a_j)$;
- if t is the constant symbol c, then the left side is $h(\bar{c}^{\mathcal{M}})$ and the right side is $\bar{c}^{\mathcal{N}}$ and these are equal since h is a homomorphism;
- if $t = ft_1 t_2 \ldots t_p$, where f is a p-ary function symbol and $t_1, t_2, \ldots, t_p$ are terms that satisfy $h(\bar{t_i}^{\mathcal{M}}[a_0, a_1, \ldots, a_{m-1}]) = \bar{t_i}^{\mathcal{N}}[h(a_0), h(a_1), \ldots, h(a_{m-1})]$ for $1 \le i \le p$ (this is the induction hypothesis), then

$$
\begin{aligned}
h\big(\bar{t}^{\mathcal{M}}&[a_0, a_1, \ldots, a_{m-1}]\big) \\
&= h\big(\bar{f}^{\mathcal{M}}\big(\bar{t_1}^{\mathcal{M}}[a_0, a_1, \ldots, a_{m-1}], \ldots, \bar{t_p}^{\mathcal{M}}[a_0, a_1, \ldots, a_{m-1}]\big)\big) \\
&= \bar{f}^{\mathcal{N}}\big(h(\bar{t_1}^{\mathcal{M}}[a_0, a_1, \ldots, a_{m-1}]), \ldots, h(\bar{t_p}^{\mathcal{M}}[a_0, a_1, \ldots, a_{m-1}])\big) \quad (*) \\
&= \bar{f}^{\mathcal{N}}\big(\bar{t_1}^{\mathcal{N}}[h(a_0), h(a_1), \ldots, h(a_{m-1})], \ldots, \\
&\qquad\qquad \bar{t_p}^{\mathcal{N}}[h(a_0), h(a_1), \ldots, h(a_{m-1})]\big) \qquad\qquad\qquad (**) \\
&= \bar{t}^{\mathcal{N}}[h(a_0), h(a_1), \ldots, h(a_{m-1})],
\end{aligned}
$$

where (*) holds because h is a homomorphism and (**) follows from the induction hypothesis. ■

Theorem 3.72 *Assume that L is a language, that $\mathcal{M} = \langle M, \ldots \rangle$ and $\mathcal{N} = \langle N, \ldots \rangle$ are two L-structures, that $h : M \longmapsto N$ is an isomorphism from $\mathcal{M}$*

onto $\mathcal{N}$, that $F = F[v_0, v_1, \ldots, v_{m-1}]$ is an arbitrary formula of L and that $a_0, a_1, \ldots, a_{m-1}$ are elements of the set M. Then F is satisfied in $\mathcal{M}$ by the sequence $(a_0, a_1, \ldots, a_{m-1})$ if and only if F is satisfied in $\mathcal{N}$ by the sequence $(h(a_0), h(a_1), \ldots, h(a_{m-1}))$.

Proof The proof is by induction on F.

- If F is atomic, there is a k-ary ($k \geq 1$) relation symbol R and terms $t_1, t_2, \ldots, t_k$ whose variables are among $v_0, v_1, \ldots, v_{m-1}$ such that $F = Rt_1t_2\ldots t_k$. So $\langle \mathcal{M} : v_0 \to a_0, v_1 \to a_1, \ldots, v_{m-1} \to a_{m-1} \rangle \models F$ if and only if

$$(\overline{t_1}^{\mathcal{M}}[a_0, a_1, \ldots, a_{m-1}], \ldots, \overline{t_k}^{\mathcal{M}}[a_0, a_1, \ldots, a_{m-1}]) \in \overline{R}^{\mathcal{M}}. \quad (*)$$

Because h is an isomorphism, (*) is equivalent to

$$(h(\overline{t_1}^{\mathcal{M}}[a_0, a_1, \ldots, a_{m-1}]), \ldots, h(\overline{t_k}^{\mathcal{M}}[a_0, a_1, \ldots, a_{m-1}])) \in \overline{R}^{\mathcal{N}},$$

or again, according to the lemma, to

$$(\overline{t_1}^{\mathcal{N}}[h(a_0), h(a_1), \ldots, h(a_{m-1})], \ldots, \overline{t_p}^{\mathcal{N}}[h(a_0), h(a_1), \ldots, h(a_{m-1})])$$
$$\in \overline{R}^{\mathcal{N}},$$

which means precisely that

$$\langle \mathcal{N} : v_0 \to h(a_0), v_1 \to h(a_1), \ldots, v_{m-1} \to h(a_{m-1}) \rangle \models F.$$

- The subsequent induction steps present no problems (in view of Remark 3.57, we may restrict our attention to the cases involving $\neg$, $\vee$ and $\exists$). For example, let us treat the case of existential quantification: so we suppose that $F = \exists v_k G$ and, as in the proof of Theorem 3.70, that $k \geq m$ and $G = G[v_0, v_1, \ldots, v_{m-1}, v_k]$; then

$$\langle \mathcal{M} : v_0 \to a_0, v_1 \to a_1, \ldots, v_{m-1} \to a_{m-1} \rangle \models F$$

means that we can find an element $a \in M$ such that

$$\langle \mathcal{M} : v_0 \to a_0, v_1 \to a_1, \ldots, v_{m-1} \to a_{m-1}, v_k \to a \rangle \models G;$$

by the induction hypothesis, this is equivalent to

$$\langle \mathcal{N} : v_0 \to h(a_0), v_1 \to h(a_1), \ldots, v_{m-1} \to h(a_{m-1}), v_k \to h(a) \rangle \models G,$$

which proves that

$$\langle \mathcal{N} : v_0 \to h(a_0), v_1 \to h(a_1), \ldots, v_{m-1} \to h(a_{m-1}) \rangle \models F.$$

Conversely, if this condition is satisfied, we can find an element $b \in N$ such that

$$\langle \mathcal{N} : v_0 \to h(a_0), v_1 \to h(a_1), \ldots, v_{m-1} \to h(a_{m-1}), v_k \to b \rangle \models G,$$

and, since h is a bijection, an element $a \in M$ such that $b = h(a)$; the conclusion now follows, as before, from the induction hypothesis.   ∎

3.6.2 Elementary equivalence

An immediate consequence of the previous theorem is that two isomorphic L-structures satisfy exactly the same closed formulas of the language L. This leads us to a notion that is absolutely fundamental in model theory, that of **elementary equivalence**.

Definition 3.73 *An L-structure M is **elementarily equivalent** to another L-structure N (we will denote this by $M \equiv N$) if and only if every closed formula of L that is satisfied in M is also satisfied in N.*

We immediately conclude that $M \equiv N$ if and only if M and N satisfy the same closed formulas of L: indeed, if a closed formula F is not satisfied in M, then $M \models \neg F$, so $N \models \neg F$ and F is not satisfied in N. Thus we see that $\equiv$ is an equivalence relation on the class of L-structures. So we will indifferently say 'M is elementarily equivalent to N' or 'M and N are elementarily equivalent'.

The notation $M \not\equiv N$ means that M and N are not elementarily equivalent.

Therefore, the following is a consequence of Theorem 3.72:

Proposition 3.74 *If two L-structures are isomorphic, then they are elementarily equivalent.*

We will have countless opportunities to observe that the converse of this is far from being true. The existence of elementarily equivalent models that are not isomorphic is clear evidence for the fact that the expressive power of first order languages is limited: in the usual practice of mathematics, when two structures are not isomorphic, we can generally discern some property that is satisfied by one and not by the other; but if the structures are elementarily equivalent, then such a distinguishing property will not be expressible as a first order formula of the language, nor even by a set of such formulas. If we permit ourselves to anticipate, we can discuss an example: it will turn out that the structures $\langle \mathbb{R}, \leq \rangle$ and $\langle \mathbb{Q}, \leq \rangle$ are elementarily equivalent (naturally, the language consists of a single binary relation symbol); they cannot possibly be isomorphic since the second is countable while the first is not; consequently, none of the properties that distinguish them can be expressible by a theory of the language. In particular, this is the case for the famous least upper bound property (every non-empty subset that has an upper bounded has a least upper bound) which is true in $\mathbb{R}$ but not in $\mathbb{Q}$.

The remarks we have just made lead us to the following definition:

Definition 3.75 *Let L be a first order language and let $X(M)$ be a property that an L-structure M may or may not satisfy.*
*The property $X(M)$ is **axiomatizable** (respectively, **finitely axiomatizable**) if there exists a theory T of L (respectively, a closed formula F of L) such that for every L-structure M, $X(M)$ is verified if and only if M is a model of T*

*(respectively, of F). When this happens, we say that the theory T (respectively, the closed formula F) **axiomatizes** the property $\mathcal{X}(\mathcal{M})$.*

*We say that $\mathcal{X}(\mathcal{M})$ is **pseudo-axiomatizable** if there exists a language L^* that extends L and a theory T of L^* such that for every L-structure $\mathcal{M}$, the property $\mathcal{X}(\mathcal{M})$ is satisfied if and only if $\mathcal{M}$ is the reduct to the language L of an L^*-structure that is a model of T.*

Obviously, every axiomatizable property is pseudo-axiomatizable.

Instead of 'axiomatizable property', we often say 'first order property'.

What we noted above can then be paraphrased as follows: the property (for a set and a binary relation on it) of 'being a totally ordered set in which every non-empty subset that has an upper bound has a least upper bound' is not axiomatizable (it is not even pseudo-axiomatizable).

Lemma 3.76 *If a property is finitely axiomatizable, then so is its negation.*

Proof This is immediate from the definition: if a property is axiomatized by the closed formula F, its negation is axiomatized by the formula $\neg F$.   ■

Lemma 3.77 *If a property is not axiomatizable, then its negation is not finitely axiomatizable.*

Proof This follows from the contrapositive of the previous lemma.   ■

To show that a property is axiomatizable, it obviously suffices to find a set of closed formulas whose models are precisely the structures that have the property in question. One suspects that it is a more delicate matter to show that a property is not axiomatizable. The fact that you have not found a theory that works clearly does not imply that such does not exist. The example of $\mathbb{R}$ and $\mathbb{Q}$ suggests a possible route: find two elementarily equivalent structures, one of which satisfies the property while the other does not. Many of the exercises will focus on this type of question. We will treat some simple examples a bit further on.

Beforehand, let us develop a very efficient tool for resolving not only these problems of non-axiomatizability but also numerous other problems of model theory. This is the **compactness theorem for predicate calculus**, which we should consider as one of the 'grand' theorems of mathematical logic. Only in Chapter 4 will we see the proof, but it would be unfortunate not to allow ourselves the possibility of using it right away (which will be done in many of the exercises), especially since its statement is very simple, all the more for those who have previously studied the analogous theorem for propositional calculus.

Theorem 3.78 *(with the axiom of choice) For a theory in a first order language to be consistent, it is necessary and sufficient that it be finitely consistent.*

As for the propositional calculus, this compactness theorem has several equivalent versions:

For a first order theory to be contradictory, it is necessary and sufficient that some finite subset be contradictory.

Or again:

For a closed formula F of a first order language L to be a semantic consequence of a theory T of L, it is necessary and sufficient that there exists a finite subset T_0 of T such that F is a semantic consequence of T_0.

The equivalence of these three versions follows directly from Theorem 3.51. Also, the 'necessary' part of the first version and the 'sufficient' part of the other two are evident.

We will now examine the question of the axiomatizability of some standard properties, relative to structures for the simplest possible language: the language whose only symbol is the equality symbol. Clearly, these structures are none other than the non-empty sets.

For every integer $n \geq 2$, the property 'is a set with at least n elements' is finitely axiomatizable thanks to the formula:

$$F_n = \exists v_1 \exists v_2 \ldots \exists v_n \bigwedge_{1 \leq i \leq j \leq n} \neg v_i \simeq v_j.$$

The property 'is a set with at most n elements' is axiomatizable by the formula $\neg F_{n+1}$.

The property 'is an infinite set' is axiomatizable by the following theory:

$$\{F_n : n \in \mathbb{N} \text{ and } n \geq 2\}.$$

Two natural questions now arise. Is the property 'is an infinite set' finitely axiomatizable? Is the property 'is a finite set' axiomatizable? The answer in both cases is 'no'.

Theorem 3.79 *The property 'is a finite set' is not pseudo-axiomatizable.*

Proof The proof is by contradiction. Suppose T is a theory in a language L that conforms with the properties required by Definition 3.75. Then $T \cup \{F_n : n \in \mathbb{N} \text{ and } n \geq 2\}$ is a set of closed formulas of L that is contradictory (for to satisfy it, a set would have to be both finite and infinite!). By the compactness theorem, we can find a finite subset $T' \subseteq T \cup \{F_n : n \in \mathbb{N} \text{ and } n \geq 2\}$ that is contradictory. There certainly exists an integer p such that

$$T' \subseteq T_p = T \cup \{F_n : 2 \leq n \leq p\}.$$

So the theory T_p itself is contradictory. But we can see right away that this is false: for the finite set $\{2, 3, \ldots, p\}$ is, according to our hypothesis, the underlying set of at least one L-structure $\mathcal{M}$ that is a model of T, and $\mathcal{M}$ obviously satisfies F_2, $F_3, \ldots, F_p$, so is a model of T_p. ■

From Lemma 3.77, we also immediately obtain:

Theorem 3.80 *The property 'is an infinite set' is not finitely axiomatizable.*

However, the infinite sets are precisely those that can support a dense total ordering. So we might say that 'being an infinite set' is a 'pseudo-finitely-axiomatizable' property.

Here is another absolutely fundamental notion:

Definition 3.81 *A theory T in a language L is **complete** if and only if*
(1) *T is consistent; and*
(2) *all the models of T are elementarily equivalent.*

Lemma 3.82 *For a theory T in a language L to be complete, it is necessary and sufficient that*
(1) *T is consistent; and*
(2) *for every closed formula F of L, we have that either $T \vdash^* F$ or $T \vdash^* \neg F$.*

Proof If the second condition is not satisfied, we can find a closed formula F of L such that $T \not\vdash^* F$ and $T \not\vdash^* \neg F$, which means that there are two models $\mathcal{M}$ and $\mathcal{N}$ of T such that $\mathcal{M} \not\models F$ and $\mathcal{N} \not\models \neg F$, or in other words, $\mathcal{M} \not\models F$ and $\mathcal{N} \models F$. So we see that $\mathcal{M} \not\equiv \mathcal{N}$, which contradicts condition (2) of the definition and so T is not complete. Conversely, if T is not complete but is consistent, we can find two models $\mathcal{A}$ and $\mathcal{B}$ of T such that $\mathcal{A} \not\equiv \mathcal{B}$, which proves that there is some closed formula F that is satisfied in $\mathcal{A}$ but not in $\mathcal{B}$; so neither $T \vdash^* F$ nor $T \vdash^* \neg F$ can hold. ∎

Example 3.83 In the language whose only symbol is equality, the theory consisting of the single formula $\forall v_0 \forall v_1 v_0 \simeq v_1$ is complete. Indeed, the models of this theory are the sets with only one element; these are all isomorphic, hence elementarily equivalent. In this same language, the empty theory is not complete: indeed, every L-structure is a model of this theory and it is not difficult to find two L-structures that are not elementarily equivalent; thus a set with only one element satisfies the formula $\forall v_0 \forall v_1 v_0 \simeq v_1$ but a set with at least two elements does not satisfy it.

Remark 3.84 *The fact that a theory is complete or not depends in an essential way on the chosen language: if we consider the formula $\forall v_0 \forall v_1 v_0 \simeq v_1$ as a formula of a language that has, in addition to the symbol $\simeq$, a unary relation symbol P, we see immediately that it is no longer a complete theory, for some of its models will satisfy $\exists v_0 P v_0$ and others will not.*

Having some more interesting examples in mind, let us consider another definition:

Definition 3.85 *Given an L-structure $\mathcal{M}$, the set of closed formulas of L that are satisfied in $\mathcal{M}$ is called the **theory of** $\mathcal{M}$ and is denoted by $Th(\mathcal{M})$:*

$$Th(\mathcal{M}) = \{F \in \mathcal{F}(L) : F \text{ is closed and } \mathcal{M} \models F\}.$$

Theorem 3.86 *For any L-structure $\mathcal{M}$, $Th(\mathcal{M})$ is a complete theory of L.*

Proof On the one hand, $Th(\mathcal{M})$ is consistent since $\mathcal{M}$ is, evidently, a model. On the other hand, for any closed formula F of L, we have

- either $\mathcal{M} \models F$, in which case $F \in Th(\mathcal{M})$, so $Th(\mathcal{M}) \vdash^* F$;
- or else $\mathcal{M} \nvDash F$, in which case $\mathcal{M} \models \neg F$, so $\neg F \in Th(\mathcal{M})$ and $Th(\mathcal{M}) \vdash^* \neg F$.

Invoking Lemma 3.82 now completes the proof. ∎

We will find many examples of complete and incomplete theories in the exercises.

3.6.3 The language associated with a structure and formulas with parameters

Consider a language L and an L-structure $\mathcal{M} = \langle M, \ldots \rangle$. We will enrich the language L to a language, denoted by L_M and called the **language associated with the L-structure** $\mathcal{M}$, in the following way: with each element $a \in M$, we associate a new constant symbol denoted by $\underline{a}$; we assume that these new symbols are really new (i.e. are distinct from all the symbols of L) and that they are pairwise distinct (if $a \neq b$, then $\underline{a} \neq \underline{b}$). We then set

$$L_M = L \cup \{\underline{a} : a \in M\}.$$

It is then not very difficult to enrich $\mathcal{M}$ into an L_M-structure. We must decide on an interpretation for each of the new symbols. No one will be surprised when we decide to interpret the symbol $\underline{a}$ by the element a. If we denote the enriched structure by $\mathcal{M}^*$, we then have for every $a \in M$:

$$\underline{a}^{\mathcal{M}^*} = a,$$

and we assign to the symbols of L the same interpretations in $\mathcal{M}^*$ that they had in $\mathcal{M}$.

With every formula $F = F[v_0, v_1, \ldots, v_{m-1}]$ of L and with every m-tuple $(a_0, a_1, \ldots, a_{m-1})$ of elements of M, we can, in a natural way, associate a closed formula of the language L_M: specifically, the formula $F_{\underline{a_0}/v_0, \underline{a_1}/v_1, \ldots, \underline{a_{m-1}}/v_{m-1}}$ which, according to our conventions, could also be written

$$F[\underline{a_0}, \underline{a_1}, \ldots, \underline{a_{m-1}}].$$

We then obtain the following result:

Theorem 3.87 *The following properties are equivalent:*

(1) $\langle \mathcal{M} : v_0 \to a_0, v_1 \to a_1, \ldots, v_{m-1} \to a_{m-1} \rangle \models F$

(2) $\mathcal{M}^* \models F[\underline{a_0}, \underline{a_1}, \ldots, \underline{a_{m-1}}]$.

Proof Recall that $F[\underline{a_0}, \underline{a_1}, \ldots, \underline{a_{m-1}}]$ is the following formula of L_M:

$$F_{\underline{a_0}/v_0, \underline{a_1}/v_1, \ldots, \underline{a_{m-1}}/v_{m-1}}.$$

Before giving the proof, we should note that it will provide the justification that we promised in Section 3.3.2 for the notation:

$$\mathcal{M} \models F[a_0, a_1, \ldots, a_{m-1}] \qquad (*)$$

that we have already suggested as a possible abbreviation for property (1) above (and of which we have already made use). Indeed, we get from property (2) to (*) by forgetting to place the asterisk on $\mathcal{M}$ and to underline the a_i. The abuses that consist in identifying, on the one hand, the elements of a model with the constant symbols that represent them in the enriched language, and, on the other hand, identifying the model with its natural enrichment to this language, present no real danger and we will often adopt them. Once the theorem is proved, an assertion such as (*) can have two distinct meanings but the theorem tells us precisely that these meanings may legitimately be treated as one. ∎

In fact, this theorem is a special case of the following more general result:

Lemma 3.88 *For any integers p and q, any pairwise distinct variables $x_0, x_1, \ldots, x_{p-1}, y_0, y_1, \ldots, y_{q-1}$, any L-formula $G = G[x_0, x_1, \ldots, x_{p-1}, y_0, y_1, \ldots, y_{q-1}]$ and any $(p+q)$-tuple $(a_0, a_1, \ldots, a_{p-1}, b_0, b_1, \ldots, b_{q-1})$ of elements of M, the following two properties are equivalent:*

(1) $\langle \mathcal{M} : x_0 \to a_0, x_1 \to a_1, \ldots, x_{p-1} \to a_{p-1},$
$\qquad y_0 \to b_0, y_1 \to b_1, \ldots, y_{q-1} \to b_{q-1} \rangle \models G;$

(2) $\langle \mathcal{M}^* : y_0 \to b_0, y_1 \to b_1, \ldots, y_{q-1} \to b_{q-1} \rangle \models G_{\underline{a_0}/x_0, \underline{a_1}/x_1, \ldots, \underline{a_{p-1}}/x_{p-1}}.$

Proof The proof is by induction on G. If G is the atomic formula $Rt_1 t_2 \ldots t_k$ (where $t_1, t_2, \ldots, t_k$ are terms of the language whose variables are among $x_0, x_1, \ldots, x_{p-1}, y_0, y_1, \ldots, y_{q-1}$), then property (1) is equivalent to

$$(\overline{t_1}^{\mathcal{M}}[a_0, \ldots, a_{p-1}, b_0, \ldots, b_{q-1}], \ldots, \overline{t_k}^{\mathcal{M}}[a_0, \ldots, a_{p-1}, b_0, \ldots, b_{q-1}]) \in \overline{R}^{\mathcal{M}},$$

and property (2) is equivalent to

$$(\overline{t_1}^{\mathcal{M}^*}[\underline{a_0}^{\mathcal{M}^*}, \ldots, \underline{a_{p-1}}^{\mathcal{M}^*}, b_0, \ldots, b_{q-1}], \ldots,$$
$$\overline{t_k}^{\mathcal{M}^*}[\underline{a_0}^{\mathcal{M}^*}, \ldots, \underline{a_{p-1}}^{\mathcal{M}^*}, b_0, \ldots, b_{q-1}]) \in \overline{R}^{\mathcal{M}^*}.$$

Now, $\overline{R}^{\mathcal{M}^*} = \overline{R}^{\mathcal{M}}$ and $\overline{t_j}^{\mathcal{M}^*} = \overline{t_j}^{\mathcal{M}}$ for $1 \leq j \leq k$ since these are symbols for a relation and for terms of L. The desired equivalence is now a consequence of the definition of the $\overline{a_j}^{\mathcal{M}^*}$.

The other steps in the induction are easy. Let us examine existential quantification: if $G = \exists z H[x_0, x_1, \ldots, x_{p-1}, y_0, y_1, \ldots, y_{q-1}, z]$ (for reasons mentioned in Theorem 3.70 and in Theorem 3.72, we may suppose that z is distinct from the

x_i and the y_j), then property (1) means that there exists an element $a \in M$ such that

$$\langle \mathcal{M} : x_0 \to a_0, \ldots, x_{p-1} \to a_{p-1}, y_0 \to b_0, \ldots, y_{q-1} \to b_{q-1}, z \to a \rangle \models H,$$

which is equivalent, by the induction hypothesis, to the existence of an element $a \in M$ such that

$$\langle \mathcal{M}^* : y_0 \to b_0, y_1 \to b_1, \ldots, y_{q-1} \to b_{q-1}, z \to a \rangle \models H_{\underline{a_0/x_0, a_1/x_1, \ldots, a_{p-1}/x_{p-1}}};$$

but, by the definition of satisfaction, this is equivalent in turn to

$$\langle \mathcal{M}^* : y_0 \to b_0, y_1 \to b_1, \ldots, y_{q-1} \to b_{q-1}, \rangle \models \exists z \, H_{\underline{a_0/x_0, a_1/x_1, \ldots, a_{p-1}/x_{p-1}}};$$

and this last formula is none other than $G_{\underline{a_0/x_0, a_1/x_1, \ldots, a_{p-1}/x_{p-1}}}$. This proves the equivalence of properties (1) and (2). ∎

To obtain the theorem, it obviously suffices to take $q = 0$ in the lemma. Observe that there was no way to bypass the generalization that this lemma provides: indeed, in the induction proof, it is not possible to only consider closed formulas of the language L_M.

The formulas of the language L_M are often called **formulas with parameters in** $\mathcal{M}$, the parameters being precisely the elements of M which have 'become' constant symbols. We will use this concept extensively in Chapter 8. At that time, we will have a need for the following two definitions that relate to a model $\mathcal{M}$ and a language L:

Definition 3.89 *The set of closed quantifier-free formulas of L_M that are satisfied in $\mathcal{M}^*$ is called the* **simple diagram of** $\mathcal{M}$ *and is denoted by* $\Delta(\mathcal{M})$.

Definition 3.90 *The set of closed formulas of the language L_M that are satisfied in $\mathcal{M}^*$ (i.e. the set $Th(\mathcal{M}^*)$) is called the* **complete diagram of** $\mathcal{M}$ *and is denoted by* $D(\mathcal{M})$.

Some authors use the phrase **elementary diagram** instead of complete diagram. They have an excellent reason for doing this that will appear in Chapter 8. As it is not necessarily simple to make the distinction between simple and elementary, we prefer to insist on our terminology (an elementary precaution...).

We will state right away a result whose proof will be given in Chapter 8. It allows us to characterize, up to isomorphism, the extensions of a structure.

Theorem 3.91 *Given an L-structure $\mathcal{M}$, for an L_M-structure $\mathcal{N}$ to be a model of the simple diagram of $\mathcal{M}$, it is necessary and sufficient that $\mathcal{M}$ be isomorphic to a sub-structure of the reduct of $\mathcal{N}$ to the language L.*

3.6.4 Functions and relations definable in a structure

Consider a first order language L and an L-structure $\mathcal{M} = \langle M, \ldots \rangle$.

Definition 3.92

- *For any integer $k \geq 1$ and any subset A of M^k, A is **definable in $\mathcal{M}$ by a formula of** L if and only if there exists a formula $F = F[w_1, w_2, \ldots, w_k]$ of L with at most k free variables such that for all elements $a_1, a_2, \ldots, a_k$ of M,*

$$(a_1, a_2, \ldots, a_k) \in A \quad \textit{if and only if } \mathcal{M} \models F[a_1, a_2, \ldots, a_k].$$

  *When this happens, we say that such a formula F is a **definition of A in** $\mathcal{M}$.*

- *An element $a \in M$ is said to be **definable in** $\mathcal{M}$ **by a formula of** L if the subset $\{a\}$ is. Any definition of $\{a\}$ is then called a **definition of the element** a.*

- *For any integer $k \geq 1$ and any map ϕ from M^k into M, ϕ is **definable in $\mathcal{M}$ by a formula of** L if and only if there exists a formula $F = F[w_1, w_2, \ldots, w_k, z]$ of L with at most $k + 1$ free variables such that for all elements $b, a_1, a_2, \ldots, a_k$ of M,*

$$\phi(a_1, a_2, \ldots, a_k) = b \quad \textit{if and only if } \mathcal{M} \models F[a_1, a_2, \ldots, a_k, b].$$

  *Such a formula is then called a **definition of ϕ in** $\mathcal{M}$.*

When we consider the graph of a map ϕ from M^k into M to be the set

$$\{(a_1, a_2, \ldots, a_k, b) \in M^{k+1} : b = \phi(a_1, a_2, \ldots, a_k)\},$$

we immediately see that such a map is definable in $\mathcal{M}$ by a formula of L if and only if its graph is also, and that the formulas that define the map are the same ones as those that define its graph.

It is important to note that the definability of a subset of M^k or of map from M^k into M depends in an essential way on the language under consideration and on the structure that accompanies the set M. For example, it may happen that a subset that is not definable in $\mathcal{M}$ by a formula of L becomes definable in an enrichment of $\mathcal{M}$ thanks to a formula of the extended language.

However, when there can be no reason to fear confusion over the subjects of language or structure, we will be content to speak of a **definable** subset (or relation) or function, without further precision.

Example 3.93

- The set M^k and the empty set are always definable subsets of M^k: the first, thanks to the formula $w_1 \simeq w_1$ and the second, thanks to its negation.

- The set of even integers is definable in the structure $\langle \mathbb{Z}, + \rangle$ by a formula of the language $\{g\}$ (g is a binary function symbol): it suffices to consider the formula $\exists w_0 g w_0 w_0 = w_1$.

Theorem 3.94 *For any integer $k \geq 1$, the set of subsets of M^k that are definable in $\mathcal{M}$ by a formula of L is a sub-algebra of the Boolean algebra of all subsets of M^k.*

Proof We have just seen that $\emptyset$ and M^k are definable. Now, if A and B are definable subsets of M^k and if $F = F[w_1, w_2, \ldots, w_k]$ and $G = G[w_1, w_2, \ldots, w_k]$ are, respectively, definitions of A and B, it is clear that $\neg F$, $(F \wedge G)$ and $(F \vee G)$ are, respectively, definitions for the complement of A in M^k, the intersection of A and B and the union of A and B. ∎

Theorem 3.95 *If k is an integer greater than or equal to 1 and A is a subset of M^k that is definable in $\mathcal{M}$ by a formula of L and if h is an automorphism of the structure $\mathcal{M}$, then the set A is invariant under h (this means that for all elements $a_1, a_2, \ldots, a_k$ of M, if $(a_1, a_2, \ldots, a_k) \in A$, then $(h(a_1), h(a_2), \ldots, h(a_k)) \in A$).*

Proof Suppose that $F = F[w_1, w_2, \ldots, w_k]$ is a formula of L that defines $A \subseteq M^k$ and that $a_1, a_2, \ldots, a_k$ are elements of M. If $(a_1, a_2, \ldots, a_k) \in A$, then $\mathcal{M} \models F[a_1, a_2, \ldots, a_k]$; in this case, then for any automorphism h of the structure $\mathcal{M}$, we also have $\mathcal{M} \models F[h(a_1), h(a_2), \ldots, h(a_k)]$ (by Theorem 3.72), which proves that $(h(a_1), h(a_2), \ldots, h(a_k)) \in A$ (because F is a definition of A). ∎

This theorem is useful when we want to show that a set is not definable: to do this, it suffices to find an automorphism of the structure under consideration that does not leave the set in question invariant.

To illustrate this, let us prove that no subsets of $\mathbb{R}$ other than $\mathbb{R}$ and $\emptyset$ are definable in the structure $\langle \mathbb{R}, \leq \rangle$ by a formula of the language $\{R\}$ (R is a binary relation symbol). We will argue by contradiction; suppose that there is a subset $A \subseteq \mathbb{R}$, distinct from $\mathbb{R}$ and $\emptyset$, that is definable by a formula $F = F[w]$ of this language. Then choose an element $a \in A$ (A is not empty) and an element $b \in \mathbb{R} - A$ ($\mathbb{R} - A$ is not empty). Observe that the map h from $\mathbb{R}$ into $\mathbb{R}$ which, with every real x, associates the real $x + b - a$, is an automorphism of $\langle \mathbb{R}, \leq \rangle$ which does not leave A invariant (since $h(a) = b$), which contradicts the previous theorem.

Next, we will introduce the notion of **definability with parameters**, which generalizes the notion we just studied.

We still consider a first order language L and an L-structure $\mathcal{M} = \langle M, \ldots \rangle$.

Definition 3.96 *For any integer k greater than or equal to 1 and any subset A of M^k, A is **definable with parameters from $\mathcal{M}$** if and only if there exist an integer $m \geq 1$, a formula $F = F[w_1, w_2, \ldots, w_k, z_1, z_2, \ldots, z_m]$ of L with at most $k + m$ free variables, and m elements $b_1, b_2, \ldots, b_m$ of M such that for all elements $a_1, a_2, \ldots, a_k$ of M,*

$(a_1, a_2, \ldots, a_k) \in A$ *if and only if $\mathcal{M} \models F[a_1, a_2, \ldots, a_k, b_1, b_2, \ldots, b_m]$.*

*When this happens, the formula $F[w_1, w_2, \ldots, w_k, b_1, b_2, \ldots, b_m]$ is called a **definition of A in $\mathcal{M}$ with parameters** $b_1, b_2, \ldots, b_m$.*

The notion of a map from M^k into M that is **definable with parameters from $\mathcal{M}$** has an analogous definition.

For example, every finite subset $\{\alpha_1, \alpha_2, \ldots, \alpha_p\}$ of M is definable in $\mathcal{M}$ with the parameters $\alpha_1, \alpha_2, \ldots, \alpha_p$ thanks to the formula

$$\bigvee_{1 \leq i \leq p} w \simeq \alpha_i.$$

We immediately see that to be definable with parameters from $\mathcal{M}$ is equivalent to being definable (in the sense of Definition 3.92) by a formula of the language L_M in the structure $\mathcal{M}^*$, the natural enrichment of $\mathcal{M}$ to L_M (see Theorem 3.87).

3.7 Models that may not respect equality

Our excursion into this topic will be as brief as possible.

We consider a language that includes the symbol for equality, $\simeq$. Let E denote the theory of L that consists of the following closed formulas (called the **axioms of equality**):

- the formula: $\forall v_0 v_0 \simeq v_0$;
- the formula: $\forall v_0 \forall v_1 (v_0 \simeq v_1 \Rightarrow v_1 \simeq v_0)$;
- the formula: $\forall v_0 \forall v_1 \forall v_2 ((v_0 \simeq v_1 \wedge v_1 \simeq v_2) \Rightarrow v_0 \simeq v_2)$;
- for every integer $k \geq 1$ and every k-ary function symbol f of L, the formula:

$$\forall v_1 \ldots \forall v_k \forall v_{k+1} \ldots \forall v_{2k} \Big(\bigwedge_{1 \leq i \leq k} v_i \simeq v_{k+i} \Rightarrow f v_1 \ldots v_k \simeq f v_{k+1} \ldots v_{2k} \Big);$$

- for every integer $k \geq 1$ and every k-ary relation symbol R of L, the formula:

$$\forall v_1 \ldots \forall v_k \forall v_{k+1} \ldots \forall v_{2k} \Big(\Big(R v_1 \ldots v_k \wedge \bigwedge_{1 \leq i \leq k} v_i \simeq v_{k+i} \Big) \Rightarrow R v_{k+1} \ldots v_{2k} \Big).$$

It is quite clear that all of these formulas are satisfied in any L-structure that respects equality.

Consider an arbitrary L-structure $\mathcal{M} = \langle M, \ldots \rangle$ in which the symbol for equality, $\simeq$, is interpreted by some binary relation on M that we will denote by θ. We are going to show that if $\mathcal{M}$ is a model of the theory E, then we can define from $\mathcal{M}$, in a natural way, another model that does respect equality and possesses some interesting properties.

So suppose that $\mathcal{M} \models E$. Then, according to the first three formulas of E, the relation θ is an equivalence relation on M that is, according to the remaining formulas of E, compatible with the functions and relations of the structure. Let A denote the quotient M/θ (the set of equivalence classes relative to θ). The equivalence class of the element $a \in M$ will be denoted by $\mathsf{cl}(a)$. We can make A into an L-structure $\mathcal{A}$ by defining the interpretations of the symbols of L in the following way:

- for each constant symbol c of L, $\bar{c}^{\mathcal{A}} = \mathsf{cl}(\bar{c}^{\mathcal{M}})$;

- for every integer $k \geq 1$ and every k-ary function symbol f of L, $\overline{f}^{\mathcal{A}}$ is the map from A^k into A which, to each k-tuple $(\mathsf{cl}(a_1), \mathsf{cl}(a_2), \ldots, \mathsf{cl}(a_k)) \in A^k$, assigns the element $\mathsf{cl}(\overline{f}^{\mathcal{M}}(a_1, a_2, \ldots, a_k))$; this makes sense because θ is compatible with $\overline{f}^{\mathcal{M}}$;

- for every integer $k \geq 1$ and every k-ary relation symbol R of L, $\overline{R}^{\mathcal{A}}$ is the k-ary relation on A defined by: $(\mathsf{cl}(a_1), \mathsf{cl}(a_2), \ldots, \mathsf{cl}(a_k)) \in \overline{R}^{\mathcal{A}}$ if and only if $(a_1, a_2, \ldots, a_k) \in \overline{R}^{\mathcal{M}}$ (again, this makes sense because θ is compatible with $\overline{R}^{\mathcal{M}}$).

We see immediately that the L-structure $\mathcal{A}$ defined in this way respects equality: indeed, the interpretation in $\mathcal{A}$ of the symbol $\simeq$ is the set of pairs $(\mathsf{cl}(a), \mathsf{cl}(b)) \in A^2$ such that $(a, b) \in \simeq^{\mathcal{M}}$, i.e. such that $(a, b) \in \theta$, or equivalently, such that $\mathsf{cl}(a) = \mathsf{cl}(b)$; this is precisely the diagonal of A^2.

Lemma 3.97 *For every formula $F = F[v_0, v_1, \ldots, v_{n-1}]$ of L and for all elements $a_0, a_1, \ldots, a_{n-1}, b_0, b_1, \ldots, b_{n-1}$ of M, we have:*

(1*) *if $(a_i, b_i) \in \theta$ for $0 \leq i \leq n - 1$, then*
$$\mathcal{M} \models F[a_0, a_1, \ldots, a_{n-1}] \text{ if and only if } \mathcal{M} \models F[b_0, b_1, \ldots, b_{n-1}];$$
(2*) *$\mathcal{M} \models F[a_0, a_1, \ldots, a_{n-1}]$ if and only if $\mathcal{A} \models F[\mathsf{cl}(a_0), \mathsf{cl}(a_1), \ldots, \mathsf{cl}(a_{n-1})]$.*

Proof We prove these two properties by induction on F. The case for atomic formulas is settled by the very definition of $\mathcal{A}$. Next, Remark 3.57 allows us to limit our attention to the induction steps that relate to $\neg$, $\wedge$ and $\exists$. It is only this last one that deserves any effort: so suppose, then, that

$$F = \exists v_m G[v_0, v_1, \ldots, v_{n-1}, v_m].$$

We may suppose that $m > n - 1$ in view of a remark that we have made on many earlier occasions. Under these conditions, for $\mathcal{M}$ to satisfy $F[a_0, a_1, \ldots, a_{n-1}]$, it is necessary and sufficient that there exist an element $b \in M$ such that $\mathcal{M} \models G[a_0, a_1, \ldots, a_{n-1}, b]$. Given that G satisfies the lemma (by the induction hypothesis) and that for every $b \in M$, $(b, b) \in \theta$ (the first formula of E), if $(a_i, b_i) \in \theta$ for all i, then we may conclude that $\mathcal{M} \models F[a_0, a_1, \ldots, a_{n-1}]$ if and only if there exists an element $b \in M$ such that $\mathcal{M} \models G[b_0, b_1, \ldots, b_{n-1}, b]$. In other words, $\mathcal{M} \models F[a_0, a_1, \ldots, a_{n-1}]$ if and only if $\mathcal{M} \models F[b_0, b_1, \ldots, b_{n-1}]$, which proves (1*). In the same way, we see by the induction hypothesis that $\mathcal{M} \models F[a_0, a_1, \ldots, a_{n-1}]$ if and only if there exists an element $b \in M$ such that

$$\mathcal{A} \models G[(\mathsf{cl}(a_0), \mathsf{cl}(a_1), \ldots, \mathsf{cl}(a_{n-1}), \mathsf{cl}(b)],$$

which is equivalent to

$$\langle \mathcal{A} : v_0 \to \mathsf{cl}(a_0), v_1 \to \mathsf{cl}(a_1), \ldots, v_{n-1} \to \mathsf{cl}(a_{n-1}) \rangle \models \exists v_m G,$$

or again, to

$$\mathcal{A} \models F[\mathsf{cl}(a_0), \mathsf{cl}(a_1), \dots, \mathsf{cl}(a_{n-1})];$$

this proves property (2*) for F.   ∎

Theorem 3.98 *For a theory T of a language L containing $\simeq$ to have a model that respects equality, it is necessary and sufficient that the theory $T \cup E$ have an (arbitrary) model.*

Proof If T has a model that respects equality, then E is satisfied in such a model, as we have already noted just after the definition of E; hence, $T \cup E$ has a model.

If $T \cup E$ has a model $\mathcal{M} = \langle M, \dots \rangle$, then since $\mathcal{M}$ is a model of E, we can, as above, construct the model $\mathcal{A}$ on the quotient of M by the interpretation in $\mathcal{M}$ of the symbol $\simeq$. It is a consequence of the preceding lemma that every formula of T (recall that these are closed formulas) that is satisfied in $\mathcal{M}$ will also be satisfied in $\mathcal{A}$. So the structure $\mathcal{A}$ is a model of T that respects equality.   ∎

The reader who is interested can show that the second and third formulas in the list of axioms of equality (the ones that express the symmetry and transitivity of equality) could have been omitted for they are derivable from the ones that express compatibility with the relations of the structure (among which can be found, naturally, the interpretation of $\simeq$).

EXERCISES FOR CHAPTER 3

1. The language L consists of a unary function symbol f and a binary function symbol g. Consider the following closed formulas:

$$F_1 : \quad \exists x \exists y f g x y \simeq f x$$
$$F_2 : \quad \forall x \forall y f g x y \simeq f x$$
$$F_3 : \quad \exists y \forall x f g x y \simeq f x$$
$$F_4 : \quad \forall x \exists y f g x y \simeq f x$$
$$F_5 : \quad \exists x \forall y f g x y \simeq f x$$
$$F_6 : \quad \forall y \exists x f g x y \simeq f x.$$

Consider the four structures whose underlying set is $\mathbb{N}^*$, where g is interpreted by the map $(m, n) \longmapsto m + n$, and where f is interpreted respectively by
(a) the constant map equal to 103;
(b) the map which, with each integer n, associates the remainder after division by 4;
(c) the map $n \longmapsto \inf(n^2 + 2, 19)$;
(d) the map which, with each integer n, associates 1 if $n = 1$ and the smallest prime divisor of n if $n > 1$.
For each of the six formulas, determine whether it is satisfied or not in each of the four structures.

2. The language consists of a unary predicate symbol P and a binary predicate symbol R. Consider the following six formulas:

$$G_1 : \quad \exists x \forall y \exists z ((Px \Rightarrow Rxy) \wedge Py \wedge \neg Ryz)$$
$$G_2 : \quad \exists x \exists z ((Rzx \Rightarrow Rxz) \Rightarrow \forall y Rxy)$$
$$G_3 : \quad \forall y (\exists z \forall t Rtz \wedge \forall x (Rxy \Rightarrow \neg Rxy))$$
$$G_4 : \quad \exists x \forall y ((Py \Rightarrow Ryx) \wedge (\forall u (Pu \Rightarrow Ruy) \Rightarrow Rxy))$$
$$G_5 : \quad \forall x \forall y ((Px \wedge Rxy) \Rightarrow ((Py \wedge \neg Ryx) \Rightarrow \exists z (\neg Rzx \wedge \neg Ryz)))$$
$$G_6 : \quad \forall z \forall u \exists x \forall y ((Rxy \wedge Pu) \Rightarrow (Py \Rightarrow Rzx)).$$

For each of these formulas, determine whether or not it is satisfied in each of the three L-structures defined below:
(a) The base set is $\mathbb{N}$, the interpretation of R is the usual order relation $\leq$, the interpretation of P is the set of even integers.

(b) The base set is $\wp(\mathbb{N})$ (the set of subsets of $\mathbb{N}$), the interpretation of R is the inclusion relation $\subseteq$, the interpretation of P is the collection of finite subsets of $\mathbb{N}$.

(c) The base set is $\mathbb{R}$, the interpretation of R is the set of pairs $(a, b) \in \mathbb{R}^2$ such that $b = a^2$, the interpretation of P is the subset of rational numbers.

3. The language L has two unary function symbols f and g.

   (a) Find three closed formulas F, G, and H of L such that for every L-structure $\mathcal{M} = \langle M, \overline{f}, \overline{g} \rangle$, we have

   $$\mathcal{M} \models F \text{ if and only if } \overline{f} = \overline{g} \text{ and } \overline{f} \text{ is a constant map;}$$

   $$\mathcal{M} \models G \text{ if and only if } \mathsf{Im}(\overline{f}) \subseteq \mathsf{Im}(\overline{g});$$

   $$\mathcal{M} \models H \text{ if and only if } \mathsf{Im}(\overline{f}) \cap \mathsf{Im}(\overline{g}) \text{ contains a single element.}$$

   (b) Consider the following five closed formulas of L:

   $$F_1 : \quad \forall x f x \simeq g x$$
   $$F_2 : \quad \forall x \forall y f x \simeq g y$$
   $$F_3 : \quad \forall x \exists y f x \simeq g y$$
   $$F_4 : \quad \exists x \forall y f x \simeq g y$$
   $$F_5 : \quad \exists x \exists y f x \simeq g y$$

   Find a model for each of the following six formulas:

   $$F_1 \wedge \neg F_2 \qquad F_2 \qquad \qquad \neg F_1 \wedge F3$$
   $$\neg F_1 \wedge F_4 \qquad \neg F_3 \wedge \neg F_4 \wedge F_5 \qquad \neg F_5.$$

4. Let L be a first order language. For every formula $F[v_0, v_1, \ldots, v_k]$ of L, the expression $\exists! v_0 F$ denotes the following formula:

   $$\exists v_0 (F[v_0, v_1, \ldots, v_k] \wedge \forall v_{k+1}(F[v_{k+1}, v_1, \ldots, v_k] \Rightarrow v_{k+1} \simeq v_0)).$$

   $\exists! v_0 F$ is read: 'there exists one and only one v_0 such that F'.

   Note that in any L-structure $\mathcal{M} = \langle M, \ldots \rangle$, $\exists! v_0 F$ is satisfied by a k-tuple $(a_1, a_2, \ldots, a_k)$ if and only if there exists a unique object $a \in M$ such that the $(k+1)$-tuple $(a, a_1, a_2, \ldots, a_k)$ satisfies F.

   Let $F[v_0, v_1]$ be a formula of L. Find a closed formula G of L which is satisfied in an L-structure $\mathcal{M} = \langle M, \ldots \rangle$ if and only if there exists a unique pair $(a, b) \in M^2$ such that $\langle \mathcal{M} : v_0 \to a, v_1 \to b \rangle \models F$. Are the formulas

   $$G, \quad \exists! v_0 \exists! v_1 F, \quad \exists! v_1 \exists! v_0 F$$

   equivalent?

5. Let L be a first order language and let $A[x, y]$ be an arbitrary formula of L with two free variables.

(a) Is the formula

$$\forall x \exists y A[x, y] \Rightarrow \exists y \forall x A[x, y]$$

satisfied in any L-structure?

(b) Repeat question (a) for the formula

$$\exists y \forall x A[x, y] \Rightarrow \forall x \exists y A[x, y].$$

(c) Show that for arbitrary formulas with two free variables $A[x, y]$ and $B[x, y]$, the following formula is universally valid:

$$(\forall x \forall y A[x, y] \Rightarrow \exists x \exists y B[x, y]) \Leftrightarrow \exists x \exists y (A[x, y] \Rightarrow B[x, y]).$$

(d) Let F be the formula

$$\forall x \forall y (A[x, y] \Rightarrow A[y, x]) \Rightarrow$$
$$((\forall u \forall v (A[u, v] \Rightarrow B[u, v]) \Rightarrow \exists x \exists y (A[x, y] \Rightarrow C[x, y])))$$

where A, B and C are arbitrary formulas with two free variables.

Show that there exists a quantifier-free formula $G = G[x, y]$ with two free variables such that F is universally equivalent to $\exists x \exists y G$.

6. Show that if two formulas are universally equivalent, then so are their universal closures.

7. In all of the languages considered in this exercise, R is a binary relation symbol, $*$ and $\oplus$ are binary function symbols, c and d are constant symbols.

   We will write $x \oplus y$ and $x * y$ respectively, rather than $\oplus xy$ and $*xy$ (with a reminder that this necessitates the use of parentheses when writing terms). x^2 will be an abbreviation for $x * x$.

(a) In each of the following six cases ($1 \leq i \leq 6$), a language L_i and two L_i-structures $\mathcal{A}_i$ and $\mathcal{B}_i$ are given and you are asked to find a closed formula of L_i that is true in $\mathcal{A}_i$ and false in $\mathcal{B}_i$.

(1) $L_1 = \{R\}$	$\mathcal{A}_1 = \langle \mathbb{N}, \leq \rangle$	$\mathcal{B}_1 = \langle \mathbb{Z}, \leq \rangle$
(2) $L_2 = \{R\}$	$\mathcal{A}_2 = \langle \mathbb{Q}, \leq \rangle$	$\mathcal{B}_2 = \langle \mathbb{Z}, \leq \rangle$
(3) $L_3 = \{*\}$	$\mathcal{A}_3 = \langle \mathbb{N}, \times \rangle$	$\mathcal{B}_3 = \langle \wp(\mathbb{N}), \cap \rangle$
(4) $L_4 = \{c, *\}$	$\mathcal{A}_4 = \langle \mathbb{N}, 1, \times \rangle$	$\mathcal{B}_4 = \langle \mathbb{Z}, 1, \times \rangle$
(5) $L_5 = \{c, d, \oplus, *\}$	$\mathcal{A}_5 = \langle \mathbb{R}, 0, 1, +, \times \rangle$	$\mathcal{B}_5 = \langle \mathbb{Q}, 0, 1, +, \times \rangle$
(6) $L_6 = \{R\}$	$\mathcal{A}_6 = \langle \mathbb{Z}, \equiv_2 \rangle$	$\mathcal{B}_6 = \langle \mathbb{Z}, \equiv_3 \rangle$

   ($\times$ and $+$ are the usual operations of multiplication and addition, $\cap$ is the operation of intersection, $\equiv_p$ is the relation of congruence modulo p.)

(b) For each of the following closed formulas of the language $\{c, \oplus, *, R\}$, find a model of the formula as well as a model of its negation.

$$F_1 : \quad \forall u \forall v \exists x (\neg v \simeq c \Rightarrow u \oplus (v * x) \simeq c)$$
$$F_2 : \quad \forall u \forall v \forall w \exists x (\neg w \simeq c \Rightarrow u \oplus (v * x) \oplus (w * x^2) \simeq c)$$

F_3 : $\forall x \forall y \forall z (Rxx \wedge ((Rxy \wedge Ryz) \Rightarrow Rxz) \wedge (Rxy \Rightarrow Ryx))$

F_4 : $\forall x \forall y \forall z (Rxx \Rightarrow R x * z \; y * z)$

F_5 : $\forall x \forall y (Rxy \Rightarrow \neg Ryx)$.

8. The language L consists of a single binary predicate symbol, R.

Consider the L-structure $\mathcal{M}$ whose base set is $M = \{n \in \mathbb{N} : n \geq 2\}$ and in which R is interpreted by the relation 'divides', i.e. $\overline{R}$ is defined for all integers m and $n \geq 2$ by the condition: $(m, n) \in \overline{R}$ if and only if m divides n.

(a) For each of the following formulas of L (with one free variable x), describe the set of elements of M that satisfy it.

F_1 : $\forall y (Ryx \Rightarrow x \simeq y)$
F_2 : $\forall y \forall z ((Ryx \wedge Rzx) \Rightarrow (Ryz \vee Rzy))$
F_3 : $\forall y \forall z (Ryx \Rightarrow (Rzy \Rightarrow Rxz))$
F_4 : $\forall t \exists y \exists z (Rtx \Rightarrow (Ryt \wedge Rzy \wedge \neg Rtz))$.

(b) Write a formula $G[x, y, z, t]$ of L such that for all a, b, c and d of M, the structure $\mathcal{M}$ satisfies $G[a, b, c, d]$ if and only if d is the greatest common divisor of a, b and c.

(c) Let H be the following closed formula of L:

$$\forall x \forall y \forall z ((\exists t (Rtx \wedge Rty) \wedge \exists t (Rty \wedge Rtz)) \Rightarrow \exists t \forall u (Rut \Rightarrow (Rux \wedge Ruz))).$$

(1) Find a prenex form of H.
(2) Is the formula H satisfied in $\mathcal{M}$?
(3) Give an example of a structure $\mathcal{M}' = \langle M', \overline{R} \rangle$ such that when $\mathcal{M}$ is replaced by $\mathcal{M}'$ in the previous question, the answer is different.

9. Let L be the first order language which has one unary relation symbol Ω and two binary relation symbols I and R.

Consider the following formulas of L:

F_1 : $\forall x \neg Rxx$
F_2 : $\forall x (\Omega x \Rightarrow \neg Rxx)$
F_3 : $\forall x \forall y \forall z ((\Omega x \wedge \Omega y \wedge Ixz \wedge Izy) \Rightarrow \Omega z)$
F_4 : $\forall x \forall y \forall z ((\Omega x \wedge \Omega y \wedge \Omega z \wedge Rxy \wedge Ryz) \Rightarrow Rxz)$
F_5 : $\forall x \forall y ((\Omega x \wedge \Omega y) \Rightarrow (\neg Rxy \vee \neg Ryx))$
F_6 : $\forall x \forall y ((\Omega x \wedge Rxy) \Rightarrow \Omega y)$
F_7 : $\forall x \forall y (\Omega x \wedge Ryx) \Rightarrow \Omega y)$
F_8 : $\forall x \exists y \exists z (Ryx \wedge Rxz)$
F_9 : $\forall x \exists y \exists z (\Omega x \Rightarrow (Ryx \wedge Rxz \wedge \Omega y \wedge \Omega z))$
F_{10} : $\forall x \forall y \exists z ((\Omega x \wedge \Omega y \wedge Rxy) \Rightarrow (Rxz \wedge Rzy \wedge \Omega z))$.

(a) Consider the L-structure $\mathcal{M}$ whose base set is $\wp(\mathbb{N})$, in which the interpretation of Ω is the unary relation '... is infinite and its complement is infinite', the interpretation of I is is the inclusion relation, and the pairs

(A, B) that satisfy the the interpretation of the binary relation R are those for which $A \subseteq B$ and $\mathsf{card}(A) = \mathsf{card}(B - A)$ (the notation $\mathsf{card}(X)$ denotes the cardinality of a set X; see Chapter 7).

For each of the formulas above, determine whether it is satisfied or not in the structure $\mathcal{M}$.

(b) We add a new unary predicate symbol D to the language. Is it possible to enrich the structure $\mathcal{M}$ with an interpretation of D such that the following four formulas are satisfied?

G_1 : $\forall x \forall y((Dx \wedge Dy) \Rightarrow (Ixy \vee Iyx))$

G_2 : $\forall x \forall y \exists z((Dx \wedge Dy \wedge Ixy \wedge \neg x \simeq y) \Rightarrow$
$(Dz \wedge Ixz \wedge Izy \wedge \neg x \simeq z \wedge \neg y \simeq z))$

G_3 : $\forall x \exists y \exists z(Dx \Rightarrow (Dy \wedge Dz \wedge Ixy \wedge Izx \wedge \neg x \simeq y \wedge \neg x \simeq z))$

G_4 : $\exists x \, Dx$

10. Let L be a language and let F be a formula of L.

The **spectrum** of F is the set of cardinalities of finite models of F; i.e. it is the set of natural numbers n such that F has a model whose base set has exactly n elements; it is denoted by $Sp(F)$.

(a) For each of the following subsets of $\mathbb{N}$, exhibit, when this is possible, an example of a language L and a closed non-contradictory formula F of L whose spectrum is the subset in question.

(1) $\emptyset$;   (2) $\mathbb{N}$;   (3) $\mathbb{N}^*$;   (4) $\{n \in \mathbb{N}^* : (\exists p \in \mathbb{N})(n = 2p)\}$;
(5) $\{n \in \mathbb{N}^* : (\exists p \in \mathbb{N})(n = p^2)\}$;   (6) $\{3\}$;   (7) $\{1, 2, 3, 4\}$;
(8) $\mathbb{N} - \{0, 1, \ldots, k\}$;
(9) the set of non-zero composite natural numbers;
(10) the set of prime numbers.

(b) Show that any formula whose spectrum is infinite has at least one infinite model.

11. Show that if all the models of a non-contradictory theory are isomorphic, then the theory is complete.

12. Let L be a first order language, let $\mathcal{M}$ be an L-structure and let A be a subset of the base set of $\mathcal{M}$.

(a) Show that if A is not empty, there exists a unique sub-structure $\mathcal{A}$ of $\mathcal{M}$ such that:

(1) the base set of $\mathcal{A}$ includes A;

(2) every sub-structure of $\mathcal{M}$ whose base set includes A is an extension of $\mathcal{A}$.

$\mathcal{A}$ is called **the substructure of $\mathcal{M}$ generated by A**.

(b) Show that, when $A = \emptyset$, there may not be a substructure generated by A. Give an example, however, in which one does exist.

(c) Suppose that L contains no function symbols of arity ≥ 1. What is the substructure generated by a subset A in this case?

(d) A sub-structure $\mathcal{N}$ of $\mathcal{M}$ is said to be **of finite type** if and only if $\mathcal{N}$ is generated by a non-empty finite subset of $\mathcal{M}$.

   Let F be a closed universal formula of L. Show that F is satisfied in $\mathcal{M}$ if and only if F is satisfied in every sub-structure of $\mathcal{M}$ of finite type.

(e) Give a counterexample to (d) for a formula that is not universal.

13. The language L consists of one constant symbol c and two unary function symbols f and g. The theory T consists of the following formulas:

$$H_1 : \quad \forall v_0 f f v_0 \simeq f v_0$$
$$H_2 : \quad \forall v_0 g g v_0 \simeq g v_0$$
$$H_3 : \quad \forall v_0 (f g v_0 \simeq c \wedge g f v_0 \simeq c)$$
$$H_4 : \quad \forall v_0 \forall v_1 ((f v_0 \simeq f v_1 \wedge g v_0 \simeq g v_1) \Rightarrow v_0 \simeq v_1)$$
$$H_5 : \quad \forall v_1 \forall v_2 ((f v_1 \simeq v_1 \wedge g v_2 \simeq_2) \Rightarrow \exists v_0 (f v_0 \simeq v_1 \wedge g v_0 \simeq v_2)).$$

(a) Show that for any term t of L, at least one of the following cases applies:

   ○ $T \vdash^* t \simeq c$;
   ○ there exists a variable x such that $T \vdash^* t \simeq x$;
   ○ there exists a variable x such that $T \vdash^* t \simeq fx$;
   ○ there exists a variable x such that $T \vdash^* t \simeq gx$.

(b) Let A and B be two non-empty sets with $a_0 \in A$ and $b_0 \in B$. We denote by $\mathcal{M}(A, B, a_0, b_0)$ the L-structure whose underlying set is $A \times B$, and in which c is interpreted by (a_0, b_0), f by the map $(a, b) \longmapsto (a, b_0)$ and g by the map $(a, b) \longmapsto (a_0, b)$. Show that $\mathcal{M}(A, B, a_0, b_0)$ is a model of T.

(c) Show that the following formulas are consequences of T:

$$H_6 : \quad fc \simeq c$$
$$H_7 : \quad gc \simeq c$$
$$H_8 : \quad \forall v_0 (f v_0 \simeq v_0 \Leftrightarrow g v_0 \simeq c)$$
$$H_9 : \quad \forall v_0 (g v_0 \simeq v_0 \Leftrightarrow f v_0 \simeq c)$$
$$H_{10} : \quad \forall v_0 (f v_0 \simeq v_0 \Leftrightarrow \exists v_1 f v_1 \simeq v_0)$$
$$H_{11} : \quad \forall v_0 (g v_0 \simeq v_0 \Leftrightarrow \exists v_1 g v_1 \simeq v_0)$$
$$H_{12} : \quad \forall v_0 ((f v_0 \simeq v_0 \wedge g v_0 \simeq v_0) \Leftrightarrow v_0 \simeq c)$$
$$H_{13} : \quad \forall v_0 \forall v_1 (f v_0 \simeq g v_1 \Rightarrow f v_0 \simeq c).$$

(d) Given four non-empty sets A, B, C and D and four elements $a_0 \in A, b_0 \in B$, $c_0 \in C$, and $d_0 \in D$, show that if A and C are equipotent and if B and D are equipotent (two sets are called equipotent if there exists a bijection between them), then the structures $\mathcal{M}(A, B, a_0, b_0)$ and $\mathcal{M}(C, D, c_0, d_0)$ are isomorphic.

(e) Let $\mathcal{M} = \langle M, \alpha, \varphi, \psi \rangle$ be a model of T. Set

$$A = \{x \in M : \varphi(x) = x\}, \quad B = \{x \in M : \psi(x) = x\}, \quad a_0 = b_0 = \alpha.$$

Show that $\mathcal{M}$ is isomorphic to the structure $\mathcal{M}(A, B, a_0, b_0)$.

For every integer $n \geq 1$, write a closed formula F_n (respectively G_n) of L that is true in $\mathcal{M}$ if and only if the set A (respectively, B) contains at least n elements.

Show, for all integers n and $p \geq 1$, that the theory

$$T_{np} = T \cup \{F_n, G_p, \neg F_{n+1}, \neg G_{p+1}\}$$

is complete.

(f) Let F be a closed formula of L that is satisfied in every infinite model of T. Show that there exists at least one integer n such that $T \cup \{F_n \vee G_n\} \vdash^* F$.

Is the theory $T \cup \{F_k \vee G_k : k \in \mathbb{N}^*\}$ complete?

The last two questions make use of concepts that will be introduced only in Chapters 7 and 8.

(g) Describe all the countable models of T.

(h) Is the theory $T'' = T \cup \{F_k : k \in \mathbb{N}^*\} \cup \{G_k : k \in \mathbb{N}^*\}$ complete? (To answer this question, we will need Vaught's theorem (see Volume 2).)

14. The language L consists of one unary function symbol f.

We let A denote the following formula:

$$\forall x (fffx \simeq x \wedge \neg fx \simeq x).$$

(a) Show that the following formula is a consequence of A:

$$\forall x \exists y \forall z (\neg ffx \simeq x \wedge \neg ffx \simeq fx \wedge fy \simeq x \wedge (fz \simeq x \Rightarrow z \simeq y)).$$

For every integer $n \in \mathbb{N}^*$, we let F_n denote the following formula;

$$\exists x_1 \exists x_2 \ldots \exists x_n \forall x \left(\left(\bigwedge_{1 \leq i < j \leq n} \neg x_i \simeq x_j \right) \wedge \left(\bigvee_{1 \leq i \leq n} x \simeq x_i \right) \right).$$

(b) Show that, for every integer $n \in \mathbb{N}^*$, the formula $A \wedge F_n$ has a model if and only if n is a multiple of 3.

(c) Show that, for every $p \in \mathbb{N}^*$, the theory $\{A, F_{3p}\}$ is complete.

(d) Exhibit a countable model of the formula A (i.e. a model of A whose base set is in bijection with the set of natural numbers).

(e) Show that all the countable models of A are isomorphic.

15. The language L consists of two unary function symbols r and l. For every term t of L, set $r^0 t = l^0 t = t$ and, for every integer k, $r^{k+1} t = r r^k t$ and $l^{k+1} t = l l^k t$.

Let F be the formula

$$\forall x \forall y \exists u \exists v (((rx \simeq ry \vee lx \simeq ly) \Rightarrow x \simeq y) \wedge$$
$$(x \simeq ru \wedge x \simeq lv) \wedge (\neg rx \simeq lx \wedge rlx \simeq lrx)),$$

and for every pair (m, n) of natural numbers, let F_{mn} be the formula

$$\forall x (\neg r^m l^n x \simeq x \wedge \neg r^m x \simeq l^n x).$$

Let T be the theory consisting of F together with the set of all the F_{mn} such that $(m, n) \neq (0, 0)$.

(a) Show that for every term t of L, there exists a variable x and integers m and n such that

$$T \vdash^* \forall x (t \simeq r^m l^n x).$$

(b) Show that the structure $\mathcal{M}_0$ whose underlying set is $\mathbb{Z} \times \mathbb{Z}$ and where r and l are interpreted respectively by the maps

$$s_r : (i, j) \mapsto (i, j + 1) \quad \text{(right successor)}$$
$$s_l : (i, j) \mapsto (i + 1, j) \quad \text{(left successor)}$$

is a model of T; it will be called the **standard model** of T.

(c) Show that for any two integers A and B, the map h_{ab} from $\mathbb{Z} \times \mathbb{Z}$ into $\mathbb{Z} \times \mathbb{Z}$ defined by $h_{ab}(i, j) = (i + a, j + b)$ is an automorphism of $\mathcal{M}_0$.

(d) Which subsets of $\mathbb{Z} \times \mathbb{Z}$ are definable in the structure $\mathcal{M}_0$ by a formula of L?

16. We will allow the following abuse of notation: for every integer $n \geq 1$, we will let $0, 1, \ldots, n - 1$ denote the elements of $\mathbb{Z}/n\mathbb{Z}$ (i.e. the respective equivalence classes modulo n of $0, 1, \ldots, n - 1$) and will let $+$ denote addition in this set.

(a) The language L has a single unary function symbol f.
    Consider the following L-structures:

$$\mathcal{M}_1 = \langle \mathbb{Z}/n\mathbb{Z}, x \mapsto x + 1 \rangle ;$$
$$\mathcal{M}_2 = \langle \mathbb{Z}/n\mathbb{Z}, x \mapsto x + 2 \rangle .$$

For each structure, determine which subsets of the underlying set are definable.

(b) The language L' has a single binary function symbol g.
    Consider the following L'-structures:

$$\mathcal{N}_1 = \langle \mathbb{Z}/3\mathbb{Z}, (x, y) \mapsto x + y \rangle ;$$
$$\mathcal{N}_1 = \langle \mathbb{Z}/6\mathbb{Z}, (x, y) \mapsto x + y \rangle ;$$
$$\mathcal{N}_1 = \langle \mathbb{R}, (x, y) \mapsto xy \rangle .$$

For each structure, determine which subsets of the underlying set are definable.

(c) The language L'' has a single binary relation symbol R.

Consider the L''-structure $\langle \mathbb{R}, \leq \rangle$.

(1) Which subsets of $\mathbb{R}$ are definable in this structure?

(2) Which subsets of $\mathbb{R}^2$ are definable in this structure?

17. Given an integer $n \geq 2$ and a binary relation S on a set E, a **cycle of order** n (or n-**cycle**) **for** S is an n-tuple $(a_1, a_2, \ldots, a_n)$ of elements of E satisfying: $(a_1, a_2) \in S$, $(a_2, a_3) \in S$, $\ldots$, $(a_{n-1}, a_n) \in S$ and $(a_n, a_1) \in S$. For example, the usual strict ordering on $\mathbb{R}$ has no n-cycles, whereas the binary relation on the set $\{1, 2, 3\}$ whose graph is $\{(1, 2), (2, 3), (3, 1)\}$ has 3-cycles but no 2-cycles.

In the first order language L that has a single binary relation symbol R, consider, for every $n \geq 2$, the following formula F_n:

$$\forall x_1 \forall x_2 \ldots \forall x_n \neg (R x_1 x_2 \wedge R x_2 x_3 \wedge \cdots \wedge R x_{n-1} x_n \wedge R x_n x_1).$$

Set $T = \{F_n : n \in \mathbb{N}, n \geq 2\}$.

We say that an L-structure $\langle M, \overline{R} \rangle$ is **cycle-free** if for every $n \geq 2$, $\overline{R}$ has no n-cycles; in the opposite case, we say that the structure **has cycles**. It is clear that the models of T are the cycle-free L-structures.

(a) For every $n \geq 2$, exhibit a model of of the formula

$$F_2 \wedge F_3 \wedge \cdots \wedge F_n \wedge \neg F_{n+1}.$$

(b) Show that if G is a closed formula of L that is a consequence of T, there exists at least one integer $p \geq 2$ such that G is satisfied in every L-structure in which the interpretation of R has no cycles of order less than or equal to p.

(c) Show that every closed formula that is a consequence of T has at least one model that has cycles.

(d) Show that T is not equivalent to any finite theory. (Hence the concept of a cycle-free binary relation, which T axiomatizes, is not finitely axiomatizable).

18. Recall that an order relation on a set E is a **well-ordering** if and only if every non-empty subset of E has a smallest element with respect to this order. This exercise shows that this concept is not pseudo-axiomatizable.

Let L_0 be the language whose only symbol is a binary relation symbol R and let L be a language that enriches L_0. Show that there does not exist a theory T of L with the following property: for every L_0-structure $\mathcal{M} = \langle M, \rho \rangle$, ρ is a well-ordering of M if and only if $\mathcal{M}$ can be enriched to an L-structure that is a model of T.

To do this, use the language L' obtained from L by adding a countably infinite set of new constant symbols, $c_0, c_1, \ldots, c_n, \ldots$ (pairwise distinct) and, for every integer n, consider the following closed formula F_n of L':

$$R c_{n+1} c_n \wedge \neg c_{n+1} \simeq c_n.$$

19. Let L be a first order language and let L' be the language obtained from L by adding new constant symbols $c_1, c_2, \ldots, c_k$.

Consider a theory T and a formula $F[x_1, x_2, \ldots, x_k]$ of L.

Show that if the closed formula $F[c_1, c_2, \ldots, c_k]$ of L' is a consequence of T (considered as a theory of L'), then $T \vdash^* \forall x_1 \forall x_2 \ldots \forall x_k F[x_1, x_2, \ldots, x_k]$ (this conclusion concerns only the language L).

20. We are given a first order language L, an L-structure $\mathcal{M} = \langle M, \ldots \rangle$ and a theory T of L.

Recall that $\Delta(\mathcal{M})$ denotes the simple diagram of $\mathcal{M}$ (Definition 3.89).

(a) Assume that no extension of $\mathcal{M}$ is a model of T. Show that there exists a quantifier-free formula $G[x_1, x_2, \ldots, x_n]$ of L such that

$$T \vdash^* \forall x_1 \forall x_2 \ldots \forall x_n G[x_1, x_2, \ldots, x_n] \text{ and}$$
$$\mathcal{M} \nvDash \forall x_1 \forall x_2 \ldots \forall x_n G[x_1, x_2, \ldots, x_n].$$

(Consider the theory $T \cup \Delta(\mathcal{M})$; use Theorem 3.91 and Exercise 19).

(b) Let $U(T)$ denote the set of closed universal formulas of L that are consequences of T.

Show that for the existence of an extension of $\mathcal{M}$ that is a model of T, it is necessary and sufficient that $\mathcal{M}$ be a model of $U(T)$.

(This result is a special case of what is called the **extension theorem**).

(c) A sub-structure of $\mathcal{M}$ that is generated by a non-empty finite subset of M is called a sub-structure **of finite type**. (The sub-structure of $\mathcal{M}$ generated by a non-empty subset $A \subseteq M$ is the smallest sub-structure of $\mathcal{M}$ whose base set includes A: see Exercise 12.)

Show that for the existence of an extension of $\mathcal{M}$ that is a model of T, it is necessary and sufficient that every sub-structure of $\mathcal{M}$ of finite type have this same property.

21. Let L be a first order language and let $\mathcal{F}_1$ denote the set of formulas of L that have at most one free variable. Given an L-structure $\mathcal{M} = \langle M, \ldots \rangle$ and an element $a \in M$, the set $\theta(a)$ of formulas in $\mathcal{F}_1$ that are satisfied in $\mathcal{M}$ by a is called the **type of a in $\mathcal{M}$** (or simply, the **type of a** if there is no ambiguity; this vocabulary will be taken up again in Chapter 8). In other words:

$$\theta(a) = \{F[v] \in \mathcal{F}_1 : v \text{ is a variable and } \mathcal{M} \vDash F[a]\}.$$

(a) Show that if all the elements of a subset $A \subseteq M$ have the same type in an L-structure $\mathcal{M} = \langle M, \ldots \rangle$, then every subset of M that is definable in $\mathcal{M}$ by a formula of L must either include A or be disjoint from A (see Definition 3.92).

(b) Let h be an automorphism of an L-structure $\mathcal{M} = \langle M, \ldots \rangle$, and let a be an element of M. Show that a and $h(a)$ have the same type.

(c) R, f, g and c are, respectively, a binary relation symbol, a unary function symbol, a binary function symbol and a constant symbol. In each of the following examples, find two elements a and b of the proposed model whose types are distinct, or show that this is not possible.

$$L_1 = \{R\} \quad \mathcal{M}_1 = \langle \mathbb{R}, \leq \rangle \, ;$$
$$L_2 = \{f\} \quad \mathcal{M}_2 = \langle \mathbb{N}, n \mapsto n + 1 \rangle \, ;$$
$$L_3 = \{f\} \quad \mathcal{M}_3 = \langle \mathbb{Z}, n \mapsto n + 1 \rangle \, ;$$
$$L_4 = \{g, c\} \quad \mathcal{M}_4 = \langle \mathbb{Z}, +, \mathbf{0} \rangle \, ;$$
$$L_5 = \{g\} \quad \mathcal{M}_5 = \langle \mathbb{Z}, + \rangle \, .$$

(d) Let T be a theory of L and let $F_1, F_2, \ldots, F_n$ be n formulas of $\mathcal{F}_1$ ($n \geq 1$). Suppose that the formula

$$G = \forall v_0 \forall v_1 \Big(\bigwedge_{1 \leq i \leq n} (F_i[v_0] \Leftrightarrow F_i[v_1]) \Rightarrow v_0 \simeq v_1 \Big)$$

is a consequence of T.

Show that every model of T has at most 2^n elements.

(e) Let S be a theory of L that has at least one infinite model.

Show that there exists a model $\mathcal{M} = \langle M, \ldots \rangle$ of S such that M contains at least two elements that have the same type.

(*Hint for an argument by contradiction*: enrich the language with two new distinct constant symbols and apply the compactness theorem to an appropriate theory in the enriched language, then make use of Exercise 19 and the preceding question.)

(f) Give an example of a language L and a theory T such that:
   • there exists at least one model of T of cardinality ≥ 2;
   • there is no model of T that contains two distinct elements of the same type.

(g) Give an example of an infinite model of a finite language L that contains no distinct elements of the same type.

4 The completeness theorems

The formalization that we have implemented thus far allows us to represent mathematical statements, or at least some of them, in the form of sequences of symbols. We will now pursue further in this direction and formalize proofs. There are many ways of doing this and, let us admit it from the start, the one we have chosen comes with a certain number of inconveniences; in particular, it does not properly reflect the manner in which proofs are conceived in the minds of mathematicians. Moreover, it does not lend itself well to the analysis of proofs, a topic that is known as proof theory and that we will discuss only briefly in this text. In its favour, our method is a bit closer to the way in which proofs are actually written and, primarily, it requires the introduction of few prerequisites. That is why it appeared to us that this approach is the easiest to understand upon first exposure.

A formal proof (or derivation) is a sequence of formulas in which each is justified either because it is an axiom or because it can be deduced from formulas that precede it in the sequence. It is quite clear that, provided we do things correctly, a formal proof can lead only to formulas that are universally valid. The converse of this assertion, namely that every universally valid formula has a formal proof, is what we will call a completeness theorem; indeed, such a result will show that the axioms and rules that we have allowed ourselves are sufficiently strong, or, in other words, that they are complete. In Section 4.2 we will present a proof of this that uses a method due to Henkin, and from this we will extract an important, purely semantic consequence, the compactness theorem. The purpose of Section 4.3 is to describe Herbrand's method which reduces the satisfiability of a formula of predicate calculus to the satisfiability of an infinite set of propositional formulas.

An essential aspect of these notions is their effective character. For example, here is a natural question: can we find an algorithm that produces proofs of theorems? We will see later, in Part 2, Chapter 6, how to think about this question in general terms. In Section 4.4, our interest turns to a restricted class of universal formulas (the universal clauses) and we will introduce a new type of proof; this is the method of resolution. This method is better suited to implementation on computers (it is the basis for the language PROLOG). We will be content to sketch the required algorithms without giving the details of a potential realization.

In this chapter, we will not speak of equality; thus, we will not assume that the equality symbol is part of our language and, when it is, we will not assume that the models we construct necessarily respect equality. However, thanks to Theorem 3.97, we can reduce to a model that does respect equality.

To avoid misunderstandings, the word 'derivation' will be reserved, throughout this chapter, to mean formal proof. The word 'proof' by itself will be used to denote what is necessary to prove the stated theorems; so we could, to conform with what was said in the introduction, call these 'meta-derivations'.

4.1 Formal proofs (or derivations)

4.1.1 Axioms and rules

In mathematics, to prove a theorem is to derive it from propositions given in advance, called **axioms**, according to exactly specified **rules**. It is this notion of proof that we will formalize in this section. What the axioms are and what the rules are must, therefore, be made absolutely precise. Let us begin with the rules. These are rules that allow us to deduce a formula from one or more other formulas. For the notion of derivation that will be presented here, there are two deduction rules:

Definition 4.1 THE DEDUCTION RULES

- *Modus ponens: from the two formulas F and $F \Rightarrow G$, modus ponens allows us to derive G.*

- *The **rule of generalization**: if F is a formula and v is a variable, the rule of generalization allows us to derive $\forall v F$ from F.*

Despite its Latin name, with which you may not be familiar, the idea behind modus ponens is totally banal.

The rule of generalization is a bit more troubling but its justification is simple: if we know, without any particular assumptions about v, how to derive $F(v)$, then we know that $\forall v F(v)$ is also true. It is used all the time in mathematics. Imagine, for example, that you wish to prove that every positive integer is the sum of four perfect squares. You would say: let n be a positive integer, and you would develop an argument that concluded with: hence, there exist a, b, c, and d such that $n = a^2 + b^2 + c^2 + d^2$, and you would, justifiably, consider the proof to be finished. This is nothing but an application of the rule of generalization in which n plays the role of the free variable. Exercise 5 will help us to appreciate the usefulness of this rule.

Definition 4.2 THE LOGICAL AXIOMS. *These are the following formulas:*

- *The **tautologies**. Recall that the tautologies of predicate calculus are obtained in the following way: we start with a tautology F of the propositional calculus whose propositional variables are, say, $A_1, A_2, \ldots, A_n$; we also have n formulas $G_1, G_2, \ldots, G_n$ of the language under consideration. The formula H obtained* by replacing in F all occurrences of A_1 by G_1, all occurrences of A_2 by G_2, and so on, is, by definition, a tautology of the predicate calculus (see Definition 3.49).

- The **quantification axioms.** These are grouped into three infinite sets (these infinite sets are usually called **axiom schemes**):

  * formulas of the form:

$$\exists v F \Leftrightarrow \neg \forall v \neg F \tag{a}$$

  where F is an arbitrary formula and v is an arbitrary variable;

  * formulas of the form:

$$\forall v (F \Rightarrow G) \Rightarrow (F \Rightarrow \forall v G) \tag{b}$$

  where F and G are arbitrary formulas and v is a variable that has no free occurrence in F;

  * formulas of the form:

$$\forall v F \Rightarrow F_{t/v} \tag{c}$$

  where F is a formula, t is a term and there is no free occurrence of v in F that lies in the scope of a quantification that binds a variable of t.

For each of these three axiom schemes, we will show that we are dealing only with universally valid formulas and we will also justify the restrictions that we have placed on the variables.

(a) These axioms present no difficulty. Their purpose is to give a syntactical definition of the existential quantifier in terms of the universal quantifier (see part (1) of Theorem 3.55.

(b) This one is not very difficult either (see part (16) of Theorem 3.55). Clearly, the restriction on the variable v is necessary: for example, consider a language that has only one unary predicate symbol P and set $F = G = Pv$. Then $\forall v(Pv \Rightarrow Pv)$ is always true, while this is not the case for $Pv \Rightarrow \forall v Pv$ which means that if something satisfies P then everything satisfies P.

(c) This scheme is the hardest one to understand because the condition that is attached to it is not simple and also, perhaps, because a superficial analysis might lead one to believe that $\forall v F \Rightarrow F_{t/v}$ is always satisfied without the need for any restrictions. Let us show that this is an illusion.

In the language consisting of the single binary relation symbol R, consider the formula $F = \exists v_1 \neg R v v_1$ and the term $t = v_1$. So that $F_{t/v} = \exists v_1 \neg R v_1 v_1$ and

$$\forall v F \Rightarrow F_{t/v} \text{ is the formula } \forall v \exists v_1 \neg R v v_1 \Rightarrow \exists v_1 \neg R v_1 v_1;$$

this formula is false, for example, in a structure whose base set contains more than one element and in which R is interpreted by equality.

We can see what is happening: contrary to what we may have naively expected, the formula $F_{t/v}$ does not, in any way, say that the object represented by t possesses the property expressed by the formula F; the reason is that the term t, which, here, is a variable, finds itself quantified in $F_{t/v}$.

Now we will show that all the formulas from scheme (c) are universally valid. Let $u_1, u_2, \ldots, u_n$ be the variables in t and let $w_1, w_2, \ldots, w_p$ be the free variables of $\forall v F$ other than $u_1, u_2, \ldots, u_n$. We insist: the variables w_i are distinct from the u_j, but the u_j may well occur, either free or bound, in $\forall v F$, and it is possible that the variable v is among the u_j. This hypothesis places us precisely in a situation where Proposition 3.45 is applicable.

Let $\mathcal{M}$ be a structure for the language of F; we will show that the formula $\forall v F \Rightarrow F_{t/v}$ is true in $\mathcal{M}$. Consider elements $a_1, a_2, \ldots, a_n, b_1, b_2, \ldots, b_p$ of the underlying set M of $\mathcal{M}$ such that

$$\left(\mathcal{M} : u_1 \to a_1, \ldots, u_n \to a_n, w_1 \to b_1, \ldots, w_p \to b_p \right) \models \forall v F.$$

This means, by definition, that for every element $a \in M$, we have

$$\left(\mathcal{M} : v \to a, u_1 \to a_1, \ldots, u_n \to a_n, w_1 \to b_1, \ldots, w_p \to b_p \right) \models F.$$

When a is taken to be the element $\bar{t}^{\mathcal{M}}[a_1, a_2, \ldots, a_n]$, we may conclude, thanks to Proposition 3.45, that

$$\left(\mathcal{M} : u_1 \to a_1, \ldots, u_n \to a_n, w_1 \to b_1, \ldots, w_p \to b_p \right) \models F_{t/v},$$

which finishes the proof.

In fact, schema (c) will be used most frequently in situations where no variable of t is bound in F (which is obviously a stronger condition than the one imposed by schema (c)), and, in particular, when t is a closed term. We will also use it in the fourth example of the next section; that example will itself be used several times during the proof of the completeness theorem.

4.1.2 Formal proofs

We can now give the definition of a derivation (or formal proof). Recall that the formulas that constitute a theory are all closed formulas.

Definition 4.3 *Let T be a theory and F be a formula of L; an L-**derivation of** F **in** T is a finite sequence of formulas $\mathcal{D} = (F_0, F_1, \ldots, F_n)$ of L that ends with F and is such that each F_i (for $0 \le i \le n$) satisfies at least one of the following conditions:*

- *$F_i \in T$;*
- *F_i is a logical axiom;*
- *F_i is derivable from one or two of the formulas that precede it in the sequence $\mathcal{D}$ by one of the two deduction rules.*

*If there exists a L-derivation of F in T, we say that F is L-**derivable in** T or that F is an L-**syntactic consequence** of T or that F is an L-**theorem** of T and we write $T \vdash_L F$. In the case where T is empty, we say that F is L-**derivable** and we write $\vdash_L F$.*

Remark 4.4 *For a formula F and a theory T in a language L, it could happen, at least a priori, that F is not derivable from T in L but that it is derivable from T if we allow a richer language. The completeness theorem will show, however, that this is not the case. Once this theorem is proved, we will simply say 'derivable' rather than 'L-derivable' and 'syntactic consequence' instead of 'L-syntactic consequence'. In the meantime, if we do not specifically mention the language L, you should consider that the language is fixed and is sufficiently rich that it contains all the formulas and theories determined by the context.*

Example 4.5 Suppose that F and G are two closed formulas and set $T = \{F, G\}$; we are going to prove that $T \vdash F \wedge G$. Here is a sequence of formulas that constitutes a derivation of $F \wedge G$ in T:

(1)	F	(F is in T)
(2)	G	(G is in T)
(3)	$F \Rightarrow (G \Rightarrow (F \wedge G))$	(a tautology)
(4)	$G \Rightarrow (F \wedge G)$	(by modus ponens from (1) and (3))
(5)	$F \wedge G$	(by modus ponens from (2) and (4)).

Example 4.6 Let F be a formula and t be a term and suppose that no free occurrence of v in F is within the scope of a quantification that binds a variable of t. We will show (Table 4.1) that $\vdash F_{t/v} \Rightarrow \exists v F$.

Table 4.1

(1)	$\forall v \neg F \Rightarrow \neg F_{t/v}$	quantifier axiom of type (c)
(2)	$(\forall v \neg F \Rightarrow \neg F_{t/v}) \Rightarrow (F_{t/v} \Rightarrow \neg \forall v \neg F)$	tautology obtained from $(A \Rightarrow \neg B) \Rightarrow (B \Rightarrow \neg A)$
(3)	$F_{t/v} \Rightarrow \neg \forall v \neg F$	by modus ponens from (1) and (2)
(4)	$\exists v F \Leftrightarrow \neg \forall v \neg F$	quantifier axiom of type (a)
(5)	$(F_{t/v} \Rightarrow \neg \forall v \neg F)$ $\Rightarrow ((\exists v F \Leftrightarrow \neg \forall v \neg F) \Rightarrow (F_{t/v} \Rightarrow \exists v F))$	tautology obtained from $(A \Rightarrow B) \Rightarrow ((C \Leftrightarrow B) \Rightarrow (A \Rightarrow C))$
(6)	$(\exists v F \Leftrightarrow \neg \forall v \neg F) \Rightarrow (F_{t/v} \Rightarrow \exists v F)$	by modus ponens from (3) and (5)
(7)	$F_{t/v} \Rightarrow \exists v F$	by modus ponens from (4) and (6)

Example 4.7 If w is a variable that has no occurrence in F (neither free nor bound), then $\vdash \forall w F_{w/v} \Rightarrow \forall v F$:

The justification for step (1) in the derivation that follows involves a somewhat acrobatic use of schema (c): in the formula $F_{w/v}$, the only free occurrences of w are those that replaced the free occurrences of v in F and, quite obviously, these do not lie within the scope of a quantifier that binds v (otherwise, they would not be free!). So formula (1) in Table 4.2 does belong to axiom schema (c).

Table 4.2

(1)	$\forall w\, F_{w/v} \Rightarrow (F_{w/v})_{v/w}$	quantifier axiom of type (c)
(1')	$\forall w\, F_{w/v} \Rightarrow F$	since w does not occur in F $(F_{w/v})_{v/w} = F$ so (1') is a rewriting of (1)
(2)	$\forall v(\forall w\, F_{w/v} \Rightarrow F)$	from (1') by generalization
(3)	$\forall v(\forall w\, F_{w/v} \Rightarrow F)$ $\Rightarrow (\forall w\, F_{w/v} \Rightarrow \forall v F)$	quantifier axiom of type (b) since v is not free in $\forall w\, F_{w/v}$
(4)	$\forall w\, F_{w/v} \Rightarrow \forall v F$	by modus ponens from (2) and (3)

Example 4.8 Let us prove that $\vdash \forall v F \Rightarrow F$.
Again, this is a use of axiom schema (c). Indeed, $F = F_{v/v}$, and it is certain that the free occurrences of v in F do not lie in the scope of a quantification that binds v.

Example 4.9 Finally, let us prove (Table 4.3) $\vdash \forall v_0 \forall v_1 F \Rightarrow \forall v_1 \forall v_0 F$.

Table 4.3

(1)	$\forall v_0 \forall v_1 F \Rightarrow \forall v_1 F$	Example 4.8
(2)	$\forall v_1 F \Rightarrow F$	Example 4.8
(3)	$(\forall v_0 \forall v_1 F \Rightarrow \forall v_1 F)$ $\Rightarrow ((\forall v_1 F \Rightarrow F) \Rightarrow (\forall v_0 \forall v_1 F \Rightarrow F))$	a tautology
(4)	$\forall v_0 \forall v_1 F \Rightarrow F$	from (1), (2) and (3) by modus ponens, twice
(5)	$\forall v_0(\forall v_0 \forall v_1 F \Rightarrow F)$	from (4) by generalization
(6)	$\forall v_0(\forall v_0 \forall v_1 F \Rightarrow F) \Rightarrow (\forall v_0 \forall v_1 F \Rightarrow \forall v_0 F)$	axiom schema (b)
(7)	$\forall v_0 \forall v_1 F \Rightarrow \forall v_0 F$	from (5) and (6) by modus ponens
(8)	$\forall v_1(\forall v_0 \forall v_1 F \Rightarrow \forall v_0 F)$	from (7) by generalization
(9)	$(\forall v_1(\forall v_0 \forall v_1 F \Rightarrow \forall v_0 F))$ $\Rightarrow (\forall v_0 \forall v_1 F \Rightarrow \forall v_1 \forall v_0 F)$	axiom schema (b)
(10)	$\forall v_0 \forall v_1 F \Rightarrow \forall v_1 \forall v_0 F$	from (8) and (9) by modus ponens

Remark 4.10 *Suppose that $F \Rightarrow G$ is a tautology. Then $\forall v F \Rightarrow \forall v G$ is a theorem.*

$\forall v F \Rightarrow F$	*Example 4.8*
$F \Rightarrow G$	*a tautology, by hypothesis*
$(\forall v F \Rightarrow F) \Rightarrow ((F \Rightarrow G) \Rightarrow (\forall v F \Rightarrow G))$	*a tautology*
$\forall v F \Rightarrow G$	*by modus ponens, twice*
$\forall v (\forall v F \Rightarrow G)$	*by generalization*
$\forall v (\forall v F \Rightarrow G) \Rightarrow (\forall v F \Rightarrow \forall v G)$	*axiom schema (b)*
$\forall v F \Rightarrow \forall v G$	*by modus ponens*

Remark 4.11 *Suppose that F is a closed formula, that $T \vdash F$ and that $T \cup \{F\} \vdash G$; then $T \vdash G$.*

To see this, note that if $(F_0, F_1, \ldots, F_n)$ is a derivation of F in T (so $F_n = F$) and $(G_0, G_1, \ldots, G_m)$ is a derivation of G in $T \cup \{F\}$ (so $G_m = G$), then $(F_0, F_1, \ldots, F_n, G_0, G_1, \ldots, G_m)$ is a derivation of G in T.

Definition 4.12 *A theory T is called **non-contradictory in** L if there is no formula F such that we have both $T \vdash_L F$ and $T \vdash_L \neg F$ simultaneously. If this does happen, T is called **contradictory in** L.*

Remark 4.13 *If T is contradictory, then every formula is derivable in T.*

Indeed, suppose that both $T \vdash F$ and $T \vdash \neg F$ and let G be an arbitrary formula. Begin a sequence by placing the derivations of F and $\neg F$ end-to-end. To produce a derivation of G from this, it suffices to add to the sequence the following formulas from Table 4.4.

Table 4.4

$F \Rightarrow (\neg F \Rightarrow G)$	a tautology
$\neg F \Rightarrow G$	by modus ponens, since F occurs earlier in the sequence
G	again by modus ponens, since $\neg$F occurs earlier in the sequence.

The converse ('if every formula is derivable in T, then T is contradictory') is obvious from the definition. Moreover, we may immediately verify that:

Remark 4.14

- *if T is contradictory, then for every formula F, $T \vdash F \wedge \neg F$;*
- *for T to be contradictory, it is necessary and sufficient that there exist one formula F such that $T \vdash F \wedge \neg F$.*

4.1.3 The finiteness theorem and the deduction theorem

The next simple but extremely important theorem is known as the **finiteness theorem**.

Theorem 4.15 *For any theory T and any formula F, if $T \vdash F$, then there is a finite subset T_0 of T such that $T_0 \vdash F$.*

Proof Let $\mathcal{D}$ be a derivation of F in T; it is a finite sequence of formulas. Only a finite number of formulas of T can appear in it. If T_0 is the finite subset of T consisting of these formulas, then $\mathcal{D}$ is also a derivation of F in T_0. ∎

Corollary 4.16 *If T is a theory all of whose finite subsets are non-contradictory, then T is non-contradictory.*

Proof If not, then from T we can derive $F \wedge \neg F$ (here, F is an arbitrary formula) and we conclude from the finiteness theorem that $F \wedge \neg F$ is derivable from some finite subset T_0 of T; so T_0 is contradictory (by Remark 4.13) and this violates our hypothesis. ∎

Let us be more precise:

Proposition 4.17 *Let I be an ordered set and for every $i \in I$, let T_i be a theory in a language L_i which is not contradictory in L_i. Suppose that if i is less than j, then $L_i \subseteq L_j$ and $T_i \subseteq T_j$. Set $T = \bigcup_{i \in I} T_i$ and $L = \bigcup_{i \in I} L_i$. Then the theory T is not contradictory in L.*

Proof The proof is almost the same as the proof of Corollary 4.16: an L-derivation D of a formula F in T is in fact an L_i-derivation in T_i for some $i \in I$ provided that L_i and T_i contains all the formula in D. Such an index i exists, thanks to our hypothesis. ∎

It is more or less obvious that if the formula $F \Rightarrow G$ is derivable in T, then the formula G is derivable in $T \cup \{F\}$ (by modus ponens). The converse of this is a very useful tool and is known as the **deduction theorem**.

Theorem 4.18 *Assume that F is a closed formula and that $T \cup \{F\} \vdash G$. Then $T \vdash F \Rightarrow G$.*

Proof Let $\mathcal{D} = (G_0, G_1, \ldots, G_n)$ be a derivation of G in $T \cup \{F\}$. We will construct a derivation $\mathcal{D}'$ of $F \Rightarrow G$ in T by inserting various formulas into the sequence $(F \Rightarrow G_0, F \Rightarrow G_1, \ldots, F \Rightarrow G_n)$.

- If G_i is a tautology, there is no problem since $F \Rightarrow G_i$ is then also a tautology.
- If G_i is a quantification axiom or is an element of T, we need to insert the two formulas G_i and $G_i \Rightarrow (F \Rightarrow G_i)$ (which is a tautology) between $F \Rightarrow G_{i-1}$ and $F \Rightarrow G_i$ (or, more simply, in front of $F \Rightarrow G_i$ in case $i = 0$); $F \Rightarrow G_i$ is then derivable from these two preceding formulas by modus ponens.
- If $G_i = F$, there is no problem since $F \Rightarrow F$ is a tautology.

- Now suppose that G_i is obtained by modus ponens, which is to say that there exist integers j and k, strictly less than i, such that G_k is the formula $G_j \Rightarrow G_i$. We then insert the following formulas between $F \Rightarrow G_{i-1}$ and $F \Rightarrow G_i$:

$(F \Rightarrow G_j) \Rightarrow ((F \Rightarrow (G_j \Rightarrow G_i)) \Rightarrow (F \Rightarrow G_i))$ (this is a tautology)

$(F \Rightarrow (G_j \Rightarrow G_i)) \Rightarrow (F \Rightarrow G_i)$ (this is derived by modus ponens from the previous formula together with $F \Rightarrow G_j$ which appears earlier in the sequence)

$F \Rightarrow G_i$ (this follows by modus ponens from the previous formula together with $F \Rightarrow (G_j \Rightarrow G_i)$ which is equal to $F \Rightarrow G_k$ and has therefore also appeared earlier).

- Suppose G_i is derived by generalization from G_j with $j < i$ (thus, G_i is $\forall v G_j$). Then the following formulas should be inserted between $F \Rightarrow G_{i-1}$ and $F \Rightarrow G_i$:

$\forall v(F \Rightarrow G_j)$ (obtained by generalization from $F \Rightarrow G_j$)

$\forall v(F \Rightarrow G_j) \Rightarrow (F \Rightarrow \forall v G_j)$ (this is quantification axiom because, since F is closed, v is certainly not free in F)

$F \Rightarrow G_i$ (this follows by modus ponens from the two preceding formulas). ∎

The next corollary justifies the general use of proofs by contradiction.

Corollary 4.19 $T \vdash F$ *if and only if* $T \cup \{\neg F\}$ *is contradictory.*

Proof It is clear that if $T \vdash F$, then $T \cup \{\neg F\}$ is contradictory. Conversely, if $T \cup \{\neg F\}$ is contradictory, any formula is derivable from it, and, in particular, F; by applying the deduction theorem, we then see that $T \vdash \neg F \Rightarrow F$. Now, $(\neg F \Rightarrow F) \Rightarrow F$ is a tautology, so we may conclude that $T \vdash F$. ∎

The least of the demands that we might make on derivations is that they should lead to formulas that are true (more precisely, a derivation from a theory should lead to formulas that are semantic consequences of the theory). That is what we will now prove.

Theorem 4.20 *Let T be a theory, F a formula and F' a universal closure of F. Then*

(1) *If $T \vdash F$, then every model of T is a model of F' (in other words, $T \vdash^* F$). and*

(2) *If $\vdash F$, then F' is universally valid.*

(Recall that the universal closures of a formula are obtained by universally quantifying all the free variables in the formula; it may have several universal closures because there is a choice in the order of the quantifications: see Definition 3.20).

Proof Since (2) is a special case of (1) (take $T = \emptyset$), it suffices to prove (1). We have already observed that all the axioms are universally valid; their universal closures are therefore true in every model of T. We will now prove the following property by induction on n:

> Let $(F_0, F_1, \ldots, F_n)$ be a derivation of F_n in T and let F_n' be a universal closure of F_n; then $T \vdash^* F_n'$.

- Case $n = 0$: then the formula F_0 is either an axiom or else is an element of T. In both instances, F_0' is true in every model of T.
- For the passage from n to $n + 1$, we distinguish several subcases:
  * if F_{n+1} is an axiom or an element of T, then, as in the previous case, every model of T is a model of F_{n+1}'.
  * if F_{n+1} is obtained by modus ponens, there exist integers i and j less than or equal to n such that F_i is $F_j \Rightarrow F_{n+1}$. Suppose that the free variables of F_i are $v_0, v_1, \ldots, v_p$ (and so the free variables of F_j and of F_{n+1} are among these); F_j and $F_j \Rightarrow F_{n+1}$ have derivations whose length is at most n. So by the induction hypothesis, if $\mathcal{M}$ is a model of T, then

$$\mathcal{M} \models \forall v_0 \forall v_1 \ldots \forall v_p F_j$$

and

$$\mathcal{M} \models \forall v_0 \forall v_1 \ldots \forall v_p (F_j \Rightarrow F_{n+1});$$

hence,

$$\mathcal{M} \models \forall v_0 \forall v_1 \ldots \forall v_p F_{n+1}.$$

Because the formulas $\forall v_0 \forall v_1 \ldots \forall v_p F_{n+1}$ and F_{n+1}' are universally equivalent, we may now conclude that $\mathcal{M} \models F_{n+1}'$.
  * if F_{n+1} is obtained by generalization, there exists an integer $i \leq n$ and a variable v such that $F_{n+1} = \forall v F_i$. F_{n+1}' is then a universal closure of F_i and the conclusion now follows from the induction hypothesis. ∎

Corollary 4.21 *If T has a model, then T is non-contradictory.*

Proof If T were contradictory, then $F \wedge \neg F$ would be derivable from it, yet this formula has no model. ∎

4.2 Henkin models

We will now prove the converse of the corollary that concluded the previous section: if T is non-contradictory, then T has a model. The proof must necessarily proceed by constructing a model and, for this reason, is certainly much more difficult. We will need a certain number of preliminary definitions and lemmas.

4.2.1   Henkin witnesses

Definition 4.22   *Let T be a theory in a language L. We say that T is **syntactically complete** (in L) if it is not contradictory in L and if, for every closed formula F of L, we have $T \vdash_L F$ or $T \vdash_L \neg F$.*

It is worth noting that the syntactical completeness of T depends on the language in which it is viewed; in principle, if we do not specify the language, it will be because the context already makes it clear.

It will follow from the completeness theorem (4.29) and from Theorem 4.20 that a theory is syntactically complete if and only it is complete (see Definition 3.80). There will be no need, from that point on, to distinguish between these two notions.

The next item is the one that will allow us to construct models:

Definition 4.23   *Let T be a theory in a language L. We say that T possesses **Henkin witnesses** if for every formula F[v] with a single free variable v, there is a constant symbol c in L such that the formula*

$$\exists v\, F[v] \Rightarrow F[c]$$

belongs to T.

The proof of the completeness theorem splits into two parts: first we show that a syntactically complete theory which possesses Henkin witnesses has a model; then we show that any non-contradictory theory can be enriched to a syntactically complete theory that possesses Henkin witnesses.

Proposition 4.24. (A)   *If T is a syntactically complete theory in a language L that possesses Henkin witnesses, then T has a model.*

Proof   Starting from a theory T that satisfies the hypotheses of the theorem, we will construct a model of T from virtually nothing.

Let $\mathcal{T}$ be the set of closed terms of L. Since the theory has Henkin witnesses, the language contains constant symbols and the set $\mathcal{T}$ is non-empty. Here is the L-structure $\mathcal{M}$ that will turn out to be a model of T:

- The underlying set of $\mathcal{M}$ is the set $\mathcal{T}$.

- If c is a constant symbol, its interpretation in $\mathcal{M}$ is c.

- If R is an n-ary relation symbol, the interpretation $\overline{R}^{\mathcal{M}}$ of R is defined by: for all $t_1, t_2, \ldots, t_n$ in $\mathcal{T}$,

$$(t_1, t_2, \ldots, t_n) \in \overline{R}^{\mathcal{M}} \text{ if and only if } T \vdash Rt_1 t_2 \ldots t_n.$$

- If f is an n-ary function symbol, the interpretation $\overline{f}^{\mathcal{M}}$ of f in $\mathcal{M}$ has to be a map from $\mathcal{T}^n$ into $\mathcal{T}$: for $t_1, t_2, \ldots, t_n$ in $\mathcal{T}$, let $\overline{f}^{\mathcal{M}}(t_1, t_2, \ldots, t_n)$ be the term $f t_1 t_2 \ldots t_n$.

With this definition, we see that for any closed atomic formula F, we have $T \vdash F$ if and only if $\mathcal{M} \models F$. We will now prove that this equivalence continues to hold for all closed formulas of L. The proof is by induction on F. Since the case for atomic formulas has been treated, we may assume that the formula F has one of the following forms:

(1) $F = \neg G$;

(2) $F = G \wedge H$;

(3) $F = G \vee H$;

(4) $F = G \Rightarrow H$;

(5) $F = G \Leftrightarrow H$;

(6) $F = \forall v G$;

(7) $F = \exists v G$.

We will only treat cases (1), (3), (6), and (7), leaving the others to the reader.

(1) By the induction hypothesis, we know that $T \vdash G$ if and only if $\mathcal{M} \models G$. Now $\mathcal{M} \models F$ if and only if it is false that $\mathcal{M} \models G$. Because T is a syntactically complete theory, $T \vdash F$ if and only if it is false that $T \vdash G$; hence the result.

(3) Suppose first that $\mathcal{M} \models F$; then $\mathcal{M} \models G$ or $\mathcal{M} \models H$; if, for example, $\mathcal{M} \models G$, then, by the induction hypothesis, $T \vdash G$; but the formula $G \Rightarrow G \vee H$ is a tautology, and so $T \vdash F$.

In the other direction, suppose $T \vdash G \vee H$. If $T \vdash G$, then, by the induction hypothesis, $\mathcal{M} \models G$, so $\mathcal{M} \models F$. If not, then, again because T is syntactically complete, $T \vdash \neg G$; and since $G \vee H \Rightarrow (\neg G \Rightarrow H)$ is a tautology, it follows that $T \vdash H$. So by the induction hypothesis, $\mathcal{M} \models H$, and so $\mathcal{M} \models F$.

(6) If $T \vdash \forall v G$ and if $t \in T$, then because the formula $\forall v G \Rightarrow G_{t/v}$ is an axiom (t is a closed term), we see that $T \vdash G_{t/v}$ and, by the induction hypothesis, $\mathcal{M} \models G_{t/v}$. Now, since every element of $\mathcal{M}$ is the interpretation of a closed term, it follows that $\mathcal{M} \models \forall v G$.

Now suppose that $\forall v G$ is not derivable in T. Invoking Remark 4.10 (noting that $\neg\neg G \Rightarrow G$ is a tautology), we see that the formula $\forall v \neg\neg G$ is not derivable in T either, and since T is syntactically complete, $T \vdash \neg\forall v \neg\neg G$. But the formula $\exists v \neg G \Leftrightarrow \neg\forall v \neg\neg G$ is an instance of axiom scheme (a); this allows us to conclude $T \vdash \exists v \neg G$ thanks to modus ponens and some tautologies. Because T has Henkin witnesses, there is a constant symbol c such that $T \vdash \neg G_{c/v}$ (v is the only free variable of G since F is closed). As T is non-contradictory, $G_{c/v}$ is not derivable in T; by the induction hypothesis, $\mathcal{M} \models \neg G_{c/v}$ so $\forall v G$ is false in $\mathcal{M}$.

(7) If $\mathcal{M} \models \exists v G$, this must be because there is an element of $t \in T$ such that $\mathcal{M} \models G_{t/v}$ so, by the induction hypothesis, $T \vdash G_{t/v}$. It is easy to find a derivation of $\exists v G$ from $G_{t/v}$ (see Example 4.6) and hence $T \vdash \exists v G$.

Conversely, suppose that $T \vdash \exists v G$ thanks to the Henkin witnesses, it follows that there exists a constant symbol c of T such that $T \vdash G_{c/v}$ so by the induction hypothesis, $\mathcal{M} \models G_{c/v}$; hence $\mathcal{M} \models \exists v G$.

This completes the proof of Proposition A.   ∎

4.2.2 The completeness theorem

We will now see how to obtain, from a non-contradictory theory, a theory that satisfies the conditions of Proposition A.

Proposition 4.25. (B) *Let T be a non-contradictory theory in a language L; then there exists a language L' that includes L and a theory T' that includes T, that is syntactically complete and possesses Henkin witnesses in L'.*

Here, since we are working with several languages, we have to be careful. We will need the following two lemmas.

Lemma 4.26 *Let S be a theory in a language L and let F be a formula in L with at most one free variable v and let c be a constant symbol that does belongs to L; if $S \vdash_{L \cup \{c\}} F_{c/v}$, then $S \vdash_L \forall v F$.*

Proof Let $(H_1, H_2, \ldots, H_n)$ be an $(L \cup \{c\})$-derivation of $F_{c/v}$ in S (thus $H_n = F_{c/v}$). Choose a variable w that does not occur in any of the formulas H_i ($1 \le i \le n$) and let K_i denote the formula obtained from H_i by substituting w for c. A straightforward examination of the rules and axioms shows that:

- if H_i is a logical axiom, then so is K_i;
- if H_i is derived either by generalization or by modus ponens from one or two preceding formulas, then K_i is derived in exactly the same way from the corresponding formulas;
- furthermore, if H_i belongs to S, then $K_i = H_i$ so K_i belongs to S.

It follows from all this that $(K_1, K_2, \ldots, K_n)$ is an L-derivation in S of $F_{w/v}$. By generalization, we now obtain $S \vdash_L \forall w F_{w/v}$ and, thanks to the quantification axioms, $S \vdash_L \forall v F$ (see Example 4.7). ∎

Lemma 4.27 *Let T be a theory in some language L, and assume that T is not L-contradictory. Let L' be the language obtained from L by adding a finite set C of constant symbols. Then T is not L'-contradictory.*

If C contains just one new constant, then it is an immediate corollary of the preceding lemma (with F a closed formula of L). In the general case, we just have to apply this several times. ∎

Following the same pattern as for the compactness theorem of propositional calculus (Theorem 1.39) we are going to give two proofs (in fact, two versions of the same proof) of Proposition B. In the second, we will appeal to the axiom of choice; in the first, which is easier to understand, we will add as a supplementary hypothesis that the set of symbols of L is finite or countable. We will see later (in Chapter 7 of Part 2) that under these conditions, we can find an enumeration $(F_0, F_1, \ldots, F_n, \ldots)$ of all the formulas of L.

First proof We add to L a countably infinite set $C = \{c_0, c_1, \ldots, c_n, \ldots\}$ of new constant symbols (this means that none of the c_i already belong to L). Denote the resulting language by L'. The language L' is also countable and we can find an enumeration $(F_0, F_1, \ldots, F_n, \ldots)$ of all the closed formulas of L.

By induction on n, we will define a language L_n and a theory T_n, starting with $L_0 = L$ and $T_0 = T$, in such a way that the following conditions are satisfied:

- T_n is not L_n contradictory;
- $L_n \subseteq L_{n+1}$ and $T_n \subseteq T_{n+1}$;
- $L_n - L$ is finite;
- $F_n \in T_{n+1}$ or $\neg F_n \in T_{n+1}$;
- if F_n is of the form $\exists v H$ and belongs to T_{n+1}, then there exists a constant symbol c such that $H_{c/v} \in T_{n+1}$.

Here is the procedure for constructing L_{n+1} and T_{n+1} from L_n and T_n: it operates on F_n which is the $(n + 1)$st formula in the enumeration that we fixed at the beginning of the proof. Let L'_n be the language obtained from L_n by adding the constants of C that occur in F_n. We see from Lemma 4.27 and by the induction hypothesis that T_n is not contradictory in L'_n. If $T_n \cup \{F_n\}$ is not contradictory in L'_n, then set $G_n = F_n$; if it is contradictory in L'_n, the reason is because $T_n \vdash_{L'_n} \neg F_n$, and we then set $G_n = \neg F_n$. In both cases, the theory $T \cup \{G_n\}$ is not contradictory in L'_n.

If G_n is not of the form $\exists v H$, we stop here and set $L_{n+1} = L'_n$ and $T_{n+1} = T_n \cup \{G_n\}$.

If G_n is of the form $\exists v H$, we choose a symbol $c \in C$ which is not in L'_n; this is possible because $L'_n - L$ is finite. We then set

$$L_{n+1} = L'_n \cup \{c\}$$

and

$$T_{n+1} = T_n \cup \{G_n, \ H_{c/v}\}.$$

The last four conditions required of T_{n+1} are clearly satisfied. It remains to prove that the theory T_{n+1} is not contradictory in L_{n+1}; we show that the opposite cannot hold: suppose that T_{n+1} is contradictory in L_{n+1}. From Corollary 4.19, we see that

$$T_n \cup \{\exists v H\} \vdash_{L_{n+1}} \neg H_{c/v}.$$

From Lemma 4.26 and by our choice of the constant c, we deduce that

$$T_n \cup \{\exists v H\} \vdash_{L_n} \forall v \neg H,$$

which is not possible since $T_n \cup \{G_n\}$ is not contradictory in L_n.

We have thus completed the construction of the theories T_n; now set

$$L' = \bigcup_{n \in \mathbb{N}} L_n$$

and
$$T' = \bigcup_{n \in \mathbb{N}} T_n.$$

The fact that T' is not contradictory in L' follows from Corollary 4.17. To show that T' is syntactically complete, suppose that F is a closed formula of L'; then there exists an integer n such that $F = F_n$ and, by construction, $F \in T_{n+1}$ or $\neg F \in T_{n+1}$. Also, T' has Henkin witnesses: let H be a formula of L' with one free variable v; again, there exists an integer n such that $F_n = \exists v H$ and, either $\neg F_n \in T_{n+1}$ or there exists a symbol c such that $H_{c/v} \in T_{n+1}$. In both cases, $T_{n+1} \vdash_{L'} \exists v H \Rightarrow H_{c/v}$, which proves that $\exists v H \Rightarrow H_{c/v} \in T'$ (otherwise T' would contain $\neg(\exists v H \Rightarrow H_{c/v})$ and would be contradictory). ∎

Second proof This time we will use Zorn's lemma. We begin by adding Henkin constants to the language. According to Lemma 4.26, if a theory T in a language L is not L-contradictory, if F is a formula of L whose only free variable is v and if c is a constant symbol that does not belong to L then the theory $T \cup \{\exists v F[v] \Rightarrow F[c]\}$ is not contradictory in $(L \cup \{c\})$. Let us introduce a new constant symbol c_F for each formula F of L that has one free variable and let L_1 be the language thus obtained. Let n be an integer and for every integer p between 1 and n, let F_p be a formula with one free variable w_p; then by applying Lemma 4.26 n times, we conclude that the theory

$$T \cup \{\exists w_p F_p[w_p] \Rightarrow F_p[c_{F_p}] : 1 \leq p \leq n\}$$

is not contradictory in $(L \cup \{c_{F_p} : 1 \leq p \leq n\})$. It then follows from Proposition 4.17 that the theory

$$T_1 = T \cup \{\exists v F[v] \Rightarrow F[c_F] : F \text{ is a formula of } L \text{ with one free variable } v\}$$

is not contradictory in L_1.

One must not jump to the conclusion that T_1 has Henkin witnesses: it has them, but only for formulas of L, and this is insufficient since the language L_1 is richer. But it will pay us to be stubborn and to repeat the process: let us, for each new formula F of L_1 with one free variable, add a new constant symbol c_F; let L_2 be the resulting language and set

$$T_2 = T_1 \cup$$
$$\{\exists v F[v] \Rightarrow F[c_F] : F \text{ is a formula of } L_1 \text{ (not of } L\text{) with one free variable } v\}.$$

T_2 is non-contradictory in L_2 and we define, using the same pattern, L_3 and T_3, and so on. All the theories T_n obtained this way are non-contradictory and, since they extend one another, we see, using Proposition 4.17, that $T' = \bigcup_{n \in \mathbb{N}} T_n$ is not contradictory in L', where $L' = \bigcup_{n \in \mathbb{N}} L_n$ and, this time, has Henkin witnesses: for if F is a formula of L' with only one free variable v, then there is an integer n such that F is in L_n. Consequently, $\exists v F[v] \Rightarrow F[c_F]$ belongs to T_{n+1} and hence to T'.

It remains to find, while remaining in L', a theory T'' that includes T', that is non-contradictory and that is syntactically complete in L'. It is clear that T'' will still have Henkin witnesses since we are no longer changing the language. The lemma that follows provides the required conclusion.

Lemma 4.28 *Let T' be a non-contradictory theory in a language L'. Then there exists a theory T'' in L' which is non-contradictory, syntactically complete, and includes T'.*

Proof It is here that we will invoke Zorn's lemma. Consider the set

$$\Gamma = \{S : S \text{ is a non-contradictory theory in } L' \text{ that includes } T'\}.$$

Γ is not empty since it contains T'. Moreover, let

$$\mathcal{S} = \{S_i : i \in I\}$$

be a subset of Γ that is totally ordered by inclusion, meaning that if i and j are elements of I, then either $S_i \subseteq S_j$ or $S_j \subseteq S_i$. Then the theory

$$S = \bigcup_{i \in I} S_i$$

is still non-contradictory (Proposition 4.17) so is an upper bound for $\mathcal{S}$ in Γ. Thus Zorn's lemma (see Part 2, Chapter 7) applies and we can find an element T'' in Γ that is maximal for inclusion. We will now show that T'' is syntactically complete.

Let F be a closed formula of L' and suppose that $F \notin T''$. This implies that $T \cup \{F\}$ strictly includes T'', so by the maximality of T'', it is contradictory. Using Corollary 4.19, we conclude that $T'' \vdash \neg F$.   ∎

This concludes the second proof of proposition B.   ∎

We will finish this section with the **completeness theorem**.

Theorem 4.29 (*Gödel, 1930*) *Every non-contradictory theory has a model.*

Proof Using Proposition B, we produce a language L' extending L and a theory T' in L' that includes T, that is non-contradictory, syntactically complete and possesses Henkin witnesses; using Proposition A, we find a model $\mathcal{M}'$ of T'. The reduct of $\mathcal{M}'$ to L is then a model of T.   ∎

The following is an alternate way of stating the completeness theorem:

Proposition 4.30 *If T is a theory and F is a closed formula that is true in every model of T, then $T \vdash F$.*

Proof If F is not derivable in T, then $T \cup \{\neg F\}$ is non-contradictory (Corollary 4.19) so it has a model.

In particular, we see that the universally valid formulas are precisely those that are derivable from the empty theory.   ∎

Remark 4.31 *We are finally convinced, thanks to the completeness theorem, that the notions of semantic consequence and syntactic consequence coincide. So from* *now on, it is no longer indispensable to distinguish them. In particular, beyond this chapter, we will identify the two symbols ⊢* and ⊢ that were reserved until now for these two notions, respectively, and will only use the second.*

Let us conclude with a harmonious marriage, that of the finiteness theorem (Theorem 4.15) and the completeness theorem (Theorem 4.29); the offspring has been expected since Chapter 3; it is the **compactness theorem for predicate calculus**:

Theorem 4.32 *If every finite subset of a theory T has a model, then T itself has a model.*

Proof Every finite subset of T is non-contradictory, and hence (by the finiteness theorem) T is non-contradictory. The completeness theorem then tells us that T has a model. ∎

4.3 Herbrand's method

4.3.1 Some examples

In this section we will give another proof of the completeness theorem, essentially for the opportunity it offers us of explaining Herbrand's method. So forget that we have already proved the completeness theorem. Despite the fact that Herbrand's method applies to all formulas, we will be content here to restrict our attention to prenex formulas. The reader who is upset by this restriction can reduce the general case to this one by proving that every formula F is syntactically equivalent to a formula in prenex form (caution: this is not the same as Theorem 3.60; here, this means that there exists a formal proof of $F \Leftrightarrow G$, where G is a prenex form of F).

We will develop an argument that is similar in style to those familiar to high school students; it begins with: 'suppose the problem has been solved'. We are trying to determine whether a formula F has a model. So we suppose that we indeed have a model at our disposal and we observe that in it, there have to be some elements that satisfy certain quantifier-free formulas. We gather, in this way, a certain amount of information and after a while, we notice either that all this information is contradictory or else that we actually have enough to construct a model. Before presenting the general case, here are two examples.

Example 4.33 The language consists of a single ternary predicate, P. We seek to determine whether the following formula F,

$$\forall v_0 \exists v_1 \exists v_2$$
$$((\neg P v_0 v_1 v_2 \Leftrightarrow P v_0 v_2 v_1) \wedge (\neg P v_0 v_1 v_2 \Leftrightarrow P v_2 v_1 v_0)$$
$$\wedge (\neg P v_0 v_1 v_2 \Leftrightarrow P v_1 v_0 v_2)),$$

has a model.

Any eventual model cannot be empty, so assume a is one of its elements. The formula tells us that there must exist two points, call them a_0 and a_1, such that

$$(\neg Paa_0a_1 \Leftrightarrow Paa_1a_0)$$
$$\wedge\,(\neg Paa_0a_1 \Leftrightarrow Pa_1a_0a)$$
$$\wedge\,(\neg Paa_0a_1 \Leftrightarrow Pa_0aa_1)$$

is true in the model. Next in turn, for a_0, we require the existence of two other points; obviously, it may happen that one of these points is equal to a or to a_0 or to a_1; but in the absence of any other information, we will give these points new names (in any case, nothing prevents us from giving several names to the same point). So there are two points a_{00} and a_{01} such that

$$(\neg Pa_0a_{00}a_{01} \Leftrightarrow Pa_0a_{01}a_{00})$$
$$\wedge\,(\neg Pa_0a_{00}a_{01} \Leftrightarrow Pa_{01}a_{00}a_0)$$
$$\wedge\,(\neg Pa_0a_{00}a_{01} \Leftrightarrow Pa_{00}a_0a_{01})$$

is true in the model. Repeating this for a_1, we find two points a_{10} and a_{11} such that

$$(\neg Pa_1a_{10}a_{11} \Leftrightarrow Pa_1a_{11}a_{10})$$
$$\wedge\,(\neg Pa_1a_{10}a_{11} \Leftrightarrow Pa_{11}a_{10}a_1)$$
$$\wedge\,(\neg Pa_1a_{10}a_{11} \Leftrightarrow Pa_{10}a_1a_{11})$$

is also true in the model. Then we have to start all over for the points a_{00}, a_{01}, a_{10}, and a_{11} and after that, repeat the process for the eight new points, etc. Let S denote the set of finite sequences of 0s and 1s. Continuing in this way then, we define, for each $s \in S$, a point a_s in such a way that for every $s \in S$,

$$(\neg Pa_sa_{s0}a_{s1} \Leftrightarrow Pa_sa_{s1}a_{s0})$$
$$\wedge\,(\neg Pa_sa_{s0}a_{s1} \Leftrightarrow Pa_{s1}a_{s0}a_s)$$
$$\wedge\,(\neg Pa_sa_{s0}a_{s1} \Leftrightarrow Pa_{s0}a_sa_{s1})$$

is true. This information now guides us toward the actual construction of a model of F: for every $s \in S$, we choose a point c_s in such a way that if $s \neq t$, then $c_s \neq c_t$ (we could, quite simply, take $c_s = s$). The underlying set of our model $\mathcal{M}$ will be

$$M = \{c_s : s \text{ is a finite sequence of 0s and 1s}\}.$$

It remains to define the interpretation of P and it suffices to do this in such a way that for every $s \in S$,

$$(\neg Pc_sc_{s0}c_{s1} \Leftrightarrow Pc_sc_{s1}c_{s0})$$
$$\wedge\,(\neg Pc_sc_{s0}c_{s1} \Leftrightarrow Pc_{s1}c_{s0}c_s)$$
$$\wedge\,(\neg Pc_sc_{s0}c_{s1} \Leftrightarrow Pc_{s0}c_sc_{s1})$$

is true. At this point, we are reduced to a problem of the propositional calculus: indeed, let us introduce, for each triple (s, t, u) of elements of $\mathcal{S}$, a propositional variable $A_{s,t,u}$. We seek an assignment of truth values δ that makes all of the following formulas true:

$$\neg A_{s,s0,s1} \Leftrightarrow A_{s,s1,s0} \quad \text{for } s \in \mathcal{S}$$
$$\neg A_{s,s0,s1} \Leftrightarrow A_{s1,s0,s} \quad \text{for } s \in \mathcal{S}$$
$$\neg A_{s,s0,s1} \Leftrightarrow A_{s0,s,s1} \quad \text{for } s \in \mathcal{S}.$$

If we find one, it will suffice to define the interpretation $\overline{P}^{\mathcal{M}}$ of P by

for all s, t, u in $\mathcal{S}$, $(c_s, c_t, c_u) \in \overline{P}^{\mathcal{M}}$ if and only if $\delta(A_{s,t,u}) = \mathbf{1}$.

If, on the contrary, there is no such assignment, we will then know that F does not have a model.

For the case at hand, such an assignment ε is very easy to find: for example, we may take $\delta(A_{s,t,u}) = \mathbf{1}$ for all triplets of the form $(s, s0, s1)$ and $\delta(A_{s,t,u}) = \mathbf{0}$ for all the others.

We have thus found a model of F.

Example 4.34 The language contains one ternary predicate symbol P and one constant symbol c. Consider the following four formulas:

$F_1 : \quad \forall v_0 \forall v_1 \forall v_2 \forall v_3 \forall v_4 \forall v_5 ((P v_0 v_1 v_3 \wedge P v_1 v_2 v_4 \wedge P v_3 v_2 v_5) \Rightarrow P v_0 v_4 v_5)$
$F_2 : \quad \forall v_0 \forall v_1 \forall v_2 \forall v_3 \forall v_4 \forall v_5 ((P v_0 v_1 v_3 \wedge P v_1 v_2 v_4 \wedge P v_0 v_4 v_5) \Rightarrow P v_3 v_2 v_5)$
$F_3 : \quad \forall v_0 P v_0 c v_0$
$F_4 : \quad \forall v_0 \exists v_1 P v_0 v_1 c.$

We wish to prove that these four formulas imply $\forall v_0 P c v_0 v_0$. So we add the negation of this formula (or rather a formula that is equivalent to this negation),

$$F_5 : \quad \exists v_0 \neg P c v_0 v_0,$$

and we attempt to construct a model of the formulas $F_1, \ldots, F_5$ by the same method as in the previous example. The failure of this attempt will give us a proof of the fact that the set $\{F_1, \ldots, F_5\}$ is contradictory and hence that $\forall v_0 P c v_0 v_0$ follows from $F_1, \ldots, F_4$.

So suppose that $\mathcal{M}$ is a model of $F_1, \ldots, F_5$. The satisfaction of the formula F_5 requires the existence of a point d in $\mathcal{M}$ such that

(1) $\quad \mathcal{M} \models \neg P c d d;$

then F_4 commits us to the existence, for $i \in \mathbb{N}$, of points c_i and d_i in $\mathcal{M}$ starting with $c_0 = c$ and $d_0 = d$, such that

(2) $\quad \mathcal{M} \models P c_i c_{i+1} c \quad$ for all $i \in \mathbb{N}$;
(3) $\quad \mathcal{M} \models P d_i d_{i+1} c \quad$ for all $i \in \mathbb{N}$.

Set $A = \{c_i : i \in \mathbb{N}\} \cup \{d_i : i \in \mathbb{N}\}$. The interpretation of P in $\mathcal{M}$ must satisfy conditions (1), (2), and (3) as well as the following conditions (4), (5), and (6) imposed by formulas F_1, F_2, and F_3:

(4) $\mathcal{M} \models (Pxyu \wedge Pyzv \wedge Puzw) \Rightarrow Pxvw$
for all elements x, y, z, u, v, w of A;

(5) $\mathcal{M} \models (Pxyu \wedge Pyzv \wedge Pxvw) \Rightarrow Puzw$
for all elements x, y, z, u, v, w of A;

(6) $\mathcal{M} \models Pxcx$ for all $x \in A$.

So we are once again reduced to the question of the satisfiability of a set of propositional formulas (whose propositional variables are the $Pxyz$ for x, y, z in A). We will show that it is not satisfiable.

To see this, observe that when we take (4) with $x = c$ and $v = w = d$, we obtain implications whose conclusion is $Pcdd$, which is false according to (1). Hence, we must have

(7) $\neg(Pcyu \wedge Pyzd \wedge Puzd)$ for all elements y, z, u of A;

and so, by taking $u = y = c$,

(8) $\neg(Pccc \wedge Pczd \wedge Pczd)$ for all z in A.

We know from (6) that $Pccc$ must be true, and so, taking $z = d_2$,

(9) $\neg Pcd_2d$.

Condition (5) with $w = x = d$, $y = d_1$, $z = d_2$ and $u = v = c$ yields

(10) $(Pdd_1c \wedge Pd_1d_2c \wedge Pdcd) \Rightarrow Pcd_2d$.

Now Pdd_1c and Pd_1d_2c are true according to (3) and $Pdcd$ is true according to (6) but Pcd_2d is false according to (9): so our set is contradictory. Thus $\forall v_0 Pcv_0v_0$ does follow from $F_1, \ldots, F_4$.

Remark 4.35 *We will better understand this proof if we think of P as the graph of a binary operation. Then F_1 and F_2 express the fact that this operation is associative, F_3 that it has a right identity, and F_4 that every element has a right inverse. The conclusion asserts that the right identity is also a left identity. This is a fact that can be proved rather easily; we are slightly surprised here to learn that we can do without the hypothesis that P is the graph of a binary operation.*

4.3.2 The avatars of a formula

We will now approach things in full generality. We will also do things better since, in situations where the attempt to construct a model of F fails, what we will obtain is a derivation of $\neg F$. First of all, here are a few definitions.

Definition 4.36

(1°) *A formula F is said to be **propositionally satisfiable** if $\neg F$ is not a tautology.*

(2°) *A finite set E of formulas is said to be **propositionally satisfiable** if the conjunction of the formulas in E is propositionally satisfiable.*

(3°) *A set of formulas is **propositionally satisfiable** if all of its finite subsets are.*

Let L be a language and let F be a fixed closed prenex formula of L. We may assume that L is countable since it suffices to retain only the symbols that actually occur in F. We are going to take as a hypothesis that in the sequence of quantifiers appearing at the beginning of F, the universal quantifiers and existential quantifiers alternate, that the first one is universal and the last one is existential. In other words, we suppose F has the form

$$F = \forall v_1 \exists v_2 \forall v_3 \ldots \forall v_{2k-1} \exists v_{2k} B[v_1, v_2, \ldots, v_{2k}],$$

where B is a quantifier-free formula. Without this hypothesis, the proof would be exactly the same, except that it would require considerably more complicated notation. Besides, we can always artificially reduce the situation for an arbitrary formula to this case by adding quantifications over variables that have no free occurrence in the formula (and proving that the resulting formula is syntactically equivalent to the initial one, which is not too difficult to do).

Let $\mathcal{T}$ be the set of terms of L and, for $i \in \mathbb{N}$, let Θ_i be the set of sequences of length i of elements of $\mathcal{T}$. For each i between 1 and k, fix, once and for all, an injective map α_i from Θ_i into $\mathbb{N}$ so that the following conditions are satisfied:

(i) if v_n occurs in one of the terms $t_1, t_2, \ldots, t_i$, then $\alpha_i(t_1, t_2, \ldots, t_i) > n$;
(ii) if $j < i$ and $(t_1, t_2, \ldots, t_i)$ is a sequence that extends $(t_1, t_2, \ldots, t_j)$, then $\alpha_j(t_1, t_2, \ldots, t_j) < \alpha_i(t_1, t_2, \ldots, t_i)$;
(iii) if τ and σ are two distinct sequences, whose respective lengths are i and j then $\alpha_i(\tau) \neq \alpha_j(\sigma)$.

It is very easy to construct such maps: for example, using the codings that will be implemented in Chapter 6, we can set, for i between 1 and k,

$$\alpha_i(t_1, t_2, \ldots, t_i) = 2^m 3^{\tau(t_1)} 5^{\tau(t_2)} \cdots \pi(i)^{\tau(t_i)},$$

where m is the largest index of any variable that occurs in one of the t_j ($1 \leq j \leq i$), where τ is the map denoted by # in Chapter 6, and where π is the function that, with each integer n, associates the $(n+1)$st prime number.

Definition 4.37 *An **avatar** of $F = \forall v_1 \exists v_2 \forall v_3 \ldots \forall v_{2k-1} \exists v_{2k} B[v_1, v_2, \ldots, v_{2k}]$ is a formula of the form*

$$B[t_1, v_{\alpha_1(t_1)}, t_2, v_{\alpha_2(t_1,t_2)}, \ldots, t_k, v_{\alpha_k(t_1,t_2,\ldots,t_k)}]$$

where $t_1, t_2, \ldots, t_k$ are arbitrary terms of L.

Each avatar A of F is a formula without quantifiers, so is a boolean combination of atomic formulas. Let At denote the set of atomic formulas of L. If we view the elements of At as propositional variables, then the avatars behave as propositional formulas. To say that a finite set E of avatars is propositionally satisfiable (see Definition 4.36) is to say that there is an assignment of truth values, 0 or 1, to each atomic formula whose extension to boolean combinations makes each of the propositional formulas corresponding to the elements of E true. Thanks to the compactness theorem for propositional calculus (Theorem 1.39), this remains true for any set of avatars. We are now in position to state the first important result of this section:

Theorem 4.38 *If the set of avatars of F is propositionally satisfiable, then F has a model.*

Proof Let δ be a map from At into $\{0, 1\}$ such that $\overline{\delta}(A) = 1$ for all avatars A of F, where, as usual, $\overline{\delta}$ is the canonical extension of δ to the set of quantifier-free formulas of L.

We have to construct a model $\mathcal{M}$ of F. The simplest idea would be to take $\mathcal{T}$ itself as the underlying set of $\mathcal{M}$ and to use δ to define the interpretations of the various symbols of the language in such a way that all the avatars of F are true in $\mathcal{M}$; this is hardly desirable, however, for symbols such as v_0 would then at the same time denote both a variable and an element of the model; and, as a consequence of our abuses of language, there would be an ambiguity when we encountered them in a formula. We are therefore going to make a copy of $\mathcal{T}$ by adding new constant symbols c_i, for $i \in \mathbb{N}$, to the language L whose purpose is to take the place of the variables v_i; let L^* denote the language obtained in this way. For each term t of L, let t^* denote the closed term of L^* obtained by replacing, for every integer n, the occurrences of v_n in t by c_n. (In other words:

$$t^* = t_{c_0/v_0, c_1/v_1, \ldots, c_n/v_n}$$

if the only variables of t are $v_0, v_1, \ldots, v_n$.) We will do the same thing with the formulas: if G is a formula of L, G^* is the closed formula of L^* obtained by replacing, for every integer n, the free occurrences of v_n in G by c_n. Let $\mathcal{T}^*$ denote the set of closed terms of L^* (this is also the set $\{t^* : t \in \mathcal{T}\}$). Also, At^* will be the set of closed atomic formulas of L^* (this too is the set $\{G^* : G \in At\}$).

It is clear that the set $\{A^* : A \text{ is an avatar of } F\}$ is also propositionally satisfiable; let us define an assignment of truth values ε on At^* in the following way:

$$\text{for all } G \in At, \ \varepsilon(G^*) = \delta(G);$$

then, for any quantifier-free formula H of L, we will have

$$\overline{\varepsilon}(H^*) = \overline{\delta}(H),$$

and consequently, for any avatar A of F, $\overline{\varepsilon}(A^*) = 1$.

We are finally ready to define $\mathcal{M}$. Its underlying set is $\mathcal{T}^*$. If c is a constant symbol of L, then $\overline{c}^{\mathcal{M}} = c$; if f is a function symbol of L, of arity n, say, and

$u_1, u_2, \ldots, u_n \in \mathcal{T}^*$, then

$$\overline{f}^{\mathcal{M}}(u_1, u_2, \ldots, u_n) = f u_1 u_2 \ldots u_n.$$

If R is a relation symbol of L, again of arity n, say, and $u_1, u_2, \ldots, u_n \in \mathcal{T}^*$, then

$$(u_1, u_2, \ldots, u_n) \in \overline{R}^{\mathcal{M}} \text{ if and only if } \varepsilon(R u_1 u_2 \ldots u_n) = 1,$$

which can also be written

$$\mathcal{M} \models R u_1 u_2 \ldots u_n \text{ if and only if } \varepsilon(R u_1 u_2 \ldots u_n) = 1.$$

This equivalence extends to all quantifier-free formulas (the proof is by induction on the height of the formula): if $H[v_1, v_2, \ldots, v_n]$ is a quantifier-free formula and $u_1, u_2, \ldots, u_n \in \mathcal{T}^*$, then

$$\mathcal{M} \models H[u_1 u_2 \ldots u_n] \text{ if and only if } \overline{\varepsilon}(H[u_1 u_2 \ldots u_n]) = 1.$$

In this way, we are assured that if A is an avatar of F, then $\mathcal{M} \models A^*$.

It remains to verify that we do have a model of F. This is where the functions α_i reveal their true nature: they are Skolem functions in disguise. For each i between 1 and k, let us define the map f_i from $(\mathcal{T}^*)^i$ into $\mathcal{T}^*$ by

$$f_i(t_1, t_2, \ldots, t_i) = c_{\alpha_i(t_1, t_2, \ldots, t_i)}.$$

Then for every sequence $(t_1, t_2, \ldots, t_k)$ of elements of $\mathcal{M}$, we have

$$\mathcal{M} \models B[t_1, f_1(t_1), t_2, f_2(t_1, t_2), \ldots, t_k, f_k(t_1, t_2, \ldots, t_k)]$$

because

$$B[t_1, f_1(t_1), t_2, f_2(t_1, t_2), \ldots, t_k, f_k(t_1, t_2, \ldots, t_k)]$$
$$= B[t_1, v_{\alpha_1(t_1)}, t_2, v_{\alpha_2(t_1, t_2)}, \ldots, t_k, v_{\alpha_k(t_1, t_2, \ldots, t_k)}]^*.$$

If we consider the functions f_i as Skolem functions, we see that $\mathcal{M}$ satisfies a Skolem form of F, and hence $\mathcal{M}$ satisfies F (Lemma 3.65).   ∎

The previous theorem shows that if F does not have a model (which is the case, in particular, if $\neg F$ is derivable), then there is a conjunction of avatars of F whose negation is a tautology. We now wish to prove the converse of this.

Theorem 4.39 *If the set of avatars of a formula F is not propositionally satisfiable, then $\neg F$ is derivable.*

Proof We know that there are a finite number of avatars of F, say A_p for p between 1 and n, such that the formula $\bigvee_{1 \leq p \leq n} \neg A_p$ is a tautology.

Let $\mathcal{A}$ denote the set of formulas of the following form:

$$\forall w_{2i+1} \exists w_{2i+2} \ldots \forall w_{2k-1} \exists w_{2k}$$
$$B[t_1, v_{\alpha_1(t_1)}, t_2, v_{\alpha_2(t_1, t_2)}, \ldots, t_i, v_{\alpha_i(t_1, t_2, \ldots, t_i)}, w_{2i+1}, w_{2i+2}, \ldots, w_{2k}]$$

where i is an integer between 0 and k, $t_1, t_2, \ldots, t_i \in \mathcal{T}$, and where the w_j are variables that do not occur in any of the terms $t_1, \ldots, t_i$ and are different from $v_{\alpha_1(t_1)}, v_{\alpha_2(t_1,t_2)}, \ldots, v_{\alpha_i(t_1,t_2,\ldots,t_i)}$.

The avatars of F clearly belong to $\mathcal{A}$ (take $i = k$). So we know there is a finite subset I of $\mathcal{A}$ such that $\vdash \bigvee_{f \in I} \neg f$. The idea is to gradually quantify all of the free variables in this formula. Once this is done, we will have a formula equivalent to $\neg F$.

So suppose that I is a finite subset of $\mathcal{A}$ such that $\vdash \bigvee_{f \in I} \neg f$. We will find another finite subset, J, of $\mathcal{A}$ such that $\vdash \bigvee_{f \in J} \neg f$ and such that the number of free variables in $\vdash \bigvee_{f \in J} \neg f$ is at least one fewer than in $\vdash \bigvee_{f \in I} \neg f$.

Let

$$f = \forall w_{2i+1} \exists w_{2i+2} \ldots \forall w_{2k-1} \exists w_{2k}$$
$$B[t_1, v_{\alpha_1(t_1)}, t_2, v_{\alpha_2(t_1,t_2)}, \ldots, t_i, v_{\alpha_i(t_1,t_2,\ldots,t_i)}, w_{2i+1}, w_{2i+2}, \ldots, w_{2k}]$$

be a formula of $\mathcal{A}$. We associate the integer $n(f) = \alpha_i(t_1, t_2, \ldots, t_i)$ with f. This is the largest index of a variable that has a free occurrence in f; in other words, if $p > n(f)$, then v_p does not occur free in f. This follows from the properties we imposed on the functions α_i. Suppose, as well, that for some other formula $f' \in \mathcal{A}$, we have $n(f) = n(f')$; set

$$f' = \forall z_{2j+1} \exists z_{2j+2} \ldots \forall z_{2k-1} \exists z_{2k}$$
$$B[u_1, v_{\alpha_1(u_1)}, u_2, v_{\alpha_2(u_1,u_2)}, \ldots, u_j, v_{\alpha_j(u_1,u_2,\ldots,u_j)}, z_{2j+1}, z_{2j+2}, \ldots, z_{2k}].$$

By construction, the images of the different functions α_i are disjoint; it follows from this that $i = j$ and, since the α_i are injective, that $t_1 = u_1, \ldots, t_i = u_i$. In other words, f and f' differ only in the names of their bound variables and hence, $\vdash f \Leftrightarrow f'$ (see the Examples in the subsection on formal proofs).

We can begin by eliminating the repeats in I; the remarks that we have just made allow us to suppose that if f and f' are in I, then $n(f) \neq n(f')$. Now choose the formula $g \in I$ for which the integer $n(g)$ is maximum. Then the variable $v_{n(g)}$ is not free in any other formula f of I. Set

$$g = \forall w_{2i+1} \exists w_{2i+2} \ldots \forall w_{2k-1} \exists w_{2k}$$
$$B[t_1, v_{\alpha_1(t_1)}, t_2, v_{\alpha_2(t_1,t_2)}, \ldots, t_i, v_{\alpha_i(t_1,t_2,\ldots,t_i)}, w_{2i+1}, w_{2i+2}, \ldots, w_{2k}]$$

and

$$H = I - \{g\}.$$

By generalization, we obtain

$$\vdash \forall v_{n(g)} \left(\bigvee_{f \in I} \neg f \right)$$

and, since it is only in g that $v_{n(g)}$ has free occurrences (see Exercise 4),

$$\vdash \left(\bigvee_{f \in H} \neg f \right) \vee \forall v_{n(g)} \neg g.$$

Let w_{2i-1} be a variable that has no free occurrence in g and take $w_{2i} = v_{n(g)}$. Let

$$g' = \forall w_{2i-1} \exists w_{2i} \ldots \exists w_{2k}$$
$$B[t_1, v_{\alpha_1(t_1)}, t_2, v_{\alpha_2(t_1,t_2)}, \ldots, t_{i-1}, v_{\alpha_{i-1}(t_1,t_2,\ldots,t_{i-1})}, w_{2i-1}, w_{2i}, \ldots, w_{2k}].$$

So $\vdash g' \Rightarrow \exists v_{n(g)} g$ ('take' $w_{2i-1} = t_i$) and hence

$$\vdash \forall v_{n(g)} \neg g \Rightarrow \neg g'$$

and it remains only to set $J = H \cup \{g'\}$ to have $\vdash \bigvee_{f \in J} \neg f$.

Once all of the free variables have been eliminated and after suppressing the duplicates, we obtain

$$\vdash \neg \forall w_1 \exists w_2 \forall w_3 \ldots \exists w_{2k} B[w_1, w_2, \ldots, w_{2k}];$$

and this last formula differs from $\neg F$ only in the names of their bound variables.   ∎

Let us recapitulate:

Theorem 4.40 *The following three conditions are equivalent:*

(i) *F does not have a model;*

(ii) *there exist avatars $A_1, A_2, \ldots, A_n$ of F such that $\bigvee_{1 \leq p \leq n} \neg A_p$ is a tautology;*

(iii) $\vdash \neg F$.

Proof The implication (i) $\Rightarrow$ (ii) is Theorem 4.38; the implication (ii) $\Rightarrow$ (iii) is Theorem 4.39; and the implication (iii) $\Rightarrow$ (i) is Theorem 4.20.   ∎

Remark 4.41 *The proof of the completeness theorem that we have just presented applies only to a single formula, whereas Henkin's method proves it for an arbitrary theory. We will see, in Exercise 8, how to use Herbrand's method to prove the completeness theorem for a countable theory.*

Remark 4.42 *From the preceding proof, we can perfectly well extract an algorithm that allows us to construct a derivation of $\neg F$ starting from a tautology of the form $\bigvee_{1 \leq p \leq n} \neg A_p$.*

4.4 Proofs using cuts

4.4.1 The cut rule

In these last two sections, we are going to present a new kind of derivation. In it, the deduction rules, rather than the axioms, will play a privileged role. But these proofs only apply to a very restricted class of formulas, the universal clauses. Why should we prove the completeness theorem once again, especially in a much less general context? Because these are the kinds of derivations that we can most easily ask a computer to do. They form the basis of the language PROLOG. Hidden within nearly every result in this section, there is an algorithm; we will be content, however, to to give the idea behind the method, without really concerning ourselves with the algorithms.

In the first sections of this chapter, we did not bother with formal proofs in propositional calculus, quite simply because we chose to brutally include all the tautologies as axioms. That is not what we will do here because we wish to provide a method that is much faster than using truth tables for determining whether a proposition is a tautology or not. So this section relates only to propositional calculus. First, a reminder concerning a definition that appeared in Chapter 1 (see Remark 1.30).

Definition 4.43   *A **clause** is a proposition of the form*

$$(\neg A_1 \vee \neg A_2 \vee \cdots \vee \neg A_n \vee B_1 \vee B_2 \vee \cdots \vee B_m)$$

where the A_i and B_j are propositional variables.

The clause

$$(\neg A_1 \vee \neg A_2 \vee \cdots \vee \neg A_n \vee B_1 \vee B_2 \vee \cdots \vee B_m)$$

is logically equivalent, when n and m are strictly positive, to the formula

$$(A_1 \wedge A_2 \wedge \cdots \wedge A_n) \Rightarrow (B_1 \vee B_2 \vee \cdots \vee B_m)$$

and this is the way we will usually write it. The **premiss** of this clause is the conjunction

$$(A_1 \wedge A_2 \wedge \cdots \wedge A_n)$$

while its **conclusion** is the disjunction

$$(B_1 \vee B_2 \vee \cdots \vee B_m).$$

It can happen that either n or m is zero; in this situation, the clause

$$(B_1 \vee B_2 \vee \cdots \vee B_m)$$

will be written

$$\Rightarrow (B_1 \vee B_2 \vee \cdots \vee B_m)$$

and the clause

$$(\neg A_1 \vee \neg A_2 \vee \cdots \vee \neg A_n)$$

will be written

$$(A_1 \wedge A_2 \wedge \cdots \wedge A_n) \Rightarrow .$$

(We could extend the conventions adopted in Chapter 1 and define the conjunction of an empty set of formulas to be the proposition that is always true and the disjunction of an empty set of formulas to be the proposition that is always false).

If n and m are both zero, we obtain an empty disjunction which, by convention, denotes the false proposition; we will call this the empty clause and denote it by $\square$.

It has already been shown that every proposition is equivalent to a finite set of clauses (Theorem 1.29).

Proofs by cut proceed exclusively with the use of the deduction rules. The first is the **rule of simplification**: if a propositional variable A appears several times in the premiss of a clause $\mathcal{A}$, then the clause $\mathcal{A}'$ obtained from $\mathcal{A}$ by deleting all but one of the occurrences of A in the premiss of $\mathcal{A}$ is a formula that is logically equivalent to $\mathcal{A}$. Obviously, the same property holds for the conclusion of $\mathcal{A}$. In these circumstances, we say that $\mathcal{A}$ has been simplified and that $\mathcal{A}'$ **is derived from $\mathcal{A}$ by simplification**.

For example,

$$(A_1 \wedge A_2 \wedge A_1 \wedge A_4) \Rightarrow (B_1 \vee B_1 \vee B_2)$$

can be simplified to

$$(A_1 \wedge A_2 \wedge A_4) \Rightarrow (B_1 \vee B_1 \vee B_2)$$

which, in turn, can be simplified to

$$(A_1 \wedge A_2 \wedge A_4) \Rightarrow (B_1 \vee B_2).$$

Definition 4.44 (*The cut rule*) *Let*

$$\mathcal{C} = (A_1 \wedge A_2 \wedge \cdots \wedge A_n) \Rightarrow (B_1 \vee B_2 \vee \cdots \vee B_m)$$

and

$$\mathcal{D} = (C_1 \wedge C_2 \wedge \cdots \wedge C_p) \Rightarrow (D_1 \vee D_2 \vee \cdots \vee D_q)$$

be two clauses. If there exist integers i ($1 \le i \le m$) and j ($1 \le j \le p$) such that $B_i = C_j$ and if $\mathcal{E}$ is the clause

$$(A_1 \wedge A_2 \wedge \cdots \wedge A_n \wedge C_1 \wedge C_2 \wedge \cdots \wedge C_{j-1} \wedge C_{j+1} \wedge \cdots \wedge C_p)$$
$$\Rightarrow (B_1 \vee B_2 \vee \cdots \vee B_{i-1} \vee B_{i+1} \vee \cdots \vee B_m \vee D_1 \vee D_2 \vee \cdots \vee D_q),$$

*we say that $\mathcal{E}$ is **derived from** $\mathcal{C}$ and $\mathcal{D}$ (or from $\mathcal{D}$ and $\mathcal{C}$) **by cut**.*

In other words, if there is a propositional variable that occurs in both the conclusion of $\mathcal{C}$ and the premiss of $\mathcal{D}$, then we can derive from these two clauses a third clause whose premiss and conclusion are, respectively, the conjunction of the premisses and the disjunction of the conclusions of $\mathcal{C}$ and $\mathcal{D}$ from which the common variable has been deleted.

Example 4.45 Consider the two clauses:

$$\mathcal{C} = (A \wedge B \wedge C) \Rightarrow (D \vee E \vee F)$$
$$\mathcal{D} = (D \wedge A \wedge G) \Rightarrow (E \vee H).$$

We can apply the cut rule so that the variable D disappears from both the conclusion of $\mathcal{C}$ and the premiss of $\mathcal{D}$. We obtain the clause

$$(A \wedge B \wedge C \wedge A \wedge G) \Rightarrow (E \vee F \vee E \vee H)$$

which, after simplification, yields the clause

$$(A \wedge B \wedge C \wedge G) \Rightarrow (E \vee F \vee H).$$

The cut rule is justified semantically by the following proposition.

Proposition 4.46 *Suppose that $\mathcal{E}$ is derived by the cut rule from $\mathcal{C}$ and $\mathcal{D}$. Then every assignment of truth values that makes both $\mathcal{C}$ and $\mathcal{D}$ true will also make $\mathcal{E}$ true.*

Proof Take $\mathcal{C}$ and $\mathcal{D}$ as in Definition 4.44, with $B_i = C_j$, and

$$\mathcal{E} = (A_1 \wedge A_2 \wedge \cdots \wedge A_n \wedge C_1 \wedge C_2 \wedge \cdots \wedge C_{j-1} \wedge C_{j+1} \wedge \cdots \wedge C_p)$$
$$\Rightarrow (B_1 \vee B_2 \vee \cdots \vee B_{i-1} \vee B_{i+1} \vee \cdots \vee B_m \vee D_1 \vee D_2 \vee \cdots \vee D_q).$$

Let ε be an assignment of truth values for which $\varepsilon(\mathcal{E}) = \mathbf{0}$. We will show that we must then have either $\varepsilon(\mathcal{C}) = \mathbf{0}$ or $\varepsilon(\mathcal{D}) = \mathbf{0}$.

We must have

(1) $\varepsilon(A_k) = \mathbf{1}$ for all k such that $1 \leq k \leq n$;

(2) $\varepsilon(C_k) = \mathbf{1}$ for all k such that $1 \leq k \leq p$ and $k \neq j$;

(3) $\varepsilon(B_k) = \mathbf{0}$ for all k such that $1 \leq k \leq m$ and $k \neq i$;

(4) $\varepsilon(D_k) = \mathbf{0}$ for all k such that $1 \leq k \leq q$.

And since $B_i = C_j$, we have

* either $\varepsilon(C_j) = \mathbf{1}$, so $\varepsilon(\mathcal{D}) = \mathbf{0}$ (because of (2) and (4)),

* or $\varepsilon(B_i) = \mathbf{0}$, so $\varepsilon(\mathcal{C}) = \mathbf{0}$ (because of (1) and (3)). ∎

Remark 4.47

• *It is obvious that if a given variable occurs simultaneously in both the premiss and the conclusion of a clause, then this clause is a tautology. The converse is also true: if a clause is a tautology, then its premiss and its conclusion have at least one variable in common.*

• *It is useful to note that the empty clause cannot be derived from another clause by simplification.*

- *Moreover, if $\square$ is derived from clauses C and D by cut, then there must be a propositional variable A such that C is $A \Rightarrow$ and D is $\Rightarrow A$, or vice versa.*

In practice, proofs by cut are usually implemented as refutations: we show that a set of clauses is not satisfiable.

Definition 4.48 *Let C be a clause and let Γ be a set of clauses. A **derivation by cut of C from** Γ is a sequence of clauses $(D_1, D_2, \ldots, D_n)$ that ends with the clause C (i.e. $D_n = C$) and is such that for all i between 1 and n, either D_i belongs to Γ, or D_i is derived by simplification from a clause D_j where j is strictly less than i, or D_i is derived by cut from two clauses D_j and D_k where j and k are strictly less than i.*

*We say that C **is derivable by cut from** Γ if there is a derivation by cut of C from Γ.*

*A **refutation of** Γ is a derivation by cut of the empty clause $\square$ from Γ.*

*If there exists a refutation of Γ, we say that Γ **is refutable**.*

This method of proof is adequate, which means that the only things we can refute are those that are never true.

Proposition 4.49 *If Γ is refutable, then Γ is not satisfiable.*

Proof Let $(D_1, D_2, \ldots, D_n)$ be a refutation of Γ and suppose, to arrive at a contradiction, the ε is an assignment of truth values that makes all of the clauses in Γ true. We will argue by induction on i, for i between 1 and n, that $\varepsilon(D_i) = 1$. This is clear if $D_i \in \Gamma$. If D_i is derived by simplification from D_j ($j < i$), then $\varepsilon(D_j) = 1$ (by the induction hypothesis) so $\varepsilon(D_i) = 1$ because D_i and D_j are logically equivalent. Finally, if D_i is obtained by cut, we invoke Proposition 4.46.

If $\square$ belongs to Γ then, according to our conventions, Γ is not satisfiable (we may observe, by the way, that this case is of no real interest: the method that we are describing is intended to be applied to sets of 'real' propositional formulas). If not, then as we have seen, $D_n = \square$ is derivable by cut from two clauses $A \Rightarrow (= D_i)$ and $\Rightarrow A (= D_j)$ where j and i are less than n (see the last item in Remark 4.47). But this is not possible for we would have to have $\varepsilon(D_i) = 1$, which implies $\varepsilon(A) = 0$ and $\varepsilon(D_j) = 1$, which implies $\varepsilon(A) = 1$. $\blacksquare$

4.4.2 Completeness of the method

We will now show that this method is complete.

Theorem 4.50 *Any set of clauses that is not satisfiable is refutable.*

Proof Let Γ be a set of clauses that is not satisfiable. Thanks to the compactness theorem for propositional calculus (Theorem 1.39), we may assume that Γ is finite. Our argument is by induction on the number of propositional variables that appear in Γ.

For a clause C, let us denote its premiss and conclusion by C^- and C^+ respectively. We will first show that the general situation is reducible to the case where Γ satisfies the following hypotheses:

(1°) Γ does not contain any tautologies;

(2°) Γ does not contain the empty clause;

(3°) all clauses in Γ are are simplified;

(4°) for any propositional variable A that occurs in a clause of Γ, there exist two distinct clauses C and D in Γ such that A occurs in the premiss of C and in the conclusion of D.

To obtain the first three conditions, there is hardly any problem: if we delete the tautologies and replace each clause of Γ by a simplified clause, we still have a set that is not satisfiable; and if Γ contains the empty clause, then we know it is not satisfiable. To obtain condition (4°), it suffices to apply the next lemma:

Lemma 4.51 *If A is a propositional variable that does not occur in the premiss of any clause of Γ (or not in the conclusion of any clause of Γ), then:*

$$\Gamma' = \{C : C \in \Gamma \text{ and } A \text{ does not occur in } C\}$$

is not satisfiable.

Proof Suppose first that A does not occur in the premiss of any clause in Γ and, in view of arriving at a contradiction, suppose that Γ' is satisfiable. Let δ be an assignment of truth values such that $\overline{\delta}(C) = 1$ for any clause C of Γ'. Let δ' be the assignment of truth values that agrees with δ everywhere except perhaps at A, and such that $\delta'(A) = 1$. If $C \in \Gamma'$, then $\overline{\delta'}(C) = \overline{\delta}(C) = 1$ (because A does not occur in C), and if $C \in \Gamma - \Gamma'$, then $\overline{\delta'}(C) = 1$ (because A appears in the conclusion of C).

The argument is analogous when A does not appear in the conclusion of any clause of Γ; in this case, we set $\delta'(A) = 0$. ∎

The proof of the theorem continues by induction on the number n of propositional variables involved in Γ. We note that n is not zero since $\Gamma \neq \emptyset$ (the empty set is satisfiable!) and $\square \notin \Gamma$.

Suppose $n = 1$. Having eliminated tautologies and the empty clause, the only simplified clauses that involve only the variable A_1 are : $A_1 \Rightarrow$ and $\Rightarrow A_1$. Since Γ is not satisfiable, $\Gamma = \{A_1 \Rightarrow, \Rightarrow A_1\}$, so Γ is refutable.

Next, let us treat the passage from n to $n+1$. Suppose there are $n+1$ propositional variables appearing in Γ and let A be one of them. Set:

$$\Gamma_0 = \{C \in \Gamma : A \text{ does not occur in } C\};$$
$$\Gamma^- = \{C \in \Gamma : A \text{ occurs in } C^-\} = \{C_1, C_2, \ldots, C_m\};$$
$$\Gamma^+ = \{C \in \Gamma : A \text{ occurs in } C^+\} = \{D_1, D_2, \ldots, D_p\}.$$

From our hypotheses, we see that Γ is the disjoint union of Γ_0, Γ^- and Γ^+ and that neither Γ^- nor Γ^+ is empty. If i is between 1 and m and j is between 1 and p, we may apply the cut rule to $\mathcal{C}_i$ and $\mathcal{D}_j$ to eliminate the variable A; let $\mathcal{E}_{i,j}$ be the clause obtained in this way. If one of the clauses $\mathcal{E}_{i,j}$ is the empty clause, then we have found a refutation of Γ. If not, set

$$\Gamma' = \Gamma_0 \cup \{\mathcal{E}_{i,j} : 1 \le i \le m \text{ and } 1 \le j \le p\}.$$

We will show that Γ' is not satisfiable. As there are only n variables appearing in Γ' (they are all those of Γ with A omitted), the induction hypothesis will tell us that Γ' is refutable; in view of the way the $\mathcal{E}_{i,j}$ were defined, this refutation of Γ' will be immediately extendible to a refutation of Γ.

The argument is by contradiction. Suppose that δ is an assignment of truth values that satisfies Γ'. Since A does not occur in Γ', we may assume that $\delta(A) = \mathbf{0}$; it is clear that Γ' is also satisfied by the assignment λ which agrees everywhere with δ except that $\lambda(A) = \mathbf{1}$. We see that δ satisfies Γ_0 (since Γ_0 is included in Γ') and Γ^- (since $\delta(A) = \mathbf{0}$) but not Γ (which is not satisfiable). So there must exist at least one integer i between 1 and p such that $\delta(\mathcal{D}_i) = \mathbf{0}$, which implies that $\delta(\mathcal{D}_i^-) = \mathbf{1}$ and $\delta(\mathcal{D}_i^+) = \mathbf{0}$.

Fix an integer j between 1 and m. We know that δ satisfies $\mathcal{E}_{i,j}$. Hence one of the following two holds:

- $\delta(\mathcal{E}_{i,j}^-) = \mathbf{0}$; now $\mathcal{E}_{i,j}^- = \mathcal{D}_i^- \wedge \mathcal{C}'$, where $\mathcal{C}'$ is the formula obtained from $\mathcal{C}_j^-$ by suppressing the variable A; in this case, $\delta(\mathcal{C}') = \mathbf{0}$ (because $\delta(\mathcal{D}_i^-) = \mathbf{1}$) and, since δ and λ agree on all the variables except A, $\lambda(\mathcal{C}') = \mathbf{0}$, which implies $\lambda(\mathcal{C}_j^-) = \mathbf{0}$.

- $\delta(\mathcal{E}_{i,j}^+) = \mathbf{1}$; this time, $\mathcal{E}_{i,j}^+ = \mathcal{C}_j^+ \vee \mathcal{D}'$, where $\mathcal{D}'$ is obtained from $\mathcal{D}_i^+$ by suppressing the variable A; as $\delta(\mathcal{D}_i^+) = \mathbf{0}$, we must have $\delta(\mathcal{C}_j^+) = \lambda(\mathcal{C}_j^+) = \mathbf{1}$.

In both cases, $\lambda(\mathcal{C}_j) = \mathbf{1}$. Since this argument is valid for all j between 1 and m, it follows that λ satisfies Γ^-. But λ also satisfies Γ_0 (since it satisfies Γ') and Γ^+ (since $\lambda(A) = \mathbf{1}$). Hence λ satisfies Γ, which is impossible.   ∎

To determine whether a finite set Γ of clauses is satisfiable, it suffices to apply the following algorithm: first, simplify all the clauses in Γ and then eliminate all clauses that have a variable that is common to both its premise and its conclusion, then apply the cut rule systematically to all pairs of formulas in Γ in all possible ways; having done this, repeat the process. After a while, we will not obtain any new clauses (because the number of propositional variables is finite, so is the set of simplified clauses and this cannot grow forever); if we obtain □ by this process, Γ is not satisfiable; otherwise it is. Obviously, if we wish to undertake this work ourselves, or even ask a computer to do it, we will have to employ a slightly more subtle strategy than the one we just described. But that is another story.

Example 4.52 We use the cut rule to derive the clause $B \Rightarrow$ from the clauses $(A \wedge B) \Rightarrow C, \Rightarrow A$ and $C \Rightarrow$ (in other words, derive $\neg B$ from $(A \wedge B) \Rightarrow C, A$

and $\neg C$).

From $(A \wedge B) \Rightarrow C$ and $C \Rightarrow$, we derive $(A \wedge B) \Rightarrow$.

From $(A \wedge B) \Rightarrow$ and $\Rightarrow A$, we derive $B \Rightarrow$.

Example 4.53 We show that the following set of clauses is not satisfiable:

$$C_1 \text{ is } (A \wedge B) \Rightarrow (C \vee D)$$
$$C_2 \text{ is } (C \wedge E \wedge F) \Rightarrow$$
$$C_3 \text{ is } (A \wedge D) \Rightarrow$$
$$C_4 \text{ is } \Rightarrow (B \vee C)$$
$$C_5 \text{ is } \Rightarrow (A \vee C)$$
$$C_6 \text{ is } C \Rightarrow E$$
$$C_7 \text{ is } C \Rightarrow F.$$

Here is a refutation of $\{C_1, \ldots, C_7\}$:

(1)	$(C \wedge C \wedge F) \Rightarrow$	from C_2 and C_6
(2)	$(C \wedge F) \Rightarrow$	from (1) by simplification
(3)	$(C \wedge C) \Rightarrow$	from (2) and C_7
(4)	$C \Rightarrow$	from (3) by simplification
(5)	$\Rightarrow A$	from (4) and C_5
(6)	$\Rightarrow B$	from (4) and C_4
(7)	$D \Rightarrow$	from (5) and C_3
(8)	$B \Rightarrow (C \vee D)$	from (5) and C_1
(9)	$\Rightarrow (C \vee D)$	from (8) and (6)
(10)	$\Rightarrow D$	from (9) and (4)
(11)	$\square$	from (10) and (7).

4.5 The method of resolution

4.5.1 Unification

In this section, we will introduce the technique of unification which will be needed to extend the method of proofs by cut to the predicate calculus. Here is the problem: in a given language containing function symbols, we have two terms $t_1[v_1, v_2, \ldots, v_n]$ and $t_2[w_1, w_2, \ldots, w_m]$ in which the v_i and the w_j are variables; the problem is to decide whether there exist terms $a_1, a_2, \ldots, a_n$ and $b_1, b_2, \ldots, b_m$ such that the terms $t_1[a_1, a_2, \ldots, a_n]$ and $t_2[b_1, b_2, \ldots, b_m]$ are identical, and if so, to find all possible solutions. This is what is called unifying the terms t_1 and t_2. We will proceed in a rather formal way.

Let L be a language with no predicate symbols and let V be a fixed subset of the set of variables. Let $T(V)$ denote the set of terms of L whose variables all belong to V. Most of the time, V will either be determined by the context or else will be of no particular importance, in which case we will not even mention it and simply write T in place of $T(V)$. We can, in a natural way, define an L-structure whose underlying set is $T(V)$; we will denote it by $\mathfrak{T}(V)$ (or $\mathfrak{T}$): if c is a constant symbol, the interpretation of c in $\mathfrak{T}$ is c itself; if f is a function symbol of arity n, then the interpretation $f^{\mathfrak{T}}$ of f in $\mathfrak{T}$ is the function from T^n into T defined by

$$f^{\mathfrak{T}}(t_1, t_2, \ldots, t_n) = f t_1 t_2 \ldots t_n.$$

For example, if L has only a single unary function symbol (and no constant symbols), and if $V = \{v_i : i \in \mathbb{N}\}$, then

$$T = \{f^n v_i : n \in \mathbb{N}, \ i \in \mathbb{N}\},$$

where it is understood that f^n is an abbreviation for the sequence consisting of n occurrences of the symbol f (it is the empty sequence if $n = 0$).

Let us recall Definition 3.31 which, in our present situation where there are no predicate symbols in L, becomes

Definition 4.54 *Let $\mathcal{M} = \langle M, \ldots \rangle$ and $\mathcal{N} = \langle N, \ldots \rangle$ be two L-structures and let α be a map from M into N. The map α is a **homomorphism** if and only if the following conditions are satisfied:*

- *for every constant symbol c of L,*

$$\alpha(\bar{c}^{\mathcal{M}}) = \bar{c}^{\mathcal{N}};$$

- *for every natural number $n \geq 1$, for every n-ary function symbol f of L and for all elements $a_1, a_2, \ldots, a_n$ belonging to M,*

$$\alpha\big(\bar{f}^{\mathcal{M}}(a_1, a_2, \ldots, a_n)\big) = \bar{f}^{\mathcal{N}}(\alpha(a_1), \alpha(a_2), \ldots, \alpha(a_n)).$$

The structure $\mathfrak{T}(V)$ is **freely generated** by the set V, which, to be precise, means the following:

Proposition 4.55 *Let $\mathcal{M} = \langle M, \ldots \rangle$ be an L-structure and let α be an arbitrary map from V into M. Then there is a unique homomorphism from $\mathfrak{T}(V)$ into $\mathcal{M}$ which extends α.*

Proof We are going to define a map β from $T(V)$ into M by induction on terms:

(i) if $t = c$ is a constant symbol, the set $\beta(t) = \bar{c}^{\mathcal{M}}$;

(ii) if $t = v_i$ is a variable, then $\beta(t) = \alpha(v_i)$;

(iii) if $t = f t_1 t_2 \ldots t_n$, where f is an n-ary function symbol and the t_i, for i between 1 and n, are terms for which $\beta(t_i)$ has already been defined, then

$$\beta(t) = \bar{f}^{\mathcal{M}}(\beta(t_1), \beta(t_2), \ldots, \beta(t_n)).$$

The map β defined in this way clearly extends α (because of (ii)) and is a homomorphism because of conditions (i) and (iii). Moreover, if β' is another homomorphism that extends α, then for every term t, $\beta(t) = \beta'(t)$: this is easily proved by induction on t. ∎

Definition 4.56 *Homomorphisms from $\mathfrak{T}$ into itself are called **substitutions**.*

This name is entirely justified: for if α is a map from V into $\mathcal{T}(V)$, with $\alpha(v_n) = u_n$, and β is the unique homomorphism of $\mathfrak{T}$ into itself that extends α, then β is none other than the map which sends a term $t[v_1, v_2, \ldots, v_n]$ into $t_{u_1/v_1, u_2/v_2, \ldots, u_n/v_n}$. A substitution is therefore entirely determined by the values it assumes on the set V of variables. Obviously, we will not bother to use two distinct notations to denote a map from V into $\mathcal{T}$ and the unique substitution that extends it.

Definition 4.57 *Let $S = \{(t_1, u_1), (t_2, u_2), \ldots, (t_n, u_n)\}$ be a finite set of pairs of terms. A **unifier of S** is a substitution σ such that, for every i between 1 and n, $\sigma(t_i) = \sigma(u_i)$. To **unify** S is to find all the unifiers of S. A finite set of pairs of terms will also be called a **system**.*

Remark 4.58 *If σ is a unifier of S and if τ is an arbitrary substitution, then $\tau \circ \sigma$ is also a unifier of S: it is clear that if $\sigma(t_i) = \sigma(u_i)$, then $\tau \circ \sigma(t_i) = \tau \circ \sigma(u_i)$.*

Example 4.59 Suppose that $V = \{v_1, v_2, v_3, v_4\}$, that c and d are constant symbols, that h is a unary function symbol and that f and g are binary function symbols.

- $S = \{(fv_1hv_2, gv_2gv_1v_3)\}$. For any substitution σ, the term

$$\sigma(fv_1hv_2) = f\sigma(v_1)h\sigma(v_2)$$

begins with the symbol f whereas the term

$$\sigma(gv_2gv_1v_3) = g\sigma(v_2)g\sigma(v_1)\sigma(v_3)$$

begins with g: so there is no unifier.
- $S = \{(v_1, gv_1v_2)\}$. Here again, there is no unifier: indeed, for any substitution σ, $\sigma(gv_1v_2) = g\sigma(v_1)\sigma(v_2)$ is a term that is strictly longer than $\sigma(v_1)$.
- $S = \{(fv_1gv_2v_3, fhv_3v_4)\}$. Let σ be a substitution and suppose, for $v_i \in V$, that $\sigma(v_i) = u_i$. For σ to be a unifier, it is necessary and sufficient that the terms $fu_1gu_2u_3$ and fhu_3u_4 be identical, hence (Theorem 3.7) that

$$u_1 = hu_3 \text{ and } gu_2u_3 = u_4.$$

So we see that the values of u_2 and u_3 can be arbitrary, but once these are fixed, these equations determine the values for u_1 and u_4. So here is a unifier, π, that is the simplest one that comes to mind:

$$\pi(v_2) = v_2, \quad \pi(v_3) = v_3, \quad \pi(v_1) = hv_3, \quad \pi(v_4) = gv_2v_3.$$

We have already seen that any substitution of the form $\tau \circ \pi$ is also a unifier. In fact, this gives us all the unifiers (we say, in this situation, that π is a principal unifier; the next definition will treat this in detail): for suppose that σ is an arbitrary unifier of S and that τ is a substitution such that $\tau(v_2) = \sigma(v_2)$ and $\tau(v_3) = \sigma(v_3)$. We will show that $\sigma = \tau \circ \pi$. Indeed, if i is equal to 2 or to 3, then $\sigma(v_i) = \tau(v_i) = \tau \circ \pi(v_i)$. Also, $\sigma(v_1) = h\sigma(v_3)$ (because σ is a unifier) and $h\sigma(v_3) = h\tau(v_3) = \tau(hv_3)$ (because τ is a substitution) and we know that $hv_3 = \pi(v_1)$. So the conclusion is that $\sigma(v_1) = \tau \circ \pi(v_1)$. The proof that $\sigma(v_4) = \tau \circ \pi(v_4)$ is similar.

Definition 4.60 *We say that π is a **principal unifier** of a system S if it is a unifier of S and if for every unifier σ of S, there is a substitution τ such that $\sigma = \tau \circ \pi$.*

The existence of a principal unifier is not a special case.

Proposition 4.61 *Every system that has a unifier has a principal unifier.*

Proof Let $S = \{(t_1, u_1), (t_2, u_2), \dots, (t_n, u_n)\}$ be a non-empty system.

Let $V(S)$ denote the (finite) set of variables that occur at least once in some term that appears in S and let Uni(S) denote the set of unifiers of S. Two systems will be called equivalent if they have the same set of unifiers. We will say that a term t **appears efficiently** in S if S contains a pair of the form (t, u) or (u, t) with $t \neq u$. The height of a system is the smallest of the heights of the terms that appear efficiently in S.

We are going to describe an algorithm that will permit us to

- either prove that S does not have a unifier;
- or else find a system, S^1, and a substitution τ such that $V(S^1)$ has strictly fewer elements than $V(S)$ and Uni(S)=$\{\sigma \circ \tau : \sigma \in \text{Uni}(S^1)\}$.

This algorithm involves three stages: **clean-up, simplification**, and **reduction**.

(A) *Clean-up (or garbage collection).* In this stage, we first eliminate from S all pairs of the form (t, t); next, if a pair occurs more than once, we only keep a single occurrence; finally, if both pairs (t, u) and (u, t) belong to the system under consideration, we keep only one of them. We then have a new system S_1 that is equivalent to S and has the same height.

(B) *Simplification.* This stage will allow us either to find an equivalent system whose height is 1 or to conclude that unification is impossible. Let h be the height of S_1 and assume it is greater than or equal to 2 (otherwise there is nothing to prove). Choose a pair (t, u) in S_1 such that the height of t, say, is equal to h. We know that the height of u is greater than or equal to h and that, hence, the first symbol of both t and u is a function symbol. We then perform the **first compatibility test**: if t and u begin with different function symbols, there is no point in continuing since neither S_1 nor S has a unifier.

Otherwise, we may write $t = f r_1 r_2 \dots r_k$ and $u = f s_1 s_2 \dots s_k$, where k is the arity of f. The system S_2 obtained from S_1 by replacing the single pair (t, u) by the

k pairs $(r_1, s_1), (r_2, s_2), \ldots, (r_k s_k)$ is equivalent to S_1. Also note that the height of each r_i is strictly less than h. As we have assumed $t \neq u$, there exists an i between 1 and k such that $r_i \neq s_i$. So by performing another clean-up, we obtain a system S_2, equivalent to S, whose height is strictly less than h and which, moreover, satisfies $V(S_2) \subseteq V(S)$ since we have not introduced any new variables.

By iterating this process, we will, unless we are led sooner to conclude that S has no unifier, arrive at a system S_3, equivalent to S, whose height is 1 and is such that $V(S_3) \subseteq V(S)$.

So let $(x_1, y_1) \in S_3$, $x_1 \neq y_1$, with the height of x_1 equal to 1.

(C) *Reduction.*   We will now concentrate on the unification of (x_1, y_1). We begin with a few tests in order to eliminate certain cases:

- the **second compatibility test:** if x_1 is a constant symbol and y_1 is not a symbol for a variable, there is no unifier.

- the **occurrence test:** if x_1 is a variable, say v_i, and if v_i occurs in y_1 (but is not equal to y_1 since $x_1 \neq y_1$), then for any substitution σ, $\sigma(v_1)$ is a proper sub-term of $\sigma(y_1)$, so $\sigma(x_1) = \sigma(y_1)$ is impossible and there is no unifier.

Except for these cases, and by interchanging x_1 and y_1 if necessary, we obtain a pair (v_i, y_1) such that the variable v_i does not occur in y_1. The unification of x_1 and y_1 is then possible: let τ_1 be the substitution defined by

- $\tau_1(v_i) = y_1$
- if $j \neq i$, $\tau_1(v_j) = v_j$.

Since τ_1 leaves all the variables that occur in y_1 fixed , $\tau_1(y_1) = y_1$. It follows that $\tau_1(v_i) = \tau_1(y_1)$, so τ_1 is a unifier of (v_i, y_1). Better yet, it is a principal unifier: for suppose σ is a unifier of (v_i, y_1). We will show that $\sigma = \sigma \circ \tau_1$. Indeed, on the one hand, if $j \neq i$, $\tau_1(v_j) = v_j$ and $\sigma \circ \tau_1(v_j) = \sigma(v_j)$; on the other hand, $\sigma(v_i) = \sigma(y_1)$ (because σ is a unifier of (v_i, y_1)) and $\tau_1(v_i) = y_1$ and hence $\sigma \circ \tau_1(v_i) = \sigma(y_1) = \sigma(v_i)$. So we have in fact found a substitution σ', namely σ, such that $\sigma = \sigma' \circ \tau_1$.

Let us return now to the system S_3. All the unifiers of S_3 are, in particular, unifiers of (x_1, y_1) so must be of the form $\sigma \circ \tau_1$, where τ_1 is the substitution defined above. Let us list S_3:

$$S_3 = \{(x_1, y_1), (x_2, y_2), \ldots, (x_m, y_m)\}.$$

For $\sigma \circ \tau_1$ to be a unifier of S_3, it is necessary and sufficient that for all i between 1 and m, $\sigma(\tau_1(x_i)) = \sigma(\tau_1(y_i))$. We already know, from our choice of τ_1, that $\sigma(\tau_1(x_1)) = \sigma(\tau_1(y_1))$. So it is necessary and sufficient that σ be a unifier of

$$S^1 = \{(\tau_1(x_2), \tau_1(y_2)), (\tau_1(x_3), \tau_1(y_3)), \ldots, \tau_1(x_m), \tau_1(y_m)\}.$$

Now exactly which variables can occur in $\tau_1(x_k)$ or in $\tau_1(y_k)$ for $2 \leq k \leq m$? Only those variables that occur in a term $\tau_1(v_j)$ such that v_j itself occurs in one of the terms x_k or y_k for k between 2 and m; i.e. such that v_j belongs to $V(S_3)$.

Referring back to the definition of τ_1, we see that $V(S^1)$ does not contain v_i and is included in $V(S_3)$. We have thereby kept our promises.

It then suffices to repeat these three operations A, B, and C, to eliminate all the free variables: either we come upon an impossibility or we find systems S^1, S^2, ..., S^k, where $V(S^k)$ is empty, and substitutions τ_1, τ_2, ..., τ_k such that for all i between 1 and $k - 1$,

$$\text{Uni}(S^i) = \{\sigma \circ \tau_{i+1} : \sigma \in \text{Uni}(S^{i+1})\}.$$

At this point, the terms appearing in S^k are all closed terms (since $V(S^k)$ is empty); hence, they are left fixed by any substitution. So, if there exists a pair $(t, u) \in S^k$ such that $t \neq u$, then neither S^k, nor S, consequently, has a unifier. If no such pair exists, then all substitutions are unifiers of S^k and the unifiers of S are then exactly those substitutions of the form $\sigma \circ \tau_k \circ \cdots \circ \tau_2 \circ \tau_1$ where σ is any substitution; $\tau_k \circ \cdots \circ \tau_2 \circ \tau_1$ is then a principal unifier of S. ∎

This proof furnishes us with an algorithm that can be used to decide whether a system has unifiers and, if it does, to find a principal unifier. We have to alternate two sub-algorithms: on the one hand, the operations of clean-up and simplification produce an equivalent system that contains no pairs of the form (t, t) and which involves at least one term of height 1; on the other hand, the reduction operation decreases the number of variables.

Example 4.62 In this example, c is a constant symbol, f and g are binary function symbols and k is a ternary function symbol. We will proceed to unify the system S:

$$S = \{(kf cg v_4 v_3 f cg v_3 v_4 k v_3 v_4 v_2 \, , \; k v_2 v_2 v_1)\}.$$

1B: Simplification. The system S is equivalent to

$$\{(f cg v_4 v_3, \; v_2), \; (f cg v_3 v_4, \; v_2) \, , \; (k v_3 v_4 v_2, \; v_1)\}.$$

1C: Reduction. Set $\tau_1(v_2) = f cg v_4 v_3$ and $\tau_1(v_i) = v_i$ for $i \neq 2$. We obtain the system

$$S^1 = \{(f cg v_3 v_4, \; f cg v_4 v_3), \; (k v_3 v_4 f cg v_4 v_3, \; v_1)\}.$$

2A and 2B: There is nothing to do: the system is simplified.

2C: Reduction. Set $\tau_2(v_1) = k v_3 v_4 f cg v_4 v_3$ and $\tau_2(v_i) = v_i$ for $i \neq 1$. We obtain the system

$$S^2 = \{(f cg v_3 v_4, \; f cg v_4 v_3)\}.$$

3B: Simplification. We obtain, in succession, systems equivalent to S^2:

$$\{(c, c), \; (g v_3 v_4, \; g v_4 v_3)\},$$

then

$$\{(c, c), \; (v_3, v_4), \; (v_4, v_3)\};$$

and after a clean-up, we see that S^2 is equivalent to $\{(v_3, v_4)\}$.

3C: Reduction. We now set $\tau_3(v_4) = v_3$ and $\tau_3(v_i) = v_i$ for $i \neq 4$. The system S^3 is empty. The substitution $\tau = \tau_3 \circ \tau_2 \circ \tau_1$ is a principal unifier of S We may calculate

$$\tau(v_1) = kv_3v_3fcgv_3v_3, \quad \tau(v_2) = fcgv_3v_3, \quad \tau(v_3) = v_3, \quad \tau(v_4) = v_3.$$

Remark 4.63 *We could have calculated this differently (for example, by reducing the pair $(kv_3v_4v_2, v_1)$ at stage 1B or by setting $\tau_3(v_3) = v_4$ at stage 3B); we would have found a different principal unifier. Exercise 15 explains how to find all principal unifiers of a system once one is known.*

Remark 4.64 *For some systems, there may arise the possibility of performing several reductions at one time. For example, suppose S contains the pairs (v_1, t_1) and (v_2, t_2). If these pairs satisfy the occurrence and the compatibility tests and if, moreover, v_2 does not occur in t_1 nor v_1 in t_2, then we may reduce using the substitution τ defined by $\tau(v_1) = t_1$, $\tau(v_2) = t_2$, and $\tau(v_i) = v_i$ for i different from 1 and 2.*

4.5.2 Proofs by resolution

We are going to adapt the method of proof by cuts to the predicate calculus. As with the propositional calculus, we deal in practice with refutations rather than derivations and the method will only apply to a restricted class of formulas, the class of universal clauses. We fix a language L (which may contain relation symbols) and denote the set of terms of L by $\mathcal{T}$. Let us agree that when we speak of substitutions in what follows, we mean substitutions in the sense of Definition 4.56 relative to the language L without its relation symbols.

Definition 4.65 *A **universal clause** is a closed formula of the following type:*

$$\forall v_1 \forall v_2 \ldots \forall v_k(\neg A_1 \vee \neg A_2 \vee \cdots \vee \neg A_n \vee B_1 \vee B_2 \vee \cdots \vee B_m),$$

where the A_i and B_j are atomic formulas.

We will often be content to write 'clause' rather than 'universal clause'. We will also adopt several conventions which simplify the writing of clauses:

(1°) We will not write the quantifiers: this is justified by the fact that all the quantifiers are universal and all the variables are quantified. The only ambiguity is then in the order in which the variables are quantified and this, in fact, has no importance.

(2°) Along with our convention to omit writing the universal quantifiers and just as for the propositional calculus, we will use the notation:

$$(A_1 \wedge A_2 \wedge \cdots \wedge A_n) \Rightarrow (B_1 \vee B_2 \vee \cdots \vee B_m)$$

to denote the universal clause:

$$\forall v_1 \forall v_2 \ldots \forall v_k(\neg A_1 \vee \neg A_2 \vee \cdots \vee \neg A_n \vee B_1 \vee B_2 \vee \cdots \vee B_m).$$

As before, $(A_1 \wedge A_2 \wedge \cdots \wedge A_n)$ will be called the **premiss** and $(B_1 \vee B_2 \vee \ldots \vee B_m)$ the **conclusion** of the clause.

(3°) It is possible that the integers n or m or both are zero. We adopt the same notations as in the previous subsection: $(A_1 \wedge A_2 \wedge \ldots \wedge A_n) \Rightarrow$ is the clause:

$$\forall v_1 \forall v_2 \ldots \forall v_k (\neg A_1 \vee \neg A_2 \vee \ldots \vee \neg A_n)$$

and $\Rightarrow (B_1 \vee B_2 \vee \ldots \vee B_m)$ is the clause:

$$\forall v_1 \forall v_2 \ldots \forall v_k (B_1 \vee B_2 \vee \cdots \vee B_m).$$

If n and m are both zero, we have, by convention, the empty clause which will again be denoted by $\square$.

Let σ be a substitution of $\mathcal{T}$ (the set of terms of L). Then σ also acts on formulas: if F is a formula whose free variables are $v_1, v_2, \ldots, v_k$, then by definition, $\sigma(F)$ is the formula

$$A_{x_1/v_1, x_2/v_2, \ldots, x_k/v_k}$$

where, for i between 1 and k, $x_i = \sigma(v_i)$. For example, if A is an atomic formula, it has the form

$$Rt_1 t_2 \ldots t_j,$$

where R is a predicate symbol of arity j and where $t_1, t_2, \ldots, t_j$ are terms; in this case,

$$\sigma(A) = R\sigma(t_1)\sigma(t_2)\ldots\sigma(t_j).$$

If $\mathcal{C}$ is a universal clause, say $\mathcal{C} = (A_1 \wedge A_2 \wedge \cdots \wedge A_n) \Rightarrow (B_1 \vee B_2 \vee \cdots \vee B_m)$, then, by definition,

$$\sigma(\mathcal{C}) = (\sigma(A_1) \wedge \sigma(A_2) \wedge \cdots \wedge \sigma(A_n)) \Rightarrow (\sigma(B_1) \vee \sigma(B_2) \vee \cdots \vee \sigma(B_m)).$$

Remark 4.66 *From the definition of satisfaction of a formula in a structure (Definition 3.42), we immediately deduce the following very important property: if F is a universal clause and if $\mathcal{M}$ is a model of F, then for every substitution σ, $\mathcal{M}$ is also a model of $\sigma(F)$.*

Unification, in turn, applies to formulas. Suppose $A_1, A_2, \ldots, A_n$ are atomic formulas. We wish to determine all substitutions σ such that $\sigma(A_1) = \sigma(A_2) = \cdots = \sigma(A_n)$; these will be called unifiers of $(A_1, A_2, \ldots, A_n)$. Since the A_i are atomic formulas, they can be written

$$A_i = R_i t_1^i t_2^i \ldots t_{m_i}^i,$$

where the R_i are predicate symbols of arity m_i and the t_j^i are terms. We have seen that if σ is a substitution, then

$$\sigma(A_i) = R_i\sigma\left(t_1^i\right)\sigma\left(t_2^i\right)\ldots\sigma\left(t_{m_i}^i\right).$$

Thus if, for distinct integers i and j between 1 and n, R_i is different from R_j, then there are no unifiers. In the opposite case, all the m_i must equal a common value, which we will call m, and the unifiers of $(A_1, A_2, \ldots, A_n)$ are precisely the unifiers of the set

$$\left\{\left(t_k^1, t_k^i\right) : 1 \leq k \leq m,\ 1 < i \leq n\right\},$$

so we may apply the results of the previous subsection: either there is no unifier or else there is a principal unifier.

Definition 4.67 *Two clauses are said to be* **separated** *if they do not have any variable in common.*

If C and D are two clauses and the set V of variables is infinite, then we can find a permutation σ of V (i.e. a bijection of V onto itself) such that C and $\sigma(D)$ are separated. It suffices to define σ so that for every variable v_i that occurs in D, $\sigma(v_i)$ is a variable that does not occur in C.

We are now in position to describe the **rule of resolution**.

Given two clauses C and D, we will explain what we mean by '**a clause $\mathcal{E}$ is derived from C and D by resolution**'.

The situation is analogous to the one for propositional calculus where we explained the cut rule. The role played there by the propositional variables is played here by the atomic formulas. The essential difference in the method is that in the present context, we will not require that some specific atomic formula appear both in the premiss of one and in the conclusion of the other; we will only require that we can reduce the situation to this case by means of a unification.

Starting with C and D (it is quite possible that $C = D$), we begin by separating them, i.e. we replace D by a clause $D' = \tau(D)$, where τ is a permutation of variables such that C and D' are separated, as we explained above. Suppose that C and D' are written:

$$C = (A_1 \wedge A_2 \wedge \cdots \wedge A_n) \Rightarrow (B_1 \vee B_2 \vee \cdots \vee B_m) \text{ and}$$
$$D' = (C_1 \wedge C_2 \wedge \cdots \wedge C_p) \Rightarrow (D_1 \vee D_2 \vee \cdots \vee D_q).$$

Suppose that certain formulas in the conclusion of C can be unified with certain formulas in the premiss of D'; more precisely, suppose there exist a non-empty subset X of $\{1, 2, \ldots, m\}$ and a non-empty subset Y of $\{1, 2, \ldots, p\}$ such that the set $\{B_i : i \in X\} \cup \{C_j : j \in Y\}$ has a unifier. Suppose then that σ is a principal unifier of $\{B_i : i \in X\} \cup \{C_j : j \in Y\}$.

The rule of resolution permits, in this circumstance, the derivation of the universal clause $\mathcal{E}$ from the clauses $\mathcal{C}$ and $\mathcal{D}$, where:

- the premiss of $\mathcal{E}$ is the conjunction of $\sigma(A_1) \wedge \sigma(A_2) \wedge \cdots \wedge \sigma(A_n)$ and the formulas $\sigma(C_h)$ for $h \in \{1, 2, \ldots, m\} - X$;
- the conclusion of $\mathcal{E}$ is the disjunction of $\sigma(D_1) \vee \sigma(D_2) \vee \cdots \vee \sigma(D_q)$ and the formulas $\sigma(B_k)$ for $k \in \{1, 2, \ldots, p\} - Y$.

We could also say that $\mathcal{E}$ is obtained from $\mathcal{C}$ and $\mathcal{D}'$ by simplification and cut (in the sense of the previous section).

Remark 4.68 *In the rule of resolution, we insist that the unification is performed with a principal unifier. The reason for this lies in computer science: it is much easier to write an algorithm that searches for a principal unifier than to write one that searches for all unifiers.*

Example 4.69 The language consists of one unary function symbol h, a binary function symbol f, a binary predicate symbol R and a ternary predicate symbol P. Consider the following clauses:

$$\mathcal{C} \text{ is } Pv_1v_2v_3 \Rightarrow (Rfv_1v_2v_3 \vee Rfv_1v_2hfv_1v_2)$$
$$\mathcal{D} \text{ is } Rv_1hv_1 \Rightarrow Pv_1v_1v_1.$$

We begin by separating the clauses: so we replace $\mathcal{D}$ by $\mathcal{D}'$:

$$\mathcal{D}' \text{ is } Rv_4hv_4 \Rightarrow Pv_4v_4v_4.$$

It is easy to unify $Pv_1v_2v_3$ in the premiss of $\mathcal{C}$ with $Pv_4v_4v_4$ in the conclusion of $\mathcal{D}'$: a principal unifier of $\{(v_1,v_4),(v_2,v_4),(v_3,v_4)\}$ is σ:

$$\sigma(v_1) = \sigma(v_2) = \sigma(v_3) = \sigma(v_4) = v_4.$$

Then

$$\sigma(\mathcal{C}) \text{ is } Pv_4v_4v_4 \Rightarrow (Rfv_4v_4v_4 \vee Rfv_4v_4hfv_4v_4)$$

and

$$\sigma(\mathcal{D}') \text{ is } Rv_4hv_4 \Rightarrow Pv_4v_4v_4.$$

From the rule of resolution, we now derive the clause

$$Rv_4hv_4 \Rightarrow (Rfv_4v_4v_4 \vee Rfv_4v_4hfv_4v_4).$$

It is also possible to unify $Rfv_1v_2v_3$, $Rfv_1v_2hfv_1v_2$, and Rv_4hv_4. For this, we must find a principal unifier of

$$\{(fv_1v_2, fv_1v_2), (v_3, hfv_1v_2), (fv_1v_2, v_4), (v_3, hv_4)\}.$$

The reader should verify that the substitution τ that follows is a principal unifier:

$$\tau(v_1) = v_1; \quad \tau(v_2) = v_2; \quad \tau(v_3) = hfv_1v_2; \quad \tau(v_4) = fv_1v_2.$$

We then have

$$\tau(\mathcal{C}) = Pv_1v_2hfv_1v_2 \Rightarrow (Rfv_1v_2hfv_1v_2 \vee Rfv_1v_2hfv_1v_2);$$

and

$$\tau(\mathcal{D}') = Rfv_1v_2hfv_1v_2 \Rightarrow Pfv_1v_2fv_1v_2fv_1v_2.$$

The rule of resolution then allows us to derive

$$Pv_1v_2hfv_1v_2 \Rightarrow Pfv_1v_2fv_1v_2fv_1v_2.$$

The next proposition expresses the fact that the rule of resolution is semantically justified.

Proposition 4.70 *Suppose that the universal clause $\mathcal{E}$ is derivable from clauses $\mathcal{C}$ and $\mathcal{D}$ by an application of the rule of resolution. If $\mathcal{M}$ is a model of $\mathcal{C}$ and $\mathcal{D}$, then $\mathcal{M}$ is also a model of $\mathcal{E}$.*

Proof Clearly, we may assume that $\mathcal{C}$ and $\mathcal{D}$ are separated. So there is a substitution σ such that $\mathcal{E}$ is obtained from $\sigma(\mathcal{C})$ and $\sigma(\mathcal{D})$ by a cut, possibly after a simplification. Let $\mathcal{M} = \langle M, \ldots \rangle$ be a model of $\mathcal{C}$ and $\mathcal{D}$. We have seen (Remark 4.66) that $\mathcal{M}$ is also a model of $\sigma(\mathcal{C})$ and $\sigma(\mathcal{D})$. Let us be explicit: suppose

$$\sigma(\mathcal{C}) = (A_1 \wedge A_2 \wedge \cdots \wedge A_n) \Rightarrow (B_1 \vee B_2 \vee \cdots \vee B_m) \text{ and}$$
$$\sigma(\mathcal{D}) = (C_1 \wedge C_2 \wedge \cdots \wedge C_p) \Rightarrow (D_1 \vee D_2 \vee \cdots \vee D_q),$$

and suppose that the cut is performed on the formulas B_i for $i \in X$ where $X \subseteq \{1, 2, \ldots, m\}$ and C_j for $j \in Y$ where $Y \subseteq \{1, 2, \ldots, p\}$. Thus $\mathcal{E}$ is equivalent to the formula

$$\mathcal{E}' = \left(A_1 \wedge A_2 \wedge \cdots \wedge A_n \wedge \left(\bigwedge_{h \in I} C_h \right) \right)$$
$$\Rightarrow \left(D_1 \vee D_2 \vee \cdots \vee D_q \vee \left(\bigvee_{k \in J} B_k \right) \right)$$

where $I = \{1, 2, \ldots, m\} - X$ and $J = \{1, 2, \ldots, p\} - Y$.

Suppose that the variables appearing in these clauses are $v_1, v_2, \ldots, v_k$ and that $a_1, a_2, \ldots, a_k$ are elements of M. Set

$$A_i' = (A_i)_{a_1/v_1, a_2/v_2, \ldots, a_k/v_k}, \quad B_i' = (B_i)_{a_1/v_1, a_2/v_2, \ldots, a_k/v_k}, \text{ etc.}$$

(Note that we are passing to the language L_M.)

We then have

$$\mathcal{M} \models (A'_1 \wedge A'_2 \wedge \cdots \wedge A'_n) \Rightarrow (B'_1 \vee B'_2 \vee \cdots \vee B'_m) \text{ and}$$
$$\mathcal{M} \models (C'_1 \wedge C'_2 \wedge \cdots \wedge C'_p) \Rightarrow (D'_1 \vee D'_2 \vee \cdots \vee D'_q).$$

By using simplification and the cut rule in the context of propositional calculus (Definition 4.44), we obtain

$$\mathcal{M} \models \left(A'_1 \wedge A'_2 \wedge \cdots \wedge A'_n \wedge \left(\bigwedge_{h \in I} C'_h \right) \right)$$
$$\Rightarrow \left(D'_1 \vee D'_2 \vee \cdots \vee D'_q \vee \left(\bigvee_{k \in J} B'_k \right) \right)$$

and hence,

$$\mathcal{M} \models \mathcal{E}'_{a_1/v_1, a_2/v_2, \ldots, a_k/v_k}.$$

∎

Definition 4.71 *Let Γ be a set of clauses. A **refutation** of Γ is a sequence of clauses $S = \{\mathcal{D}_1, \mathcal{D}_2, \ldots, \mathcal{D}_n\}$ that ends with $\square$ and is such that for every i between 1 and n, either $\mathcal{D}_i$ belongs to Γ, or $\mathcal{D}_i$ is derivable from two clauses that precede it in S by the rule of resolution. We say that Γ is **refutable** if there exists a refutation of Γ.*

It follows from what we said, as in Proposition 4.49, that if Γ is refutable, then Γ does not have a model. It is the converse that we are concerned with here. So if Γ is not refutable, we need to construct a model of Γ. We will reduce this to the case for propositional calculus by using a simplification of Herbrand's method.

Recall that $V = \{v_k : k \in \mathbb{N}\}$ is the set of variables and $\mathcal{T}$ is the set of terms of L. Since there are no quantifiers that require manipulation, we need not be as scrupulous as we were in Section 4.3 and we will construct a model whose underlying set is precisely $\mathcal{T}$. The interpretation of the function symbols is defined as in the proof of Theorem 4.38: if f is a function symbol of arity n, then, naturally enough, the interpretation of f is the function from $\mathcal{T}^n$ into $\mathcal{T}$ that sends $(t_1, t_2, \ldots, t_n)$ to $ft_1t_2 \ldots t_n$. Let $\mathcal{P}$ be the set of atomic formulas. To completely define our L-structure, it remains to decide, for every integer n, for every n-ary predicate symbol R and for every sequence $(t_1, t_2, \ldots, t_n)$, whether $Rt_1t_2 \ldots t_n$ is true or not. All these decisions are independent from one another and may be taken arbitrarily. In other words, for every function δ from $\mathcal{P}$ into $\{0,1\}$, we construct an L-structure $\mathcal{M}_\delta$ in the following way: if R is an n-ary predicate symbol and if $t_1, t_2, \ldots, t_n \in \mathcal{T}$, then $(t_1, t_2, \ldots, t_n)$ belongs to the interpretation of R in $\mathcal{M}_\delta$ (i.e. $\mathcal{M}_\delta \models Rt_1t_2 \ldots t_n$) if and only if $\delta(Rt_1t_2 \ldots t_n) = 1$.

We will consider the elements of $\mathcal{P}$ as propositional variables and δ as an assignment of truth values. We extend δ in a canonical way to a map $\overline{\delta}$ which assigns truth values to the quantifier-free formulas of L and so, if F is one of these formulas without quantifiers,

$$\mathcal{M}_\delta \models F \text{ if and only if } \bar{\delta}(F) = 1.$$

Now let $G = \forall v_1 \forall v_2 \ldots \forall v_n F[v_1, v_2, \ldots, v_n]$ be a universal formula. Under what conditions is $\mathcal{M}_\delta$ a model of G? The response is simple: it is necessary and sufficient that for all sequences $(t_1, t_2, \ldots, t_n)$ of elements of $\mathcal{T}$, we have

$$\mathcal{M}_\delta \models F[t_1, t_2, \ldots, t_n],$$

which amounts to saying that for any substitution σ, $\mathcal{M}_\delta \models \sigma(F)$, or, alternatively, $\bar{\delta}(\sigma(F)) = 1$.

It remains to prove:

Theorem 4.72 *Let $\Gamma = \{\mathcal{C}_i : 1 \leq i \leq n\}$ be a set of universal clauses that has no model; then Γ is refutable.*

Proof The set

$$X = \{\sigma(\mathcal{C}_i) : 1 \leq i \leq n \text{ and } \sigma \text{ is a substitution}\}$$

is not propositionally satisfiable: if δ were a distribution of truth values satisfying X, then the structure $\mathcal{M}_\delta$ constructed above would be a model of Γ. By the compactness theorem for propositional calculus, we conclude that there is a finite subset X_0 of X which is not propositionally satisfiable, and, by Theorem 4.50, that there is a refutation of X_0 using simplification and the cut rule. We are going to see how to transform this refutation into a refutation of Γ (in the sense of Definition 4.71). First, we must introduce some more precise concepts.

- If Δ is a set of universal clauses and $\mathcal{D}$ is a universal clause, we will say that $\mathcal{D}$ is **derivable from** Δ **by cut** if there exists a sequence $(\mathcal{D}_1, \mathcal{D}_2, \ldots, \mathcal{D}_n)$ such that $\mathcal{D} = \mathcal{D}_n$ and, for all i between 1 and n, either $\mathcal{D}_i \in \Delta$, or there exists $j < i$ such that $\mathcal{D}_i$ is derived from $\mathcal{D}_j$ by simplification, or there exist j and k less than i such that $\mathcal{D}_i$ is derived from $\mathcal{D}_j$ and $\mathcal{D}_k$ by the cut rule.

- If Δ is a set of universal clauses and $\mathcal{D}$ is a universal clause, we will say that $\mathcal{D}$ is **derivable from** Δ **by resolution** if there exists a sequence $(\mathcal{D}_1, \mathcal{D}_2, \ldots, \mathcal{D}_n)$ such that $\mathcal{D} = \mathcal{D}_n$ and, for all i between 1 and n, either $\mathcal{D}_i \in \Delta$ or there exist j and k less than i such that $\mathcal{D}_i$ is derived from $\mathcal{D}_j$ and $\mathcal{D}_k$ by resolution.

- If $\mathcal{C}$ is a clause, $\mathcal{C}^+$ will denote the set of atomic formulas appearing in its conclusion, and $\mathcal{C}^-$ the set of those appearing in its premiss. If $\mathcal{C}$ and $\mathcal{D}$ are two clauses, we will write $\mathcal{C} \subseteq \mathcal{D}$ if $\mathcal{C}^- \subseteq \mathcal{D}^-$ and $\mathcal{C}^+ \subseteq \mathcal{D}^+$.

We already know that the empty clause $\square$ is derivable by cut from X_0. We will prove:

Lemma 4.73 *Let $\mathcal{D}$ be a clause that is derivable by cut from X. Then there exists a clause $\mathcal{E}$ that is derivable by resolution from Γ and a substitution τ such that $\tau(\mathcal{E}) \subseteq \mathcal{D}$.*

First of all, it is clear that the theorem follows from the lemma: by applying it to the empty clause, we see that there exists a clause $\mathcal{E}$ that is derivable from Γ by resolution and a substitution τ such that $\tau(\mathcal{E}) \subseteq \square$. Necessarily, $\tau(\mathcal{E}) = \square$ and $\mathcal{E} = \square$.

Proof (of the Lemma) Let $(\mathcal{D}_1, \mathcal{D}_2, \ldots, \mathcal{D}_n)$ be a derivation from X of $\mathcal{D}$ by cut (so $\mathcal{D} = \mathcal{D}_n$). We will argue by induction on the integer n. There are several cases to consider.

(a) $\mathcal{D}_n \in X$; this means that there is a substitution σ and a clause $\mathcal{C}_i \in \Gamma$ such that $\mathcal{D} = \sigma(\mathcal{C}_i)$. It then suffices to take $\mathcal{E} = \mathcal{C}_i$ and $\tau = \sigma$.

(b) There exists an $i < n$ such that $\mathcal{D}_n$ is obtained by simplification from $\mathcal{D}_i$. This implies that $\mathcal{D}_i \subseteq \mathcal{D}_n$ since $\mathcal{D}_i^- = \mathcal{D}_n^-$ and $\mathcal{D}_i^+ = \mathcal{D}_n^+$. By the induction hypothesis, there exists a clause $\mathcal{E}$ that is derivable by resolution from Γ and a substitution τ such that $\tau(\mathcal{E}) \subseteq \mathcal{D}_i$. The conclusion follows.

(c) $\mathcal{D}_n$ is obtained by cut from two clauses $\mathcal{D}_i$ and $\mathcal{D}_j$ where i and j are integers less than n. By the induction hypothesis, there exist clauses $\mathcal{E}_1$ and $\mathcal{E}_2$ derivable by resolution from Γ and substitutions τ_1 and τ_2 such that $\tau_1(\mathcal{E}_1) \subseteq \mathcal{D}_i$ and $\tau_2(\mathcal{E}_2) \subseteq \mathcal{D}_j$. Let us separate $\mathcal{E}_1$ and $\mathcal{E}_2$: let σ be a permutation of the set of variables such that, by setting $\mathcal{E}_3 = \sigma(\mathcal{E}_2)$, $\mathcal{E}_1$ and $\mathcal{E}_3$ are separated. There is then a substitution μ such that: if v_i is a variable that occurs in $\mathcal{E}_1$, then $\mu(v_i) = \tau_1(v_i)$ and if v_i is a variable that occurs in $\mathcal{E}_3$, then $\mu(v_i) = \tau_2 \circ \sigma^{-1}(v_i)$. Under these conditions,

$$\mu(\mathcal{E}_1) = \tau_1(\mathcal{E}_1) \subseteq \mathcal{D}_i;$$
$$\mu(\mathcal{E}_3) = \tau_2 \circ \sigma^{-1}(\mathcal{E}_3) = \tau_2(\mathcal{E}_2) \subseteq \mathcal{D}_j.$$

Let us write $\mathcal{E}_1$ and $\mathcal{E}_3$ explicitly:

$$\mathcal{E}_1 = (A_1 \wedge A_2 \wedge \cdots \wedge A_r) \Rightarrow (B_1 \vee B_2 \vee \cdots \vee B_s) \text{ and}$$
$$\mathcal{E}_3 = (C_1 \wedge C_2 \wedge \cdots \wedge C_t) \Rightarrow (D_1 \vee D_2 \vee \cdots \vee D_u).$$

We know that the cut rule applies to the pair $(\mathcal{D}_i, \mathcal{D}_j)$. This means that there is an atomic formula E that appears in the conclusion of $\mathcal{D}_i$ and in the premiss of $\mathcal{D}_j$ (or vice versa); we may assume, without loss of generality, that $E \in \mathcal{D}_i^+$ and $E \in \mathcal{D}_j^-$. Because $\mathcal{D}_n$ is obtained by cut from $\mathcal{D}_i$ and $\mathcal{D}_j$ and since $\mu(\mathcal{E}_1) \subseteq \mathcal{D}_i$ and $\mu(\mathcal{E}_3) \subseteq \mathcal{D}_j$, we see that

$$\{\mu(A_i) : 1 \leq i \leq r\} \cup (\{\mu(C_i) : 1 \leq i \leq t\} - \{E\}) \subseteq \mathcal{D}_n^-, \text{ and} \tag{*}$$
$$\{\mu(D_i) : 1 \leq i \leq u\} \cup (\{\mu(B_i) : 1 \leq i \leq s\} - \{E\}) \subseteq \mathcal{D}_n^+.$$

The formula E may not appear in $\mu(\mathcal{E}_1)^+$ nor in $\mu(\mathcal{E}_3)^-$; this forces us to distinguish several cases:

(1) for all k between 1 and s, $E \neq \mu(B_k)$; in this case, $\mu(\mathcal{E}_1) \subseteq \mathcal{D}_n$ and we have the desired conclusion;

(2) for all k between 1 and t, $E \neq \mu(C_k)$; in this case, $\mu(\mathcal{E}_3) = \tau_2(\mathcal{E}_2) \subseteq \mathcal{D}_n$ and again we have what we seek;

(3) The sets

$$I = \{i : 1 \le i \le s \text{ and } \mu(B_i) = E\}$$

and

$$J = \{j : 1 \le j \le t \text{ and } \mu(C_j) = E\}$$

are non-empty. Then we can unify $\{B_i : i \in I\} \cup \{C_j : j \in J\}$ and apply the rule of resolution to $\mathcal{E}_1$ and $\mathcal{E}_2$. Here, more precisely, is how this can be done. We have already separated $\mathcal{E}_1$ and $\mathcal{E}_2$ to obtain $\mathcal{E}_1$ and $\mathcal{E}_3$. Let σ be a principal unifier of $\{B_i : i \in I\} \cup \{C_j : j \in J\}$. Since μ unifies these formulas, there is a substitution τ such that $\mu = \tau \circ \sigma$. By resolution, we obtain a clause $\mathcal{E}_4$ with all the desired properties:

- It is derivable by resolution from Γ (since $\mathcal{E}_1$ and $\mathcal{E}_2$ are).
- $\mathcal{E}_4^+ = \{\sigma(B_i) : 1 \le i \le s \text{ and } i \notin I\} \cup \{\sigma(D_i) : 1 \le i \le u\}$; this shows that

$$\tau(\mathcal{E}_4^+) = \tau(\mathcal{E}_4)^+ = \{\mu(B_i) : 1 \le i \le s \text{ and } i \notin I\} \cup \{\sigma(D_i) : 1 \le i \le u\}.$$

  Invoking (*), we conclude that $\tau(\mathcal{E}_4^+) \subseteq \mathcal{D}_n^+$.
- A similar argument shows that $\tau(\mathcal{E}_4^-) \subseteq \mathcal{D}_n^-$.   ∎

Example 4.74 The language contains one constant symbol c, two unary predicate symbols S and Q, a binary predicate symbol R and a unary function symbol f. We will derive the empty clause from the following six clauses:

(1) $Qfv_0 \Rightarrow Sv_0$ (2) $\Rightarrow (Sv_0 \vee Rfv_0v_0)$
(3) $Pv_0 \Rightarrow Qv_0$ (4) $(Pv_1 \wedge Rv_0v_1) \Rightarrow Pv_0$
(5) $\Rightarrow Pc$ (6) $Sc \Rightarrow$

First, we apply the rule of resolution to (1) and (6) by unifying Sv_0 and Sc using the principal unifier $\sigma(v_0) = c$ (by convention, if we are not explicit about $\sigma(v_i)$, it is understood that $\sigma(v_i) = v_i$). We obtain

(7) $Qfc \Rightarrow$.

Then, from (3) and (7), by unifying Qfc and Qv_0 (the principal unifier is $\sigma'(v_0) = fc$), we obtain

(8) $Pfc \Rightarrow$;

and together with (4), by unifying Pfc and Pv_0 using σ', we obtain

(9) $(Pv_1 \wedge Rfcv_1) \Rightarrow$.

We may now unify $Rfcv_1$ from (9) and Rfv_0v_0 from (2). The principal unifier is $\sigma''(v_0) = \sigma''(v_1) = c$ and the rule of resolution yields

(10) $Pc \Rightarrow Sc$.

With (5), we obtain $\Rightarrow Sc$ and with (6) we obtain the empty clause.

Example 4.75 We will reconsider Example 4.34. Formulas F_4 and F_5 are not clauses. To reduce the situation to this case, we introduce Skolem functions, specifically, a constant symbol d and a unary function symbol f. We must then derive the empty clause from the following clauses:

(1) $(Pv_0v_1v_3 \wedge Pv_1v_2v_4 \wedge Pv_3v_2v_5) \Rightarrow Pv_0v_4v_5$

(2) $(Pv_0v_1v_3 \wedge Pv_1v_2v_4 \wedge Pv_0v_4v_5) \Rightarrow Pv_3v_2v_5$

(3) $\Rightarrow Pv_0cv_0$

(4) $\Rightarrow Pv_0fv_0c$

(5) $Pcdd \Rightarrow$

From (1) and (5), by unifying $Pv_0v_4v_5$ and $Pcdd$ (the unifier is obvious), we obtain

(6) $(Pcv_1v_3 \wedge Pv_1v_2d \wedge Pv_3v_2d) \Rightarrow.$

Clauses (6) and (3) are separated and we can unify Pv_0cv_0 and Pcv_1v_3 using the principal unifier $\sigma(v_0) = \sigma(v_1) = \sigma(v_3) = c$; we obtain

(7) $(Pcv_2d \wedge Pcv_2d) \Rightarrow.$

We now apply the rule of resolution to (7) and (2): they must first be separated, so we replace (7) by $(Pcv_6d \wedge Pcv_6d) \Rightarrow.$ We then unify Pcv_6d and $Pv_3v_2v_5$ (using $\sigma(v_3) = c, \sigma(v_5) = d, \sigma(v_6) = v_2$) and we obtain

(8) $(Pv_0v_1c \wedge Pv_1v_2v_4 \wedge Pv_0v_4d) \Rightarrow.$

To separate (3) and (8), we replace (3) by Pv_6cv_6 and by unifying this formula with Pv_0v_4d (using $\sigma(v_0) = \sigma(v_6) = d, \sigma(v_4) = c$), we obtain

(9) $(Pdv_1c \wedge Pv_1v_2c) \Rightarrow.$

We then apply the rule of resolution to (4) and (9) by unifying Pv_0fv_0c and Pdv_1c (using $\sigma(v_0) = d, \sigma(v_1) = fd$); from this, we derive

(10) $Pfdv_2c \Rightarrow$

and a final application of the rule of resolution to (4) and (10) (using $\sigma(v_0) = fd, \sigma(v_2) = ffd$) yields the empty clause.

Proposition 4.70 and Theorem 4.72 furnish a theoretical basis for the method of resolution. Imagine that we wish to derive a formula F from the formulas $G_1, G_2, \dots, G_n$. We would instead try to show that the set

$$\{G_1, G_2, \dots, G_n\} \cup \{\neg F\}$$

does not have a model. We begin by replacing these formulas by a set of clauses. To accomplish this, we first add Skolem functions and in this way we obtain universal formulas. We then replace the quantifier-free part of each of these formulas by their conjunctive normal forms (see Definition 1.28). Finally we distribute the universal quantifiers over the conjunctions (see Theorem 3.55) to produce a set of clauses.

It is then from this set that we try to derive the empty clause using the rule of resolution. Obviously, in practice, we cannot merely apply the rule of resolution systematically in all possible ways. It is necessary to adopt a strategy; a great number of these strategies, which we will not treat here, have been elaborated.

Note that there is an essential theoretical difference between the propositional calculus and the predicate calculus when it comes to applying the method: we have already mentioned that in the former case, the search is bounded. By this, we mean that after a certain number of applications of the cut rule, a number which can be bounded in advance (at most 3^n if there are n propositional variables: see Exercise 10), if we have not obtained the empty clause, then we will never obtain it and the original set of clauses is satisfiable. This is no longer true for the predicate calculus (except for some particularly impoverished languages). In the general case, we cannot be certain that the search is complete until the empty clause is obtained. It is only if we never obtain the empty clause and if the search is conducted in a systematic fashion that we may conclude for certain that the initial set of clauses is satisfiable (Theorem 4.72). We will express this difference in a more striking way in Chapter 6: the propositional calculus is decidable while, in general, the predicate calculus is not.

EXERCISES FOR CHAPTER 4

1. Is the formula $F \Rightarrow \forall v F$ universally valid for any formula F?

2. (This exercise is motivated by Example 4.7.)

  (a) Give an example which illustrates why the restriction 'w is not free in F' must accompany the assertion that $\forall v F \Rightarrow F_{w/v}$ is universally valid.

  (b) Give an example which illustrates why the restriction 'w is not bound in F' must accompany the assertion that $\forall v F \Rightarrow \forall w F_{w/v}$ is universally valid.

3. In a language L, we are given a theory T and two formulas F and G. Assume that

$$T \vdash \exists v_0 F \text{ and } T \vdash \forall v_0 (F \Rightarrow G).$$

Without appealing to the completeness theorem, show that

$$T \vdash \exists v_0 G.$$

4. Let F and G be two formulas and suppose that the variable v is not free in F. Give a derivation of $F \vee \forall v G$ starting from $\forall v (F \vee G)$.

5. Let L be a language and $\mathcal{F}$ be the set of formulas of L.

  (a) Show that there exists at least one map ϕ from the set $\mathcal{F}$ into $\{0,1\}$ that satisfies the following conditions:

   (1) If $F \in \mathcal{F}$ and F begins with a universal quantifier, then $\phi(F) = 0$;
   (2) If $F \in \mathcal{F}$ and F begins with an existential quantifier, then $\phi(F) = 1$;
   (3) If F is of the form $\neg G$, then $\phi(F) = 1 - \phi(G)$;
   (4) If F is of the form $(G \; \alpha \; H)$, where α is a binary connective, then $\phi(F) = \tilde{\alpha}(\phi(G), \phi(H))$, where $\tilde{\alpha}$ is the map from $\{0, 1\}^2$ into $\{0, 1\}$ corresponding to the connective α.

  (b) Show that if F is an axiom, then $\phi(F) = 1$.

  (c) Show that if $\phi(F \Rightarrow G) = 1$ and $\phi(F) = 1$, then $\phi(G) = 1$;

  (d) Show that if there exists a derivation of F that makes no appeal to the rule of generalization, then $\phi(F) = 1$. Conclude from this that we cannot do without the rule of generalization, i.e. that there exist formulas that are derivable but which are not derivable without using the rule of generalization.

6. Using a method analogous to the one just used in Exercise 5, we can show that the quantification axiom schema (c) is indispensable.

  (a) Define a map ϕ from $\mathcal{F}$ into $\{0, 1\}$ which satisfies conditions (3) and (4) from Exercise 5 as well as:

(1) If $F \in \mathcal{F}$ and F begins with a universal quantifier, then $\phi(F) = 1$;

(2) If $F \in \mathcal{F}$ and F begins with an existential quantifier, then $\phi(F) = 0$.

(b) Show that if F has a derivation in which axiom schema (c) is never used, then $\phi(F) = 1$.

(c) Find a formula that is derivable but whose derivations all use schema (c).

7. For any formula F, let F^* denote the formula obtained by replacing all occurrences of the existential quantifier by the universal quantifier. Show that if F has a derivation that makes no appeal to axiom schema (a), then

$$\vdash F^*.$$

Find a formula that is derivable but whose derivations all use schema (a).

8. Use Herbrand's method to prove the following completeness theorem.

Let $T = \{F_n : n \in \mathbb{N}\}$ be a set of closed formulas in prenex form. Show that if T is non-contradictory, then T has a model.

9. Using a derivation with cuts, exhibit a refutation of each of the following four sets of clauses:

(a) $\{(A \wedge B) \Rightarrow C, \ \Rightarrow A, \ C \Rightarrow, \ \Rightarrow B\}$;

(b) $\{(A \wedge B) \Rightarrow C, \ A \Rightarrow B, \ \Rightarrow A, \ C \Rightarrow\}$;

(c) $\{(A \wedge B) \Rightarrow, \ C \Rightarrow A, \ \Rightarrow C, \ D \Rightarrow B, \ \Rightarrow (D \vee B)\}$;

(d) $\{(A \wedge B) \Rightarrow (C \vee D), \ (C \wedge E \wedge F) \Rightarrow, \ (A \wedge D) \Rightarrow, \ \Rightarrow (B \vee C),$
$\Rightarrow (A \vee C), \ C \Rightarrow E, \ C \Rightarrow F\}$.

10. Suppose the set of propositional variables has cardinality n and is equal to $\{A_1, A_2, \ldots, A_n\}$. We wish to count the number of clauses, but subject to two conditions: first, we count only those in which no propositional variable occurs more than once (if a propositional variable occurs more than once in the premiss or more than once in the conclusion, the clause can be simplified; if a propositional variable occurs in both the premiss and the conclusion of a clause, then this clause is a tautology); second, clauses which differ only in the order of the variables that occur in either the premiss or the conclusion should be counted only once. We will say that a clause is reduced if it is of the form

$$(A_{i_1} \wedge A_{i_2} \wedge \cdots \wedge A_{i_n}) \Rightarrow (A_{j_1} \vee A_{j_2} \vee \cdots \vee A_{j_m})$$

where $(i_1, i_2, \ldots, i_n)$ and $(j_1, j_2, \ldots, j_m)$ are strictly increasing sequences. The number we seek is the number of reduced clauses. What is it?

11. Let S be a set of 7 clauses such that at least three distinct propositional variables occur in each of them. Show that S is satisfiable.

EXERCISES FOR CHAPTER 4

12. Let Γ consist of the following four clauses:

$$\mathcal{C}_1 \text{ is } (A \wedge B) \Rightarrow C \qquad \mathcal{C}_2 \text{ is } (B \wedge C) \Rightarrow$$
$$\mathcal{C}_3 \text{ is } A \Rightarrow B \qquad\qquad \mathcal{C}_4 \text{ is } \Rightarrow A.$$

(a) Exhibit a refutation of Γ using a proof with cuts.

(b) Show that the clause $\mathcal{D} = A \Rightarrow C$ is derivable from clauses $\mathcal{C}_1$ and $\mathcal{C}_3$ using the cut rule and simplification but that the set $\{\mathcal{C}_2, \mathcal{C}_4, \mathcal{D}\}$ is not refutable.

13. The language L has two constant symbols a and b, two unary function symbols h and k, and two binary function symbols f and g. Unify each of the following five pairs of formulas:

(a) $(g v_0 g h b g v_2 v_3, \ g g g v_5 v_1 h v_4 g h b v_0)$;

(b) $(g v_3 g g v_2 a g v_0 v_1, \ g g g v_0 v_2 g v_0 v_1 g g v_2 a v_3)$;

(c) $(g v_4 g g f g v_1 v_6 f v_5 v_{12} g f v_{10} a v_2 f v_7 v_8,$
$\quad g f f v_3 v_9 f v_{11} v_{11} g g v_2 g f v_{10} a f v_6 v_0 v_4)$;

(d) $(f v_2 f f f g v_4 v_1 f v_3 v_5 f g v_7 b v_0 g v_6 v_8,$
$\quad f g g v_9 v_3 g v_9 v_{10} f f v_0 f g v_7 b f v_{11} v_{12} v_2)$;

(e) $(g v_0 g g g k v_5 g v_{11} v_7 g g v_1 g h v_2 k v_8 v_6 h v_9,$
$\quad g h g v_{10} v_3 g g v_6 g g k k v_4 g h v_2 v_1 g v_{12} v_8 v_0)$.

14. Let V be the set of variables and $\mathcal{T}$ the set of terms. Let α be a permutation of V (i.e. a bijection of V onto itself). Show that the unique substitution, σ, that extends α is bijective.

Conversely, assuming that the substitution σ is a bijection from $\mathcal{T}$ into $\mathcal{T}$, show that the restriction of σ to V is a permutation of V.

15. In this exercise, we assume that the set V of variables is finite and is equal to $\{v_1, v_2, \ldots, v_n\}$. Let $\mathcal{T}$ be the set of terms.

(a) Suppose that t is a term, σ is a substitution, and that $\sigma(t) = t$. Show that if v occurs in t, then $\sigma(v) = v$.

(b) Let σ and π be two substitutions and let v be a variable. Assume that $\pi = \sigma \circ \pi$. Show that at least one of the following two assertions must be true:

(i) $\sigma(v) = v$;
(ii) for any term t, v does not occur in $\pi(t)$.

(c) Suppose now that $\pi, \pi_1, \sigma, \sigma_1$ are substitutions, that $\pi = \sigma \circ \pi_1$ and that $\pi_1 = \sigma_1 \circ \pi$. Set

$$A = \{v \in V : \text{there exists a term } t \text{ such that } v \text{ occurs in } \pi(t)\}, \text{ and}$$
$$B = \{\sigma_1(v) : v \in A\}.$$

Show that $B \subseteq V$, that σ_1 is a bijection from A onto B, that σ is a bijection from B onto A, and that $\sigma \circ \sigma_1$ is the identity on A.

Construct bijections σ' and σ_1' from V into V such that

(i) for every $v \in A$, $\sigma_1(v) = \sigma_1'(v)$;
(ii) for every $v \in B$, $\sigma(v) = \sigma'(v)$;
(iii) $\sigma' \circ \sigma_1'$ and $\sigma_1' \circ \sigma'$ are equal to the identity on V;
(iv) $\pi = \sigma' \circ \pi_1$;
(v) $\pi_1 = \sigma_1' \circ \pi$.

(d) Suppose that π is a principal unifier of a system S. Show that the set of principal unifiers of S is equal to

$$\{\sigma \circ \pi : \sigma \text{ is a substitution and a bijection from } T \text{ onto } T\}.$$

16. The language consists of three unary relation symbols P, R and S, one binary relation symbol Q, and one unary function symbol f. Apply the rule of resolution in two different ways to the following two universal clauses:

$$Sv_0 \Rightarrow (Pv_0 \vee Rv_0), \qquad (Pv_0 \wedge Pfv_1) \Rightarrow Qv_0v_1.$$

17. The language is the same as in Exercise 16. Add Skolem functions to the language to rewrite the following formulas as clauses and use the method of resolution to show that the set they form is contradictory:

$$\exists v_0 \forall v_1 (Pv_0 \wedge (Rv_1 \Rightarrow Qv_0v_1));$$
$$\forall v_0 \forall v_1 (\neg Pv_0 \vee \neg Sv_1 \vee \neg Qv_0v_1);$$
$$\exists v_1 (Rv_1 \wedge Sv_1).$$

18. The language consists of one binary relation symbol R and one unary function symbol f. Consider the following formulas:

$$F_1 \quad \forall v_0 (\exists v_1 Rv_0v_1 \Rightarrow Rv_0fv_0);$$
$$F_2 \quad \forall v_0 \exists v_1 Rv_0v_1;$$
$$F_3 \quad \exists v_0 Rffv_0v_0;$$
$$G \quad \exists v_0 \exists v_1 \exists v_2 (Rv_0v_1 \wedge Rv_1v_2 \wedge Rv_2v_0).$$

Using the method of resolution, show that G is a consequence of $\{F_1, F_2, F_3\}$.

Solutions

Solutions to the exercises for Chapter 1

1. For an arbitrary formula F, we let $b[F]$ and $n[F]$ denote the number of occurrences of symbols for binary connectives and for negation, respectively. We will prove by induction that the length of F is

$$\lg[F] = 4b[F] + n[F] + 1. \tag{*}$$

- If $F \in P$, then $b[F] = n[F] = 0$ and $\lg[F] = 1$; so (*) is verified.
- If $F = \neg G$, then $b[F] = b[G]$, $n[F] = n[G] + 1$, and $\lg[F] = \lg[G] + 1$. By the induction hypothesis, $\lg[G] = 4b[G] + n[G] + 1$. It follows that (*) is again satisfied.
- If $F = (G \, \alpha \, H)$ (where α is a binary connective), then $b[F] = b[G] + b[H] + 1$, $n[F] = n[G] + n[H]$, and $\lg[F] = \lg[G] + \lg[H] + 3$. By the induction hypothesis, $\lg[G] = 4b[G] + n[G] + 1$ and $\lg[H] = 4b[H] + n[H] + 1$. Here too, (*) is satisfied.

2. (a) Let X_n denote the set of lengths of formulas of height n. Since the formulas of height 0 are the elements of P, we have $X_0 = \{1\}$.

We have proved (Theorem 1.5) that the height of a formula is always strictly less than its length. We may conclude from this that every element of X_n is greater than or equal to $n + 1$. Now the formula $\neg\neg\ldots\neg A$ (with n occurrences of the symbol $\neg$) is of length $n + 1$ and of height n. This proves that $n + 1$ is an element of X_n and that it is the smallest element.

To convince ourselves that X_n is finite, we note that if all the formulas of height $n - 1$ have a length that is at most some integer x, then all the formulas of height n will have a length that is at most $2x + 3$: to see this, observe that if F is a formula of maximum length among the formulas of height $n - 1$, then the length of any formula of height n will be less than or equal to the length of the formula $(F \, \alpha \, F)$, where α is a symbol for a binary connective. Since $X_0 = \{1\}$, we have thereby proved by induction that for every integer n, the set X_n is bounded and hence finite (since it is a set of natural numbers). Let L_n denote the greatest element of X_n. The argument above establishes the following recurrence relation:

$$L_n = 2L_{n-1} + 3.$$

A classic computation then shows that

$$L_n = 2^{n+2} - 3.$$

The length of any formula of height n is thus between $n + 1$ and $2^{n+2} - 3$, where the minimum corresponds to formulas that use n occurrences of the negation symbol and no occurrences of symbols for binary connectives while the maximum corresponds to formulas in which the negation symbol does not occur. The reader may find it amusing to show that all values between these two extreme values are possible, except for

$$n + 2, \; n + 3, \; 2^{n+2} - 8, \; 2^{n+2} - 5 \text{ and } 2^{n+2} - 4.$$

3. (a) For any word W on the alphabet $P \cup \{\neg, \wedge, \vee, \Rightarrow, \Leftrightarrow, (\ \}$, let $b[W]$ denote the number of occurrences in W of symbols for binary connectives and recall that $o[W]$ is the number of opening parentheses in W.

In a way that is more or less analogous to the one used for formulas, we prove the following four facts by induction:

(1) Every pseudoformula ends with a propositional variable.
(2) If F is a pseudoformula, then $o[F] = b[F]$.
(3) If W is a proper initial segment of a pseudoformula F, then $o[W] \geq b[W]$.
(4) If W is a proper initial segment of a pseudoformula and if the last symbol of W is a propositional variable, then $o[W] > b[W]$.

From this, it is not difficult to show that

(5) A proper initial segment of a pseudoformula cannot be a pseudoformula.

We may then continue, as in the case for formulas, and arrive at the **unique readability theorem** for pseudoformulas:

For any pseudoformula F, one and only one of the following three situations is applicable:

- F is a propositional variable;
- there is a unique pseudoformula G such that $F = \neg G$;
- there exists a unique symbol α for a binary connective and a unique pair of pseudoformulas H and K such that $F = (H \; \alpha \; K)$.

(b) So we have proved that, for writing formulas, we can very well do without closing parentheses. We must avoid jumping to the conclusion that we could just as well eliminate opening rather than closing parentheses: the culprit here is the presence of the symbol for the unary connective, negation, which destroys all symmetry. As an illustration, consider the formula $\neg(A \Rightarrow B)$ which, in the context above, becomes the pseudoformula $\neg(A \Rightarrow B$; if opening parentheses are suppressed, this becomes $\neg A \Rightarrow B)$, but this is exactly the same word that would result if the formula $\neg(A \Rightarrow B)$ were to receive the same treatment. So this procedure is doomed to fail.

4. (a) T^* is the smallest set of words on the alphabet $P \cup \{\neg, \Rightarrow,), (\}$ that includes P and is closed under the operations $X \mapsto \neg X$ and $(X, Y) \mapsto (X \Rightarrow Y)$.

To have an equivalent definition from below, we would set $T_0^* = P$ and for every integer $n \in \mathbb{N}$,

$$T_{n+1}^* = T_n^* \cup \{\neg F : F \in T_n^*\} \cup \{(F \Rightarrow G) : F \in T_n^*, G \in T_n^*\}.$$

We would then have $T^* = \bigcup_{n \in \mathbb{N}} T_n^*$.

(b) The given conditions allow us to define a unique map $\widehat{\mu}$ from T^* into T^* by induction: $\widehat{\mu}$ is already defined (equal to μ) on $T_0^* = P$, and, if we assume it is defined on T_n^*, the given conditions determine its values on T_{n+1}^*.

(c) We argue by induction on G. If $G \in P$, then the only sub-formula of G is $F = G$ and, in this case, $\widehat{\mu}(F) = \widehat{\mu}(G)$ is a sub-formula of $\widehat{\mu}(G)$. If $G = \neg H$ and if F is a sub-formula of G, then either $F = G$ and $\widehat{\mu}(F) = \widehat{\mu}(G)$ is a sub-formula of $\widehat{\mu}(G)$, or else F is a sub-formula of H and, by the induction hypothesis, $\widehat{\mu}(F)$ is a sub-formula of $\widehat{\mu}(H)$, hence also of $\neg \widehat{\mu}(H) = \widehat{\mu}(\neg H) = \widehat{\mu}(G)$. If $G = (H \Rightarrow K)$ and if F is a sub-formula of G, then either $F = G$ and $\widehat{\mu}(F) = \widehat{\mu}(G)$ is a sub-formula of $\widehat{\mu}(G)$, or else F is a sub-formula of H or a sub-formula of K and, by the induction hypothesis, $\widehat{\mu}(F)$ is a sub-formula of $\widehat{\mu}(H)$ or a sub-formula of $\widehat{\mu}(K)$, hence also of the formula

$$\neg(\widehat{\mu}(K) \Rightarrow \widehat{\mu}(H)) = \widehat{\mu}((H \Rightarrow K)) = \widehat{\mu}(G).$$

(d) $\widehat{\mu_0}((A \Rightarrow B)) = \neg(\neg B \Rightarrow \neg A)$ and $\widehat{\mu_0}((\neg A \Rightarrow B)) = \neg(\neg B \Rightarrow \neg\neg A)$.

It would obviously be an error to write, for example:

$$\widehat{\mu_0}((A \Rightarrow B)) = \neg(A \Rightarrow B),$$

despite the fact that these formulas turn out to be logically equivalent. With each formula of T^*, the map $\widehat{\mu_0}$ associates a uniquely determined formula.

The last property is proved by induction on F. If $F \in P$, $\widehat{\mu_0}(F) = \neg F$, thus $\widehat{\mu_0}(F) \sim \neg F$. If $F = \neg G$ and if (induction hypothesis) $\widehat{\mu_0}(G) \sim \neg G$, then $\widehat{\mu_0}(F) = \neg\widehat{\mu_0}(G)$ is logically equivalent to $\neg\neg G = \neg F$. If $F = (G \Rightarrow H)$ and if (induction hypothesis) $\widehat{\mu_0}(G) \sim \neg G$ and $\widehat{\mu_0}(H) \sim \neg H$, then $\widehat{\mu_0}(F) = \neg(\widehat{\mu_0}(H) \Rightarrow \widehat{\mu_0}(G))$ is logically equivalent to $\neg(\neg H \Rightarrow \neg G)$, hence also to $\neg(G \Rightarrow G) = \neg F$.

5. We will treat question (b) directly since (a) merely presents two special cases. As F_n ($n \in \mathbb{N}$, $n \geq 2$) we take the following formula:

$$\left(\bigwedge_{1 \leq i \leq n} (A_i \Rightarrow B_i) \wedge \bigwedge_{1 \leq i < j \leq n} \neg(B_i \wedge B_j) \wedge \bigvee_{1 \leq i \leq n} A_i \right) \Rightarrow \bigwedge_{1 \leq i \leq n} (B_i \Rightarrow A_i).$$

If we set $G_n = \bigwedge_{1 \leq i \leq n} (A_i \Rightarrow B_i)$, $H_n = \bigwedge_{1 \leq i < j \leq n} \neg(B_i \wedge B_j)$, $K_n = \bigvee_{1 \leq i \leq n} A_i$, and $L_n = \bigwedge_{1 \leq i \leq n} (B_i \Rightarrow A_i)$, then we can write

$$F_n = ((G_n \wedge H_n \wedge K_N) \Rightarrow L_n).$$

Let δ be an assignment of truth values and suppose that $\bar{\delta}(L_n) = \mathbf{0}$. This means that we can find an index j between 1 and n such that $\delta(B_j) = \mathbf{1}$ and $\delta(A_j) = \mathbf{0}$. If we suppose as well that $\bar{\delta}(H_n) = \mathbf{1}$, then we must conclude that for all indices i between 1 and n and distinct from j, we have $\delta(B_i) = \mathbf{0}$. Now let us add the hypothesis that $\bar{\delta}(G_n) = \mathbf{1}$. Then for all i different from j, we must have $\delta(A_i) = \mathbf{0}$. But since $\delta(A_j)$ is also zero, the conclusion is that δ does not satisfy the formula K_n. In this way, we see that it is not possible for δ to assign the value $\mathbf{0}$ to L_n while simultaneously assigning the value $\mathbf{1}$ to G_n, to H_n and to K_n. This amounts to saying that δ cannot assign the value $\mathbf{0}$ to F_n. So this formula is a tautology.

6. (a) With the help of items 38 and 53 from the list in Section 1.2.3, we see that E is logically equivalent to

$$(B \Rightarrow (C \Rightarrow (A \Leftrightarrow (B \Rightarrow C)))),$$

which is a formula that satisfies the required conditions.

(b) Any assignment of truth values that gives the value $\mathbf{0}$ to B or to C will satisfy the formula E. If δ is as assignment of truth values such that $\delta(B) = \delta(C) = \mathbf{1}$, then $\bar{\delta}((\neg B \vee C)) = \mathbf{1}$, from which we deduce that $\bar{\delta}(E) = \mathbf{1}$ if and only if $\delta(A) = \mathbf{1}$. Thus, there is a unique assignment of truth values to the set $\{A, B, C\}$ that makes the formula E false: namely, it assigns $\mathbf{1}$ to B and to C and $\mathbf{0}$ to A. This observation provides us with the CCNF of the formula E:

$$(A \vee \neg B \vee \neg C),$$

which is, at the same time, in reduced disjunctive normal form.

(c) It follows from what has just been said that there are seven assignments of truth values to $\{A, B, C\}$ that satisfy E. So there are seven elementary conjunctions in the CDNF of E.

(d) In question (a) we noted that the formula E is logically equivalent to

$$(B \Rightarrow (C \Rightarrow (A \Leftrightarrow (B \Rightarrow C)))),$$

but this formula is also logically equivalent to

$$(C \Rightarrow (B \Rightarrow (A \Leftrightarrow (B \Rightarrow C))))$$

(again, refer to item 53 of Section 1.2.3). Also, the disjunctive normal form of E that we found in (b) is clearly logically equivalent to $(C \Rightarrow (B \Rightarrow A))$; this proves the desired result.

7. (a) Let δ be an assignment of truth values to P that satisfies F and let i be an integer between 1 and n. We see immediately that

- if $\delta(A_i) = \mathbf{1}$, then $\delta(A_{i+1}) = \delta(A_{i+2}) = \ldots \delta(A_n) = \mathbf{1}$;
- if $\delta(A_i) = \mathbf{0}$, then $\delta(A_{i-1}) = \delta(A_{i-2}) = \ldots \delta(A_1) = \mathbf{0}$.

It follows that the assignments of truth values to P that satisfy F are the $n+1$ assignments δ_p $(0 \le p \le n)$ that give the value $\mathbf{1}$ to the last p variables in P and the value $\mathbf{0}$ to the first $n - p$ variables; specifically:

$$\delta_p(A_i) = \begin{cases} \mathbf{0} & \text{if } i \le n - p; \\ \mathbf{1} & \text{if } i > n - p. \end{cases}$$

From this, we may obtain the CDNF of F:

$$(\neg A_1 \wedge \neg A_2 \wedge \cdots \wedge \neg A_n) \vee (\neg A_1 \wedge \cdots \wedge \neg A_{n-1} \wedge A_n)$$
$$\vee (\neg A_1 \wedge \cdots \wedge \neg A_{n-2} \wedge A_{n-1} \wedge A_n) \vee \ldots$$
$$\cdots \vee (\neg A_1 \wedge A_2 \wedge \cdots \wedge A_n) \vee (A_1 \wedge A_2 \wedge \cdots \wedge A_{n-1} \wedge A_n).$$

(b) Let δ be an assignment of truth values to P that satisfies G. Obviously, δ must satisfy F, so δ must be one of the assignments δ_p above. But δ must also satisfy $(A_n \Rightarrow A_1)$ so it is not possible that we have both $\delta(A_n) = \mathbf{1}$ and $\delta(A_1) = \mathbf{0}$; this excludes the possibility that δ could equal δ_p when $1 \le p \le n - 1$. We easily verify that the assignments of truth values δ_0 and δ_n (i.e. the two constant assignments) satisfy G. It follows from what was just said that these are the only ones.

We conclude from this that the CDNF of G is

$$(\neg A_1 \wedge \neg A_2 \wedge \cdots \wedge \neg A_n) \vee (A_1 \wedge A_2 \wedge \cdots \wedge A_{n-1} \wedge A_n).$$

(c) The formula $(A_i \Rightarrow \neg A_j)$ is logically equivalent to $\neg(A_i \wedge A_j)$. Consequently, a necessary and sufficient condition for an assignment of truth values to satisfy H is that there not exist a pair of distinct indices i and j such that $\delta(A_i) = \delta(A_j) = \mathbf{1}$. So the assignments of truth values to P that satisfy H are those that give the value $\mathbf{1}$ to at most one variable in P: these are the $n + 1$ assignments λ_p $(0 \le p \le n)$ defined by

$$\lambda_p(A_i) = \begin{cases} \mathbf{0} & \text{if } i \ne p \\ \mathbf{1} & \text{if } i = p. \end{cases}$$

(Thus λ_0 is the constant assignment equal to $\mathbf{0}$). We conclude from this that the canonical disjunctive normal form of H is

$$(\neg A_1 \wedge \neg A_2 \wedge \cdots \wedge \neg A_n) \vee (A_1 \wedge \neg A_2 \wedge \cdots \wedge \neg A_n)$$
$$\vee (\neg A_1 \wedge A_2 \wedge \neg A_3 \wedge \cdots \wedge \neg A_n) \vee \ldots$$
$$\cdots \vee (\neg A_1 \wedge \cdots \wedge \neg A_{n-2} \wedge A_{n-1} \wedge \neg A_n) \vee (\neg A_1 \wedge \cdots \wedge \neg A_{n-1} \wedge A_n).$$

8. (a) Set $F = \bigvee_{1 \le i < j \le n}(A_i \wedge A_j)$ and $G = \bigwedge_{1 \le i \le n}(\bigvee_{j \ne i} A_j)$.

Let δ be an assignment of truth values on P.

For δ to satisfy F, it is necessary and sufficient that δ assume the value **1** for at least two of the variables $A_1, A_2, \ldots, A_n$. In order that δ not satisfy G, it is necessary and sufficient that there exist an index i such that $\delta(A_j) = \mathbf{0}$ for every index j different from i. In other words, $\overline{\delta}(G) = \mathbf{0}$ if and only if δ assigns the value **1** to at most one of the variables $A_1, A_2, \ldots, A_n$. We may conclude that δ satisfies G if and only if δ assigns the value **1** to at most one of the variables $A_1, A_2, \ldots, A_n$. So we see that δ satisfies G if and only if δ satisfies F. The formula $(F \Leftrightarrow G)$ is thus a tautology.

(b) Set $H = \bigvee_{1 \le i \le n} A_i$ and consider an assignment of truth values δ on P. For δ to satisfy H, it is necessary and sufficient that δ assign the value **1** to at least one of the variables $A_1, A_2, \ldots, A_n$. Now we have just seen that δ satisfies G if and only if δ assigns the value **1** to at least two of the variables $A_1, A_2, \ldots, A_n$. It follows immediately that

- if δ assigns the value **0** to all the variables in P, then $\overline{\delta}(G) = \overline{\delta}(H) = \mathbf{0}$;
- if δ assigns the value **1** to one and only one of the variables in P, then $\overline{\delta}(G) = \mathbf{0}$ and $\overline{\delta}(H) = \mathbf{1}$;
- if δ assigns the value **1** to at least two of the variables in P, then $\overline{\delta}(G) = \overline{\delta}(H) = \mathbf{1}$.

Thus the assignments of truth values that make the formula $(H \Leftrightarrow G)$ false are precisely the n assignments $\delta_1, \delta_2, \ldots, \delta_n$ defined by

$$\delta_i(A_j) = \begin{cases} \mathbf{1} & \text{if } i = j \\ \mathbf{0} & \text{if } i \ne j \end{cases} \quad (\text{for } 1 \le i \le n \text{ and } 1 \le j \le n).$$

(c) Once we know the assignments of truth values that make the formula $(H \Leftrightarrow G)$ false, we immediately obtain its CCNF:

$$(\neg A_1 \vee A_2 \vee \cdots \vee A_n) \wedge (A_1 \vee \neg A_2 \vee A_3 \vee \cdots \vee A_n) \wedge \ldots$$
$$\wedge (A_1 \vee \cdots \vee A_{n-1} \vee \neg A_n).$$

For $1 \le i \le n$, the ith of these n clauses is logically equivalent to

$$\left(A_i \Rightarrow \bigvee_{j \ne i} A_j \right);$$

this provides us with the desired result.

9. Consider a set of five propositional variables:

$$P = \{A, B, C, D, E\}.$$

Intuitively, the variable A (respectively, B, C, D, E) will have the value 'true' if and only if person a (respectively, b, c, d, e) is present. The safe can be opened if and only if the following propositional formula is satisfied:

$$F = (A \wedge B) \vee (A \wedge C \wedge D) \vee (B \wedge D \wedge E).$$

Let $S_1, S_2, \ldots, S_n$ denote the keys to the safe. To be able to open the safe, it is necessary and sufficient that for every integer i between 1 and n, at least one person possessing the key S_i be present. If, for example, the holders of the key S_i are persons c and e, the ability to unlock S_i is equivalent to the satisfaction of the formula $(C \vee E)$. The ability to open the safe is thus equivalent to the satisfaction of the conjunction of formulas of this type (in which i assumes the values $1, 2, \ldots, n$), i.e. to a formula in conjunctive normal form. Now, as we saw above, the ability to open the safe is equivalent to the satisfaction of the formula F which is in disjunctive normal form. So it will suffice to find a conjunctive normal form for F that is as reduced as possible; the number of disjuncts (clauses) in this CNF will correspond to the minimum number of keys required and each clause will provide a list of those persons who should receive a key to the corresponding lock. By distributing the conjunctions over the disjunctions in F (which leads to a CNF that has eighteen clauses, each having three variables) and then simplifying (using the properties of idempotence and absorption as in items (2) and (10) from the list of tautologies: for example, $(B \vee C \vee B)$ becomes $(B \vee C)$ which then allows us to eliminate $(B \vee C \vee D)$ and $(B \vee C \vee E)$), we arrive at the following CNF for F:

$$(A \vee B) \wedge (A \vee D) \wedge (A \vee E) \wedge (B \vee C) \wedge (B \vee D).$$

Some fastidious verifications would convince us that we cannot further reduce the number of clauses. We will not attempt to treat this issue.

The CNF that we have obtained tells us that the number of locks required is five and that one possible distribution of keys consists in giving:

- to a, the keys for S_1, S_2, and S_3;
- to b, the keys for S_1, S_4, and S_5;
- to c, the key for S_4;
- to d, the keys for S_2 and S_5;
- to e, the key for S_3.

10. Let δ be an assignment of truth values to P that satisfies $\mathcal{A}$. Let H_δ denote the set of elements i of $\mathbb{Z}/15\mathbb{Z}$ such that $\delta(A_i) = 1$. We see that $0 \in H_\delta$ and that H_δ is closed under the operations $i \mapsto -i$ and $(i, j) \mapsto i + j$. This means that H_δ is necessarily a subgroup of the group $\langle \mathbb{Z}/15\mathbb{Z}, + \rangle$. Conversely, if H is a subgroup of $\langle \mathbb{Z}/15\mathbb{Z}, + \rangle$, then the assignment of truth values δ defined by

$$\delta(A_i) = \begin{cases} 1 & \text{if } i \in H \\ 0 & \text{if } i \notin H \end{cases}$$

clearly satisfies the set $\mathcal{A}$. Now $\langle \mathbb{Z}/15\mathbb{Z}, + \rangle$ has precisely the following four subgroups: $\mathbb{Z}/15\mathbb{Z}$, $\{0\}$, $\{0, 5, 10\}$, and $\{0, 3, 6, 9, 12\}$. There are thus four assignments of truth values that satisfy $\mathcal{A}$: the one that is constant and equal to $\mathbf{1}$, the one whose value is $\mathbf{1}$ on A_0 and $\mathbf{0}$ elsewhere, the one whose value is $\mathbf{1}$ on A_0, A_5, and A_{10} and $\mathbf{0}$ elsewhere and, finally, the one whose value is $\mathbf{1}$ on A_0, A_3, A_6, A_9, and A_{12} and $\mathbf{0}$ elsewhere.

11. The formulas $F_\Rightarrow$ and $G_\vee$ are tautologies. The formulas $G_\wedge, G_\Leftrightarrow, G_\not\Rightarrow$, and $G_\curlyvee$ are antilogies. The six other formulas are neutral.

  We list them here and place, below each of them, a simpler formula that is logically equivalent. The proofs present no difficulties.

$F_\wedge$	$F_\vee$	$G_\Rightarrow$	$F_\Leftrightarrow$	$F_\not\Rightarrow$	$F_\curlyvee$
$A \wedge B$	$A \vee B$	$\neg(A \Leftrightarrow B)$	B	A	$\neg A \wedge B$

12. (a) The given conditions entirely determine φ: the first tells us that φ must assume the value $\mathbf{1}$ at the points $(0, 0, 0)$, $(0, 1, 0)$, $(1, 0, 0)$, and $(1, 0, 1)$; the second tells us that φ must assume the value $\mathbf{0}$ at the points $(0, 0, 1)$, $(0, 1, 1,)$, $(1, 1, 0)$, and $(1, 1, 1)$.

  (b) A DNF for φ is given by the following formula:

$$G[A, B, C] = (\neg B \wedge \neg C) \vee (\neg A \wedge B \wedge \neg C) \vee (A \wedge \neg B \wedge C).$$

  (c) It suffices to take a formula that is logically equivalent to $G[A, A, A]$ in case 1, to $G[A, B, B]$ in case 2, to $G[A, A, B]$ in case 3, to $G[A, B, A]$ in case 4, to $G[A, G[B, B, B], A]$ in case 5, and to $(G[A, B, B] \Rightarrow G[A, B, A])$ in case 6. Here are some possible answers:

  (1) $\neg A$ (2) $\neg B$ (3) $(\neg A \wedge \neg B)$
  (4) $(\neg A \vee \neg B)$ (5) $(A \Rightarrow B)$ (6) $(\neg A \vee A)$.

  (d) As the formula $(A \vee B)$ is logically equivalent to $(\neg\neg A \vee \neg\neg B)$, we will refer to case 4 of the preceding question; it is then not difficult to verify that the connective ψ defined by

  for all x and y belonging to $\{\mathbf{0}, \mathbf{1}\}$,
$$\psi(x, y) = \varphi(\varphi(x, x, x), \varphi(y, y, y), \varphi(x, x, x))$$

  is precisely disjunction.

  (e) Using composition, the connectives disjunction (question (d)) and negation (case 1 of question (c)) can be expressed in terms of the sole connective φ. As every connective can be expressed in terms of negation and disjunction, it follows that every connective can be expressed in terms of the connective φ. Thus, $\{\varphi\}$ is a complete system of connectives.

13. The solution is

$$p = (a \wedge b \wedge d) \vee (b \wedge c \wedge d) \vee (a \wedge \neg b \wedge c) \vee (a \wedge c \wedge \neg d);$$
$$q = (b \wedge d) \Leftrightarrow (a \Leftrightarrow c)$$
$$r = \neg(b \Leftrightarrow d).$$

It is not difficult to verify this. Here, of course, $\neg$, $\wedge$, $\vee$, and $\Leftrightarrow$ denote operations on $\{0, 1\}$.

14. (a) Our concern is to express the connectives as operations in $\mathbb{Z}/2\mathbb{Z}$. For all elements x and y of this set, we have

$$\neg x = 1 + x \qquad x \wedge y = xy \qquad x \vee y = x + y + xy$$
$$x \Rightarrow y = 1 + x + xy \qquad x \Leftrightarrow y = 1 + x + y.$$

The verifications are elementary. As usual, we permit ourselves to write xy in place of $x \times y$.

(b) For all elements x and y of $\mathbb{Z}/2\mathbb{Z}$, we have

$$xy = x \wedge y;$$
$$x + y = \neg(x \Leftrightarrow y) = x \nLeftrightarrow y = (x \wedge \neg y) \vee (\neg x \wedge y).$$

(c) Obviously inspired by question (a), we define P_F by induction.

- If F is the propositional variable A_i ($1 \leq i \leq n$), set $P_F = X_i$.
- If $F = \neg G$, set $P_F = 1 + P_G$.
- If $F = (G \wedge H)$, set $P_F = P_G \times P_H$.
- If $F = (G \vee H)$, set $P_F = P_G + P_H + P_G P_H$.
- If $F = (G \Rightarrow H)$, set $P_F = 1 + P_G + P_G P_H$.
- If $F = (G \Leftrightarrow H)$, set $P_F = 1 + P_G + P_H$.

We then prove by induction that for every assignment of truth values $\delta \in \{0, 1\}^P$, we have

$$\bar{\delta}(F) = \widetilde{P}_F(\delta(A_1), \delta(A_2), \ldots, \delta(A_n)). \tag{*}$$

If F is the propositional variable A_i, we have $P_F = X_i$, and the associated polynomial function $\widetilde{P}_F$ is the ith projection, i.e. the map from $\{0, 1\}^P$ into $\{0, 1\}$ whose value on any n-tuple $(\varepsilon_1, \varepsilon_2, \ldots, \varepsilon_n)$ is ε_i. This shows that the relation (*) is satisfied.

If $F = (G \wedge H)$ and if (induction hypothesis) for all assignments of truth values δ, we have

$$\bar{\delta}(G) = \widetilde{P}_G(\delta(A_1), \delta(A_2), \ldots, \delta(A_n)) \text{ and}$$
$$\bar{\delta}(H) = \widetilde{P}_H(\delta(A_1), \delta(A_2), \ldots, \delta(A_n)),$$

then, given that $P_F = P_G P_H$, we would have $\widetilde{P}_F = \widetilde{P}_G \widetilde{P}_H$; thus, for any assignment δ,

$$\widetilde{P}_F(\delta(A_1), \delta(A_2), \ldots, \delta(A_n)) = \overline{\delta}(G)\overline{\delta}(H) = \overline{\delta}((G \wedge H))$$

(by definition of $\overline{\delta}$); this proves (*).

The other steps in the induction are treated in an analogous manner.

The definition that we have adopted defines, for every formula F, a unique polynomial P_F; but we could have chosen others while still preserving property (*): for example, the polynomial associated with the formula $(A_1 \Leftrightarrow A_2)$ is, according to our definition, $1 + X_1 + X_2$, but it is clear that the polynomial $1 + X_1^2 + X_2^5$ would work just as well. What is unique, for a given formula F, is the associated **polynomial function** (it is the truth table of F).

(d) From what we have just seen, we conclude that for a formula to be a tautology, it is necessary and sufficient that the associated polynomial (or rather, the polynomial function) assume the constant value **1**. For two formulas to be logically equivalent, it is necessary and sufficient that their associated polynomial functions coincide. To illustrate this, let us verify that the formulas

$$G = (A \Rightarrow (B \Rightarrow C)) \quad \text{and} \quad H = ((A \wedge B) \Rightarrow C)$$

are logically equivalent (to simplify, we have used A, B, and C instead of A_1, A_2, and A_3 and X, Y, and Z instead of X_1, X_2, and X_3). We have

$$P_G = 1 + X + X(1 + Y + YZ) = 1 + XY + XYZ$$

(since $X + Y$ is the null polynomial), and

$$P_H = 1 + ((XY) + (XY)Z) = 1 + XY + XYZ = P_G.$$

15. (a) We have seen (Lemma 1.34) that if the formulas F and G have no propositional variable in common, and if the formula $(F \Rightarrow G)$ is a tautology, then either the formula G is a tautology or the formula F is an antilogy. It is obvious that in the first case, the formula $\top$ is an interpolant between F and G while in the second case, the formula $\bot$ is an interpolant between F and G.

(b) The proof is by induction on F. If the height of F is 0, this is obvious. We then simply need to verify that, given two formulas G and H, if both of them are logically equivalent to one of the three formulas $\top$, $\bot$, and A, then the same is true for $(G \wedge H)$ and $(G \vee H)$; this presents no difficulty.

(c) The argument is analogous. This time, we show that if each of the formulas G and H is logically equivalent to one of the eight formulas $\top$, $\bot$, A, B, $\neg A$, $\neg B$, $(A \Leftrightarrow B)$, $\neg(A \Leftrightarrow B)$, then this is also the case for the formulas

$(G \wedge H)$ and $(G \vee H)$. There are 64 cases to examine but some elementary arguments allow us to reduce this number considerably.

(d) Since we already know that the systems $\{\neg, \vee\}$ and $\{\neg, \wedge\}$ are complete, to show that a given system of connectives is complete, it suffices to prove that the connectives $\neg$ and $\wedge$, or else the connectives $\neg$ and $\wedge$, can be obtained by composition from the system of connectives under consideration. We will apply this remark to the systems that have been proposed. For all elements x and y belonging to $\{0, 1\}$, we have

- $\neg x = x \Rightarrow 0$ and $x \vee y = (x \Rightarrow 0) \Rightarrow y$; hence $\{\Rightarrow, 0\}$ is complete.
- $\neg x = x \Leftrightarrow 0$ and $\vee \in \{0, \Leftrightarrow, \vee\}$; hence $\{0, \Leftrightarrow, \vee\}$ is complete.
- $\neg x = x \Leftrightarrow 0$ and $\wedge \in \{0, \Leftrightarrow, \wedge\}$; hence $\{0, \Leftrightarrow, \wedge\}$ is complete.
- $\neg x = x \curlyvee x$ and $x \wedge y = (x \curlyvee y) \curlyvee (x \curlyvee y)$; hence $\{\curlyvee\}$ is complete.
- $\neg x = x \curlywedge x$ and $x \vee y = (x \curlywedge y) \curlywedge (x \curlywedge y)$; hence $\{\curlywedge\}$ is complete.

(Here, the symbols $\neg, \wedge, \vee, \Rightarrow, \Leftrightarrow, \curlyvee$, and $\curlywedge$ denote operations on $\{0, 1\}$.)

(e) Let δ_1 denote the assignment of truth values to P that is constant and equal to 1 and let $\mathcal{H}$ denote the set of formulas that can be written using the symbols for connectives $\top, \Rightarrow, \wedge$, and $\vee$, to the exclusion of all others. $\mathcal{H}$ is defined inductively as the smallest set of formulas that includes $P \cup \{\top\}$ and that is closed under the operations

$$(M, N) \mapsto (M \Rightarrow N), \quad (M, N) \mapsto (M \wedge N) \text{ and } (M, N) \mapsto (M \vee N).$$

(We may compare this with Exercise 20). By induction, we show that the assignment of truth values δ_1 satisfies all the formulas that belong to $\mathcal{H}$. For the propositional variables, this is true by definition of δ_1; it is also true for the formula $\top$, hence for all formulas in $\mathcal{H}$ of height 0. Assuming that F and G are two formulas of $\mathcal{H}$ such that $\delta_1(F) = \delta_1(G) = 1$, then we have

$$\delta_1((F \Rightarrow G)) = \delta_1((F \wedge G)) = \delta_1((F \vee G)) = 1.$$

It follows from this that the formula $\neg A$ is not equivalent to any formula of $\mathcal{H}$ since $\delta_1(\neg A) = 0$. The conclusion is that $\{1, \Rightarrow, \wedge, \vee\}$ is not a complete system.

It is easy to find a formula, $\neg A$ for example, that is not logically equivalent to any of the three formulas $\top, \perp$, and A. We conclude from this, by virtue of question (b), that $\neg A$ is not logically equivalent to any formula that can be written using the variable A and only the connectives $\top, \perp, \wedge$, and $\vee$; this shows that the system $\{\wedge, \vee, 0, 1\}$ is not complete. One may object that we have not considered the possibility that $\neg A$ could be logically equivalent to a formula that involves only the connectives $\top, \perp, \wedge$, and $\vee$ but that might involve variables other than A: if G were such a formula, the formula G' obtained from A by replacing all occurrences of variables other than A by $\top$ would still be equivalent to $\neg A$ (Lemma 1.19) and this would contradict what we proved earlier.

A similar argument, using question (c), will show that the system $\{0, 1, \neg,$ $\Leftrightarrow\}$ is not complete. Consider, for example, the formula $(A \vee B)$ which is not logically equivalent to any of the eight formulas in the set E of question (c). From this, we conclude that it is not logically equivalent to any formula that can be written using only the connectives $\top$, $\bot$, $\neg$, and $\Leftrightarrow$. (If such a formula existed, we could find one that is logically equivalent, that is written with the same connectives and with the propositional variables A and B alone.)

(f) Refer to the tables of one-place and two-place connectives given in Section 1.3. No system of one-place connectives is complete: for example, the formula $(A_1 \vee A_2)$ is not logically equivalent to any formula written with only the connectives $\top$, $\bot$, and $\neg$.

Let us now prove that, with the exception of φ_9 and φ_{15}, no two-place connective constitutes, by itself, a complete system. As far as φ_2, φ_4, φ_6, φ_8, φ_{10}, φ_{12}, φ_{14}, and φ_{16} are concerned, notice that each of these connectives assumes the value 1 at the point $(1, 1)$, which proves that the connective φ_5, for example, which assumes the value 0 at the point $(1, 1)$, cannot be obtained by composition from one of these (nor from several of these, for that matter). This argument was already used in another form, in question (e) to show that the system $\{1, \Rightarrow, \wedge, \vee\}$ (which is none other than $\{\varphi_2, \varphi_8, \varphi_{14}, \varphi_{16}\}$) is not complete. A kind of dual argument applies for connectives that assume the value 0 at $(0, 0)$: the case of φ_1, φ_3, φ_5, and φ_7 are thereby settled. As for φ_{11} and φ_{13}, they really depend on only one of their arguments; so if either of these constituted a complete system by itself, the same would be true for the corresponding one-place connective and we have seen that this is not true. Since we said in question (d) that each of the connectives φ_9 and φ_{15} constitutes a complete system, we have arrived at the desired conclusion.

16. (a) We have already indicated, several times, the logical equivalence between $(A \Leftrightarrow (B \Leftrightarrow C))$ and $((A \Leftrightarrow B) \Leftrightarrow C)$ (item 58 in Section 1.2.3); we could check this with the help of the corresponding polynomials in $\mathbb{Z}/2\mathbb{Z}[X, Y, Z]$ (Exercise 14) or, as well, by verifying that these two formulas are satisfied by precisely the same assignments of truth values to $\{A, B, C\}$, namely $(0, 0, 1)$, $(0, 1, 0)$, $(1, 0, 0)$, and $(1, 1, 1)$. As for the formula $((A \Leftrightarrow B) \wedge (B \Leftrightarrow C))$, it is satisfied by the two assignments of truth values $(0, 0, 0)$ and $(1, 1, 1)$ and only these; it is therefore not logically equivalent to $(A \Leftrightarrow (B \Leftrightarrow C))$. Now it is a nearly universal practice, in contemporary mathematics, to write equivalences in a 'chain'; for example, when we say of three properties I, II, and III that they are equivalent, which we would write I$\Leftrightarrow$II$\Leftrightarrow$III, what we mean is that either all three of them are true or else that all three of them are false; and we, in effect, interpret the written expression I$\Leftrightarrow$II$\Leftrightarrow$III as (I$\Leftrightarrow$II) $\wedge$ (II$\Leftrightarrow$III) and certainly not as (I$\Leftrightarrow$(II$\Leftrightarrow$III)) which, as we have seen, has a different meaning. Naturally, this goes against the usual conventions relating to associativity which, were they applied in this case, would lead to interpreting I⇔II⇔III indifferently, as an abbreviation for either (I⇔(II⇔III)) or ((I⇔II)⇔III). That is the reason why, in this particular case, it is important not to use any abbreviations: neither the one suggested by associativity, for the formula $(A \Leftrightarrow (B \Leftrightarrow C))$, nor the one dictated by mathematical practice, for the formula $((A \Leftrightarrow B) \wedge (B \Leftrightarrow C))$.

(b) The argument is by induction on the cardinality, n, of the set $\mathcal{B}$. For $n = 2$, it is obvious. Suppose that the property is true for all sets of cardinality less than n and let us prove it if the cardinality of $\mathcal{B}$ is equal to n.

Let $F \in \mathcal{G}(\mathcal{B})$. It is appropriate to distinguish three cases:

- The formula F has the form $(H \Leftrightarrow B)$, where B is a propositional variable belonging to $\mathcal{B}$ and $H \in \mathcal{G}(\mathcal{B} - \{B\})$. For an assignment of truth values δ to satisfy F, it is necessary and sufficient that one of the following two situations holds:

   (i) δ satisfies H and B
   (ii) δ does not satisfy H nor B.

In case (i), there are an even number of elements of $\mathcal{B} - \{B\}$ that are not satisfied by δ and exactly the same number in $\mathcal{B}$. In case (ii), there are an odd number of elements of $\mathcal{B} - \{B\}$ that are not satisfied by δ and, as $\delta(B) = \mathbf{0}$, there are an even number in $\mathcal{B}$.

- The formula F has the form $(B \Leftrightarrow H)$, where B is a propositional variable in $\mathcal{B}$ and $H \in \mathcal{G}(\mathcal{B} - \{B\})$. The same analysis applies.
- There exists a partition of $\mathcal{B}$ into two sets $\mathcal{B}_1$ and $\mathcal{B}_2$, each having at least two elements, and formulas H_1 and H_2 belonging respectively to $\mathcal{G}(\mathcal{B}_1)$ and $\mathcal{G}(\mathcal{B}_2)$ such that $F = (H_1 \Leftrightarrow H_2)$. For δ to satisfy F, it is necessary and sufficient that δ satisfy both H_1 and H_2 or else that δ satisfy neither H_1 nor H_2. In the first case, the number of propositional variables in $\mathcal{B}_1$ that are not satisfied by δ is even (by the induction hypothesis) and so is the number of propositional variables in $\mathcal{B}_2$ that are not satisfied by δ, which, in all, make an even number. In the second case, the number of propositional variables in $\mathcal{B}_1$ that are not satisfied by δ is odd and so is the number of propositional variables in $\mathcal{B}_2$ that are not satisfied by δ, which, once again, make an even number in all.

(c) We prove by induction on the cardinality of the set $\mathcal{B}$ that, for every formula $G \in \mathcal{G}(\mathcal{B})$, the formula $\tilde{G}$ is satisfied by an assignment of truth values δ if and only if δ satisfies an odd number of propositional variables in $\mathcal{B}$. The proof is analogous to the one for (b). The required equivalence follows immediately.

(d) Let $x \in E$. For each integer i between 1 and k, we introduce a propositional variable A_i and define the assignment of truth values δ by

$$\delta(A_i) = \mathbf{1} \text{ if and only if } x \in X_i.$$

We easily see, by induction on the integer k, that $x \in X_1 \Delta X_2 \Delta \ldots \Delta X_k$ if and only if δ satisfies the following formula F:

$$F = (\ldots ((A_1 \Leftrightarrow A_2) \Leftrightarrow A_3) \cdots \Leftrightarrow A_k);$$

the result now follows from question (c). (Also see Exercise 2 from Chapter 2.)

17. (a) We argue by contradiction. If the formula

$$((F[A_1, A_2, \ldots, A_n, A] \wedge F[A_1, A_2, \ldots, A_n, B]) \Rightarrow (A \Leftrightarrow B))$$

is not a tautology, there exists an assignment of truth values δ that satisfies the formulas $F[A_1, A_2, \ldots, A_n, A]$ and $F[A_1, A_2, \ldots, A_n, B]$ and does not satisfy the formula $(A \Leftrightarrow B)$. But, by hypothesis, δ must satisfy the formula

$$(F[A_1, A_2, \ldots, A_n, A] \Rightarrow (G[A_1, A_2, \ldots, A_n] \Leftrightarrow A)),$$

since this is a tautology, as well as the formula obtained by substituting B for A in the preceding, namely,

$$(F[A_1, A_2, \ldots, A_n, B] \Rightarrow (G[A_1, A_2, \ldots, A_n] \Leftrightarrow B)).$$

Since $\delta(F[A_1, A_2, \ldots, A_n, A]) = \delta(F[A_1, A_2, \ldots, A_n, B]) = \mathbf{1}$, we must have

$$\delta(G[A_1, A_2, \ldots, A_n]) = \delta(A) = \delta(B),$$

which is incompatible with the fact that δ does not satisfy $(A \Leftrightarrow B)$.

(b) For each of the cases under consideration, we present formulas $G[A_1, A_2, \ldots, A_n]$ that are possible definitions of A modulo F (these need not be unique). The verifications are immediate and are left to the reader.

    (1) $G = A_1$
    (2) $G = A_1$ or $G = A_2$
    (3) $G = A_1$ or $G = A_2$ or $G = (A_1 \vee \neg A_1)$
    (4) $G = (A_1 \vee \neg A_1)$ or $G = \neg A_2$
    (5) $G = (A_1 \vee \neg A_1)$ or $G = (A_1 \Leftrightarrow (A_2 \Leftrightarrow A_3))$.

18. (a) It suffices to prove that if $\varphi_F(\varepsilon_1, \varepsilon_2, \ldots, \varepsilon_n, \mathbf{1}) = \mathbf{1}$, then we do not have $\varphi_F(\varepsilon_1, \varepsilon_2, \ldots, \varepsilon_n, \mathbf{0}) = \mathbf{1}$. We argue by contradiction. If these two equalities were verified, this would mean that the assignments of truth values λ and μ on $\{A_1, A_2, \ldots, A_n, A\}$ defined by $\lambda(A_1) = \mu(A_1) = \varepsilon_1$, $\lambda(A_2) = \mu(A_2) = \varepsilon_2, \ldots, \lambda(A_n) = \mu(A_n) = \varepsilon_n, \lambda(A) = \mathbf{1}$ and $\mu(A) = \mathbf{0}$

both satisfy the formula $F[A_1, A_2, \ldots, A_n, A]$. This amounts to saying that the assignment of truth values δ on $\{A_1, A_2, \ldots, A_n, A, B\}$ defined by

$$\delta(A_1) = \varepsilon_1, \ \delta(A_2) = \varepsilon_2, \ldots, \ \delta(A_n) = \varepsilon_n, \ \delta(A) = \mathbf{1} \text{ and } \delta(B) = \mathbf{0}$$

simultaneously satisfies the formulas

$$F[A_1, A_2, \ldots, A_n, A] \text{ and } F[A_1, A_2, \ldots, A_n, B].$$

As we have assumed that

$$((F[A_1, A_2, \ldots, A_n, A] \wedge F[A_1, A_2, \ldots, A_n, B]) \Rightarrow (A \Leftrightarrow B))$$

is a tautology, δ must also satisfy $(A \Leftrightarrow B)$; but this contradicts the definition of δ.

(b) Having chosen G as indicated, let us consider a distribution of truth values δ on $\{A_1, A_2, \ldots, A_n, A\}$ that satisfies the formula $F[A_1, A_2, \ldots, A_n, A]$ and set

$$\varepsilon_1 = \delta(A_1), \ \varepsilon_2 = \delta(A_2), \ldots, \varepsilon_n = \delta(A_n).$$

- If $\delta(A) = \mathbf{0}$, then $\varphi_F(\varepsilon_1, \varepsilon_2, \ldots, \varepsilon_n, \mathbf{0}) = \mathbf{1}$, hence $\varphi_G(\varepsilon_1, \varepsilon_2, \ldots, \varepsilon_n) = \psi(\varepsilon_1, \varepsilon_2, \ldots, \varepsilon_n) = \mathbf{0}$, which means that $\overline{\delta}(G) = \mathbf{0}$.
- If $\delta(A) = \mathbf{1}$, then $\varphi_F(\varepsilon_1, \varepsilon_2, \ldots, \varepsilon_n, \mathbf{1}) = \mathbf{1}$, hence according to the first question, $\varphi_G(\varepsilon_1, \varepsilon_2, \ldots, \varepsilon_n) = \psi(\varepsilon_1, \varepsilon_2, \ldots, \varepsilon_n) = \mathbf{1}$, which means that $\overline{\delta}(G) = \mathbf{1}$.

So in every case, $\delta(A) = \overline{\delta}(G)$, which shows that δ satisfies the formula $(G[A_1, A_2, \ldots, A_n] \Leftrightarrow A)$. Thus we have proved that the formula

$$(F[A_1, A_2, \ldots, A_n, A] \Rightarrow (G[A_1, A_2, \ldots, A_n] \Leftrightarrow A))$$

is a tautology, since every assignment that satisfies the left side of this implication also satisfies the right side.

19. (a) A formula is determined, up to logical equivalence, by its truth table or, which amounts to the same thing, by the set of assignments of truth values that satisfy it. The set of assignments of truth values on $P = \{A, B, C, D, E\}$ has $2^5 = 32$ elements. There are therefore $_{17}C_{32}$ subsets that have seventeen elements. If one prefers, there are $_{17}C_{32}$ ways to place seventeen $\mathbf{1}$s and fifteen $\mathbf{0}$s in the last column of a truth table that has 32 lines. Consequently, there are, up to logical equivalence,

$$_{17}C_{32} = \frac{32!}{17! \ 15!} = 565\ 722\ 720$$

formulas that are satisfied by seventeen assignments of truth values (whereas there are $2^{32} = 4\ 294\ 967\ 296$ equivalence classes for the equivalence relation on formulas that is logical equivalence).

(b) Let Λ denote the set of assignments of truth values δ on $\{A, B, C, D, E\}$ that satisfy $\delta(A) = \delta(B) = \mathbf{1}$. For a formula to be a consequence of $(A \wedge B)$, it is necessary and sufficient that it receive the value $\mathbf{1}$ for all the assignments that belong to Λ. There are eight such assignments (we extend each of the eight assignments from $\{C, D, E\}$ into $\{\mathbf{0}, \mathbf{1}\}$ by setting their values on A and B equal to $\mathbf{1}$). A formula that is a consequence of $(A \wedge B)$ is therefore determined, up to logical equivalence, by the set of assignments of truth values, other than the eight required, that satisfy it. So there are as many such formulas as there are subsets of the set $\{\mathbf{0}, \mathbf{1}\}^P - \Lambda$, which is to say,

$$2^{24} = 16\ 777\ 216.$$

If one prefers, in the last column of the truth table of a formula that is a consequence of $(A \wedge B)$, the value $\mathbf{1}$ must appear in the eight lines corresponding to the assignments of truth tables in Λ; in the other 24 lines, we may arbitrarily place the values $\mathbf{0}$ or $\mathbf{1}$; this leads to exactly 2^{24} possible truth tables.

20. (a) We argue by induction on F. If F is a propositional variable, we may obviously take the formula G to be F itself. Next, let H and K be two formulas with which we have associated (induction hypothesis) negation-free formulas H' and K' such that H is logically equivalent to H' or to $\neg H'$ and K is logically equivalent to K' or to $\neg K'$.

We will distinguish four possibilities:

11. $H \sim H'$ and $K \sim K'$;
10. $H \sim H'$ and $K \sim \neg K'$;
01. $H \sim \neg H'$ and $K \sim K'$;
00. $H \sim \neg H'$ and $K \sim \neg K'$.

We will prove that in each of the following five cases:

$I. \quad F = \neg H$
$II. \quad F = (H \wedge K)$
$III. \quad F = (H \vee K)$
$IV. \quad F = (H \Rightarrow K)$
$V. \quad F = (H \Leftrightarrow K)$

we can find a formula G that does not involve negation and is such that F is logically equivalent either to G or to $\neg G$.

I. If H is logically equivalent to H', F is logically equivalent to $\neg H'$; if H is logically equivalent to $\neg H'$, F is logically equivalent to $\neg\neg H'$, hence to H'; we may therefore take $G = H'$.

II. In case 11, we take $G = (H' \wedge K')$ and we have $F \sim G$. In case 10, we take $G = (H' \Rightarrow K')$ and we have $F \sim \neg G$. In case 01, we take $G = (K' \Rightarrow H')$ and we have $F \sim \neg G$. Finally, in case 00, we take $G = (H' \vee K')$ and we have $F \sim \neg G$.

III. In case 11, we take $G = (H' \vee K')$ and we have $F \sim G$. In case 10, we take $G = (K' \Rightarrow H')$ and we have $F \sim G$. In case 01, we take $G = (H' \Rightarrow K')$ and we have $F \sim G$. Finally, in case 00, we take $G = (H' \wedge K')$ and we have $F \sim \neg G$.

IV. In case 11, we take $G = (H' \Rightarrow K')$ and we have $F \sim G$. In case 10, we take $G = (H' \wedge K')$ and we have $F \sim \neg G$. In case 01, we take $G = (H' \vee K')$ and we have $F \sim G$. Finally, in case 00, we take $G = (K' \Rightarrow H')$ and we have $F \sim G$.

V. In cases 11 and 00, we take $G = (H' \Leftrightarrow K')$ and we have $F \sim G$. In cases 10 and 00, take $G = (H' \Leftrightarrow K')$ and we have $F \sim \neg G$.

(b) It is obvious that (i) implies (ii). To see that (ii) implies (i) is not difficult: as the formula $(G \Leftrightarrow H)$ is logically equivalent, for all formulas G and H, to $((G \Rightarrow H) \wedge (H \Rightarrow G))$, any formula written using the four connectives $\wedge$, $\vee$, $\Rightarrow$, and $\Leftrightarrow$ is logically equivalent to a formula that can be written using only the first three of these symbols. (Perfectionists would prove this by induction on F.) Let us now prove the equivalence of (ii) and (iii).

(ii) implies (iii). We can see this by induction on the height of formulas F written without negation: if it is a variable, $\overline{\delta_1}(F) = \mathbf{1}$ by definition of δ_1, if it is $(G \wedge H)$, $(G \vee H)$, $(G \Rightarrow H)$, or $(G \Leftrightarrow H)$, where $\overline{\delta_1}(G) = \mathbf{1}$ and $\overline{\delta_1}(H) = \mathbf{1}$, then we obviously have $\overline{\delta_1}(F) = \mathbf{1}$ also.

Conversely, let F be a formula such that $\overline{\delta_1}(F) = \mathbf{1}$. According to (a), we can find a negation-free formula G such that F is logically equivalent either to G or to $\neg G$. As G is negation-free, we may conclude, knowing that (ii) implies (iii), that $\overline{\delta_1}(G) = \mathbf{1} = \overline{\delta_1}(F)$. It is therefore not possible for F to be logically equivalent to $\neg G$; F is consequently logically equivalent to G. We have thus shown that (iii) implies (ii).

21. (a) The reflexivity, transitivity, and antisymmetry of the relation $\ll$ are proved without difficulty. This is certainly an order relation, but this is not a total ordering: if $n \geq 2$, the assignments of truth values λ and μ defined by $\lambda(A_1) = \mathbf{0}$, $\mu(A_1) = \mathbf{1}$, $\lambda(A_i) = \mathbf{1}$, and $\mu(A_i) = \mathbf{0}$ for $2 \leq i \leq n$ do not satisfy $\lambda \ll \mu$ nor $\mu \ll \lambda$.

(b) The formula $(A_1 \Rightarrow A_2)$ is an example of a formula that is not increasing and whose negation is not increasing: to see this, it is sufficient to consider the assignments of truth values λ, μ, and ν defined by

$$\lambda(A_i) = \mathbf{0} \text{ for all } i \in \{1, 2, \ldots, n\};$$
$$\mu(A_1) = \mathbf{1} \text{ and } \mu(A_i) = \mathbf{0} \text{ for all } i \in \{2, 3, \ldots, n\};$$
$$\nu(A_i) = \mathbf{1} \text{ for all } i \in \{1, 2, \ldots, n\}.$$

So we have $\lambda \ll \mu$, $\lambda(F) = \mathbf{1}$, and $\mu(F) = \mathbf{0}$; hence F is not increasing. On the other hand, $\mu \ll \nu$, $\mu(\neg F) = \mathbf{1}$, and $\nu(\neg F) = \mathbf{0}$; hence $\neg F$ is not increasing.

(c) First we prove 'if'. It is clear that if F is a tautology, or if $\neg F$ is a tautology (i.e. F is an antilogy), then F is increasing. Let $\mathcal{C}$ denote the set of formulas in which the symbols $\neg$, $\Rightarrow$, and $\Leftrightarrow$ do not occur. Because it is evident that any formula logically equivalent to an increasing formula is itself increasing, it suffices to prove that if F belongs to $\mathcal{C}$, then F is increasing. We prove this by induction on the height of the formula F: it needs to be proved that propositional formulas are increasing (this is obvious) and that if G and H are increasing, then so are $(G \wedge H)$ and $(G \vee H)$; this is not very difficult.

Next we prove 'only if'. Let F be an increasing formula that is neither a tautology nor an antilogy. We must prove that there exists a formula G belonging to $\mathcal{C}$, defined above, that is logically equivalent to F.

Let $\Delta(F)$ denote the set of assignments of truth values that satisfy F:

$$\Delta(F) = \{\delta \in \{0, 1\}^P : \overline{\delta}(F) = 1\}.$$

For an arbitrary assignment of truth values δ, set

$$V_+(\delta) = \{A \in P : \delta(A) = 1\}.$$

Note that the sets defined this way are finite and that $\Delta(F)$ is not empty (otherwise $\neg F$ would be a tautology). We also see that for every δ belonging to $\Delta F)$, $V_+(\delta)$ is not empty: for there is only one assignment of truth values δ for which $V_+(\delta) = \emptyset$, namely the assignment δ_0 defined by $\delta_0(A) = 0$ for all propositional variables A. And if δ_0 belonged to $\Delta(F)$, we would have $\delta_0(F) = 1$; but as F is increasing and since every assignment of truth values δ satisfies $\delta_0 \ll \delta$, we would also have $\delta(F) = 1$ for all $\delta \in \{0, 1\}^P$; but this is impossible since F is not a tautology.

We may thus define the formula

$$G = \bigvee_{\delta \in \Delta(F)} \left(\bigwedge_{A \in V_+(\delta)} A \right)$$

which is clearly an element of the set $\mathcal{C}$.

Now let us show that F and G are logically equivalent. Let λ be an assignment of truth values on P that satisfies F; it follows that $\lambda \in \Delta(F)$, hence the formula $\wedge_{A \in V_+(\lambda)} A$ is one of the terms in the disjunction that is the formula G. Since we have, by definition, that $\lambda(A) = 1$ for all $A \in V_+(\lambda)$, we conclude that $\overline{\lambda}(\wedge_{A \in V_+(\delta)} A) = 1$ and hence that $\overline{\lambda}(G) = 1$. Conversely, if μ is an element of $\{0, 1\}^P$ that satisfies G, there exists an assignment of truth values $\delta \in \Delta(F)$ such that $\overline{\mu}(\wedge_{A \in V_+(\delta)} A) = 1$; this means that for all A belonging to $V_+(\delta)$, $\mu(A) = 1$, which in turn means that for every propositional variable A, if $\delta(A) = 1$, then $\mu(A) = 1$; in other words, for every propositional variable A, $\delta(A) \leq \mu(A)$; this says that $\delta \ll \mu$. Since $\delta \in \Delta(F)$ (so $\overline{\delta}(F) = 1$) and F is an increasing formula, it follows that $\overline{\mu}(F) = 1$. We have established that F and G are satisfied by exactly the same assignments of truth values to P; so the two formulas are logically equivalent.

22. Given a finite set of formulas $\{F_1, F_2, \ldots, F_k\}$, to show that it is independent, it suffices to find, for each index i, an assignment of truth values that satisfies all the $F_j (j \neq i)$ and that does not satisfy F_i; on the contrary, to prove that it is not independent, we show that one of the F_i is a consequence of the others.

In the situation where the set of formulas under consideration is not independent, we invoke the following observation when we wish to find an equivalent independent subset: if $\mathcal{A}$ is a set of formulas and if the formula F is a consequence of $\mathcal{A} - \{F\}$, then the sets $\mathcal{A}$ and $\mathcal{A} - \{F\}$ are equivalent.

(a) We will only treat sets (1), (2), and (6). None of the other sets is independent.

(1) The set $\{(A \Rightarrow B), (B \Rightarrow C), (C \Rightarrow A)\}$ is independent. Note that the assignment of truth values δ defined by $\delta(A) = \mathbf{1}$, $\delta(B) = \mathbf{0}$, and $\delta(C) = \mathbf{1}$ does not satisfy the first formulas but does satisfy the other two; similarly, we note that any assignment of truth values that makes one of these formulas false will necessarily satisfy the other two.

(2) The set $\{(A \Rightarrow B), (B \Rightarrow C), (A \Rightarrow C)\}$ is not independent since

$$\{(A \Rightarrow B), (B \Rightarrow C)\} \vdash^* (A \Rightarrow C).$$

The subset $\{(A \Rightarrow B), (B \Rightarrow C)\}$ is independent and is equivalent to the given set. It is easy to see that this is the only subset with this property.

(6) The set $\{((A \Rightarrow B) \Rightarrow C), (A \Rightarrow C), (B \Rightarrow C), (C \Rightarrow (B \Rightarrow A)),$ $((A \Rightarrow B) \Rightarrow (A \Leftrightarrow B))\}$ is not independent; observe that the last formula is equivalent to $(B \Rightarrow A)$ and so the next to last formula is a consequence of it. There are two independent equivalent subsets:

$$\{((A \Rightarrow B) \Rightarrow C), (A \Rightarrow C), (C \Rightarrow (B \Rightarrow A))\} \text{ and}$$
$$\{((A \Rightarrow B) \Rightarrow C), (A \Rightarrow C), ((A \Rightarrow B) \Rightarrow (A \Leftrightarrow B))\}.$$

(b) The empty set is independent: if it were not, it would contain a formula F such that $\emptyset - \{F\} \models^* F$; this is obviously impossible. If $\mathcal{A} = \{G\}$, then $\mathcal{A} - \{G\} \models^* G$ is equivalent to $\emptyset \models^* G$ (which means that G is a tautology). Consequently, a necessary and sufficient condition for a set containing a single formula to be independent is that this formula is not a tautology.

(c) We establish this property by induction on the number of formulas in the set. It is true when this number is 0 because $\emptyset$ is an independent subset that is equivalent to $\emptyset$. Suppose that every set containing n formulas has at least one equivalent independent subset and consider a set $\mathcal{A}$ of $n + 1$ formulas. If $\mathcal{A}$ is independent, it is itself an independent subset equivalent to $\mathcal{A}$. If not, then we can find in $\mathcal{A}$ a formula F that is a consequence of $\mathcal{B} = \mathcal{A} - \{F\}$. $\mathcal{B}$, which contains n formulas, has, by the induction hypothesis, an independent subset C that is equivalent to B. But B is equivalent to A according to the remark at the beginning of this solution. As a result, C is a subset of A that is independent and equivalent to A.

Remark: Concerning this proof, there are two errors to be avoided (which, by the way, are related). The first consists in believing that if A is an independent set of formulas and if F is a formula that is not a consequence of A, then $A \cup \{F\}$ is independent. The second consists in thinking that a maximal independent subset of a set of formulas is necessarily equivalent to this set. The example that follows shows that these two ideas are incorrect.

Let $A = \{A\}$, $F = (A \wedge B)$ and $B = A \cup \{F\}$. We see immediately that A is independent, that F is not a consequence of A, that $A \cup \{F\}$ is not independent and that A is a maximal independent subset of B but is not equivalent to B.

(d) If A is an independent set of formulas and if B is a subset of A (finite or not), then B is independent. Now suppose that A is a set of formulas that is not independent. So there is at least one formula G in A such that $A - \{G\} \vdash^* G$. According to the compactness theorem, there is some finite subset B of $A - \{G\}$ such that $B \vdash^* G$. Set $C = B \cup \{G\}$; we then have $C - \{G\} \vdash^* G$, which proves that C is a finite subset of A that is not independent. Thus, for a set of formulas to be independent, it is necessary and sufficient that all its finite subsets are independent.

(e) For each integer $n \geq 1$, set $F_n = A_1 \wedge A_2 \wedge \cdots \wedge A_n$ and let A denote the set $\{F_n : n \in \mathbb{N}^*\}$. For $n \leq p$, F_n is a consequence of F_p; thus the only subsets of A that are independent consist of a single element. It is also clear that for every n, F_{n+1} is not a consequence of F_n (just take the assignment of truth values that satisfies $A_1, A_2, \ldots, A_n$ and not A_{n+1}); it follows that no independent subset of A can be equivalent to A. None the less, there exist independent sets that are equivalent to A, for example, $\{A_1, A_2, \ldots, A_n, \ldots\}$.

(f) We seek a set of formulas equivalent to $\mathcal{F} = \{F_0, F_1, \ldots, F_n, \ldots\}$. First, we obtain a set that is equivalent to $\mathcal{F}$ by removing those formulas F_n that are a consequence of $\{F_0, F_1, \ldots, F_{n-1}\}$. In other words, we may assume that for every n, the formula F_n is not a consequence of $\{F_0, F_1, \ldots, F_{n-1}\}$ and, in particular, that F_0 is not a tautology. We then consider the following set $\mathcal{G}$:

$$\mathcal{G} = \{F_0, \ F_0 \Rightarrow F_1, \ (F_0 \wedge F_1) \Rightarrow F_2, \ldots,$$
$$(F_0 \wedge F_1 \wedge \cdots \wedge F_n) \Rightarrow F_{n+1}, \ldots\}.$$

It is clear that if an assignment of truth values satisfies all the formulas F_n, then it satisfies $\mathcal{G}$; conversely, if it satisfies all the formulas in $\mathcal{G}$, then we see, by induction on n, that it satisfies all the formulas F_n. The sets $\mathcal{F}$ and $\mathcal{G}$ are thus equivalent. We will show that $\mathcal{G}$ is independent by exhibiting, for every formula G in $\mathcal{G}$, an assignment of truth values that does not satisfy G but that satisfies all the other formulas of $\mathcal{G}$.

- If $G = F_0$, we take an assignment that makes F_0 false (one exists since F_0 is not a tautology). The other formulas of $\mathcal{G}$ are then all satisfied.
- If $G = (F_0 \wedge F_1 \wedge \cdots \wedge F_n) \Rightarrow F_{n+1}$, we choose an assignment of truth values δ that satisfies $F_0, F_1, \ldots, F_n$ and that makes F_{n+1} false (one exists since F_{n+1} is not a consequence of $\{F_0, F_1, \ldots, F_{n-1}\}$); it is easy to verify that δ has the required properties.

23. (a) For each $a \in E$, consider the formula F_a:

$$F_a = \left(\bigvee_{1 \le i \le k} A_{a,i} \right) \wedge \left(\bigwedge_{1 \le i < j \le k} \neg (A_{a,i} \wedge A_{a,j}) \right),$$

and for every pair $(a, b) \in E^2$, the formula:

$$H_{a,b} = \bigwedge_{1 \le i \le k} \neg (A_{a,i} \wedge A_{b,i}).$$

We will show that G is k-colourable if and only if the set

$$\mathcal{A}(E, G) = \{F_a : a \in E\} \cup \{H_{a,b} : (a, b) \in G\}$$

is satisfiable. If f is a function from E into $\{1, 2, \ldots, k\}$ that satisfies the conditions that are required for G to be k-colourable, we define the assignment of truth values δ by

$$\delta(A_{a,i}) = \mathbf{1} \text{ if and only if } f(a) = i,$$

and we easily see that δ satisfies all the formulas in $\mathcal{A}(E, G)$.

Conversely, suppose that we have an assignment of truth values δ that satisfies all the formulas in $\mathcal{A}(E, G)$. Let $a \in E$. The fact that F_n is satisfied by δ means that there is one and only one integer i between 1 and k such that $\delta(A_{a,i}) = \mathbf{1}$; denote this integer by $f(a)$. The fact that δ satisfies the formulas $H_{a,b}$ for $(a, b) \in G$ shows that if $(a, b) \in G$, then $f(a) \neq f(b)$. G is therefore k-colourable.

(b) It is clear that if a graph is k-colourable, then all its subgraphs and, in particular, all its finite subgraphs are k-colourable. Conversely, suppose that all the finite subgraphs of G are k-colourable. We will establish that G is k-colourable by showing that $\mathcal{A}(E, G)$ is satisfiable and, to do this, we will invoke the compactness theorem. So let $\mathcal{A}$ be a finite subset of $\mathcal{A}(E, G)$. Let E' denote the subset of points a in E such that for some integer i, the variable $A_{a,i}$ occurs in some formula of $\mathcal{A}$ and let G' denote the restriction of G to E'. G' is a finite subgraph of G and $\mathcal{A} \subseteq \mathcal{A}(E', G')$. Since G' is k-colourable, $\mathcal{A}(E', G')$ is satisfiable (question (a)) and so too is $\mathcal{A}$. Hence, by the compactness theorem, $\mathcal{A}(E, G)$ is satisfiable and G is k-colourable (question (a)).

The following illustration explains the terminology used in this exercise. For E, take a set of countries and for G the relation 'have a common border that does not reduce to a single point'; in this context, k-colourability corresponds to the possibility of attributing a colour to each country (in view of producing a geographical map), the diverse colours available being $E_1, E_2, \ldots, E_k$. The obvious constraint is that bordering countries must always receive distinct colours. Cartographers rarely have to deal with an infinite set of countries, so this exercise is of no great use to them. It is rather the problem of finding the smallest possible value for k that concerned specialists in graph theory for a long time. The conjecture was that this smallest value is 4 (for a certain class of graphs that are not too complicated): this was the famous four-colour problem that remained unsolved until 1986, when two American mathematicians together with a number of powerful computers produced a 'proof' of this conjecture. The quotes are justified here by the fact that, even if the number of pages of this book were increased beyond anything the reader might imagine, and if we had the necessary competence, we would still be hard pressed to provide it here ....
This is the first example of its kind for a theorem of mathematics. The only thing that logic has taught us is that the four-colour problem is no easier for finite sets than it is for infinite sets.

24. (a) Intuitively, the propositional variable $A_{x,y}$ has the value **1** if and only if x is less than or equal to y. We set:

$$\mathcal{B}(G) = \{A_{x,x} : x \in G\};$$
$$\mathcal{C}(G) = \{(A_{x,y} \wedge A_{y,z}) \Rightarrow A_{x,z} : x \in G, \ y \in G, \ z \in G\};$$
$$\mathcal{D}(G) = \{(A_{x,y} \not\Leftrightarrow A_{y,x}) : x \in G, \ y \in G, \ x \neq y\};$$
$$\mathcal{E}(G) = \{A_{x,y} \Rightarrow A_{x\cdot z, y\cdot z} : x \in G, \ y \in G, \ z \in G\};$$

and

$$\mathcal{A}(G) = \mathcal{B}(G) \cup \mathcal{C}(G) \cup \mathcal{D}(G) \cup \mathcal{E}(G).$$

Suppose that the group G is orderable. Let $\leq$ be a total ordering on G that is compatible with the group operation. Define an assignment of truth values δ on P by setting:

for all elements x and y of G,
$$\delta((x, y)) = \mathbf{1} \text{ if and only if } x \leq y.$$

This assignment satisfies all the formulas in $\mathcal{A}(G)$: those in $\mathcal{B}(G)$ because the relation $\leq$ is reflexive, those in $\mathcal{C}(G)$ because it is transitive, those in $\mathcal{D}(G)$ because it is antisymmetric and total, and finally, those in $\mathcal{E}(G)$ because it is compatible with the group operation.
We have thus shown that the set $\mathcal{A}(G)$ is satisfiable when the group G is orderable. Conversely, suppose that $\mathcal{A}(G)$ is satisfiable. Let λ be an

assignment of truth values that satisfies it and define a binary relation R on G as follows:

for all elements x and y of G,
$$(x, y) \in R \text{ if and only if } \lambda(A_{x,y}) = 1.$$

The fact that the sets $\mathcal{B}(G)$, $\mathcal{C}(G)$, and $\mathcal{E}(G)$ are satisfied shows that R is reflexive, transitive, and compatible with the group operation. Also, R is antisymmetric and total because λ satisfies the formulas in $\mathcal{D}(G)$. This relation is therefore a total ordering that is compatible with the group operation, i.e. G is orderable.

(b) If a group is orderable, it is clear that all its subgroups (in particular, the subgroups of finite type) are orderable (by the restriction to the subgroup of the order on G). It is the converse of this property that is not obvious. Suppose that $\langle G, \cdot, 1 \rangle$ is a group all of whose subgroups of finite type are orderable. According to (a), to prove that G is orderable, it suffices to show that the set of formulas $\mathcal{A}(G)$ is satisfiable. By the compactness theorem, it is then sufficient to show that every finite subset of $\mathcal{A}(G)$ is satisfiable. Let $\mathcal{U}$ be a finite subset of $\mathcal{A}(G)$. Let M denote the set of elements of G that appear in at least one formula of $\mathcal{U}$; this is, of course, a finite subset of G, so the subgroup, H, that it generates is of finite type. According to our hypothesis, H is orderable, hence (question (a)) the set of formulas $\mathcal{A}(H)$ is satisfiable. Because $\mathcal{U}$ is included in $\mathcal{A}(H)$, $\mathcal{U}$ is therefore satisfiable.

(c) First we will show the easy part of the equivalence: if an abelian group is orderable, then it is torsion-free. Let $\langle G, \cdot, 1 \rangle$ be an abelian group that is ordered by $\leq$. Suppose that G is a torsion group, i.e. that there is an element x in G, distinct from 1, and a non-zero natural number n such that $x^n = 1$. (Such an element x is called a **torsion element**). Because the ordering $\leq$ is total, we have either $1 \leq x$ or $x \leq 1$. If we suppose that $1 \leq x$, then using the compatibility of $\leq$ with the group operation $\cdot$, we obtain successively

$$x \leq x^2, x^2 \leq x^3, \ldots, x^{n-1} \leq x^n = 1.$$

It follows that $x \leq 1$ by transitivity and hence, by antisymmetry, $x = 1$, which is not possible. In a similar way, we arrive at a contradiction from the assumption that $x \leq 1$. Thus the group is torsion-free.

Let us now tackle the other implication. Suppose that $\langle G, \cdot, 1 \rangle$ is a torsion-free abelian group and consider a subgroup, H, of G of finite type. Then H is also a torsion-free abelian group (for a torsion element in H would obviously be a torsion element in G also). If H is the trivial subgroup $\{1\}$, it is obviously orderable. If H is non-trivial, then by the theorem stated as part of the exercise, there is a non-zero natural number p such that $\langle H, \cdot, 1 \rangle$ is isomorphic to the group $\langle \mathbb{Z}^p, +, 0 \rangle$. Now the group $\langle G, \cdot, 1 \rangle$ is orderable: it suffices to consider the lexicographical ordering on $\mathbb{Z}^p$, specifically, if the elements $(a_1, a_2, \ldots, a_p)$ and $(b_1, b_2, \ldots, b_p)$ of $\mathbb{Z}^p$ are distinct and if k

is the smallest of the indices i between 1 and p such that $a_i \neq b_i$, then

$$(a_1, a_2, \ldots, a_p) < (b_1, b_2, \ldots, b_p) \text{ if and only if } a_k < b_k.$$

This is clearly a total ordering and it is compatible with the group operation (which, in this case, is coordinate-wise addition): more precisely, if

$$(a_1, a_2, \ldots, a_p) < (b_1, b_2, \ldots, b_p),$$

then for any element $(c_1, c_2, \ldots, c_p)$ of $\mathbb{Z}^p$, we also have

$$(a_1, a_2, \ldots, a_p) + (c_1, c_2, \ldots, c_p) < b_1, b_2, \ldots, b_p) + (c_1, c_2, \ldots, c_p),$$

which is to say,

$$(a_1 + c_1, a_2 + c_2, \ldots, a_p + c_p) < (b_1 + c_1, b_2 + c_2, \ldots, b_p + c_p).$$

If φ is an isomorphism of the group $\langle G, \cdot, \mathbf{1} \rangle$ onto the group $\langle \mathbb{Z}^p, +, \mathbf{0} \rangle$, then the binary relation $\prec$ defined on H by

for all elements x and y of H,
$x \prec y$ if and only if $\varphi(x) \leq \varphi(y)$

is a total ordering on H that is compatible with the group operation (what we are saying here is that any group that is isomorphic to an orderable group is orderable).

We have just proved that every subgroup of $\langle G, \cdot, \mathbf{1} \rangle$ of finite type is orderable. According to question (b), this proves that G itself is orderable.

25. Where precision matters, we will denote properties I, II, and III by $\text{I}_{E,F,R}$, $\text{II}_{E,F,R}$, and $\text{III}_{E,F,R}$ respectively.

(a) If E is the empty set, III is trivially verified (the empty map is injective). This is also the case when E consists of a single element, a: for then, I means that R_a is non-empty, so by choosing an element b in R_a and setting $f(a) = b$, we define a map f that satisfies the requirements. Let k be a natural number greater than or equal to 2. We suppose (induction hypothesis) that for all sets X and Y and relations $S \subseteq X \times Y$, if $\mathsf{card}(X) < k$ and if $\text{I}_{X,Y,S}$ is satisfied, then so is $\text{III}_{X,Y,S}$. Now suppose that E is a set with k elements.

First, we examine case 1: there exists at least one non-empty subset A of E, distinct from E, such that $\mathsf{card}(A) = \mathsf{card}(R_A)$. By the induction hypothesis, we can find an injective map f_1 from A into R_A such that for every element a in A, $f_1(a) \in R_a$. Set $B = E - A$, $C = F - R_A$, and $S = R \cap (B \times C)$. We will show that $\text{I}_{B,C,S}$ is true. To see this, consider an arbitrary subset M of B and set $N = A \cup M$. It is not difficult to verify that

$$R_N = R_A \cup R_M = R_A \cup (R_M - R_A) = R_A \cup S_M.$$

This implies (because R_A and S_M are disjoint) that

$$\mathrm{card}(R_N) = \mathrm{card}(R_A) + \mathrm{card}(S_M)$$

and, according to $I_{E,F,R}$, $\mathrm{card}(R_N) \geq \mathrm{card}(A) + \mathrm{card}(S_M)$ (since A and M are disjoint). It follows that

$$\mathrm{card}(A) + \mathrm{card}(S_M) \geq \mathrm{card}(A) + \mathrm{card}(M).$$

Since $\mathrm{card}(A)$ is finite, we may conclude that

$$\mathrm{card}(S_M) \geq \mathrm{card}(M),$$

which shows that $I_{B,C,S}$ is satisfied.

So the induction hypothesis is applicable; this yields an injective map f_2 from B into C such that, for all b in B, $f_2(b)$ belongs to S_b, which means that $(b, f_2(b)) \in R$. The desired map f is the map that is equal to f_1 on A and to f_2 on B.

Next, consider case 2: for every non-empty subset A of E, distinct from E, the cardinality of A is strictly greater than that of R_A. Choose an element u in E (which is not empty) and an element v in R_u (which is also not empty by virtue of $I_{E,F,R}$ and which even has, in the present case, at least two elements). By the induction hypothesis, there is an injective map from $E - \{u\}$ into $R_u - \{v\}$. It suffices to extend this map to the whole of E by setting $f(u) = v$.

(b) If $E = F = \mathbb{N}$ and if $R = \{(0, n) : n \in \mathbb{N}\} \cup \{(n + 1, n) : n \in \mathbb{N}\}$, we can check that property I is satisfied while II and III are not.

(c) Let X and Y be two sets and let $S \subseteq X \times Y$ be a binary relation such that properties $I_{X,Y,S}$ and $II_{X,Y,S}$ hold.

Take $P = X \times Y$ as the set of propositional variables and consider the following sets of formulas:

$$\mathcal{B}_{X,Y,S} = \left\{ \bigvee_{y \in S_x} (x, y) : x \in X \right\}$$

$$\mathcal{C}_{X,Y,S} = \{\neg((x, y) \wedge (x, z)) : x \in X, \ y \in Y, \ z \in Y \text{ and } y \neq z\}$$

$$\mathcal{D}_{X,Y,S} = \{\neg((x, y) \wedge (t, y)) : x \in X, \ t \in X, \ y \in Y \text{ and } x \neq t\}.$$

Then set:

$$\mathcal{A}_{X,Y,S} = \mathcal{B}_{X,Y,S} \cup \mathcal{C}_{X,Y,S} \cup \mathcal{D}_{X,Y,S}.$$

Observe that for each element x of X, $\bigvee_{y \in S_x} (x, y)$ is a formula because the set S_x is finite (property II) and non-empty (property I: $\mathrm{card}(S_x) \geq \mathrm{card}(\{x\})$).

If δ is an assignment of truth values that satisfies $\mathcal{A}_{X,Y,S}$, then we can define a map f from X into Y that satisfies the conditions in III in the following way: for all $x \in X$, $f(x)$ is the unique element of Y for which $\delta(A_{x,y}) = \mathbf{1}$.

Conversely, if we have a map f that satisfies the conditions in III, then we can obtain an assignment of truth values δ that satisfies $\mathcal{A}_{X,Y,S}$ by setting:

$$\text{for all } (x, y) \in X \times Y,$$
$$\delta(A_{x,y}) = \mathbf{1} \text{ if and only if } (x) = y.$$

So we see that the set of formulas $\mathcal{A}_{X,Y,S}$ is satisfiable if and only if property $\mathrm{III}_{X,Y,S}$ is verified.

Now consider the sets E and F and the relation R that we have been studying. Since they satisfy properties I and II, to show that they also satisfy III, it suffices to prove that the associated set of formulas $\mathcal{A}_{E,F,R}$ is satisfiable. By the compactness theorem, this amounts to proving that every finite subset of this set is satisfiable. Let $\mathcal{G}$ be such a finite subset. We easily see that there exists a finite subset X of E such that, if we set $Y = R_X$ and $S = R \cap (X \times Y)$, then $\mathcal{G}$ is included in $\mathcal{A}_{X,Y,S}$. Now properties $\mathrm{I}_{X,Y,S}$ and $\mathrm{II}_{X,Y,S}$ are obviously verified; hence, from (a), $\mathrm{III}_{X,Y,S}$ is also verified and, from the preceding argument, $\mathcal{A}_{X,Y,S}$ is satisfiable as well as $\mathcal{G}$.

The property proved in this exercise is known as the **marriage theorem.** We can illustrate it this way: E represents a set of men, F a set of women, $(x, y) \in R$ means that 'x knows y' and $y = f(x)$ means 'x marries y'. What we have proved is that if an arbitrary number of men chosen from E collectively know at least the same number of women, then it is possible to marry each man in E to a woman in F that he knows, with no polygamy occurring. (Let us add that if we wish to consider the improbable case in which these sets are infinite, we would have to add the hypothesis that each man knows only a finite number of women...). Such an illustration could naturally be criticized, specifically for the asymmetric roles played by E and F, or else for its conformity with the imposed rules: monogamy and marriage between people who know each other.

Solutions to the exercises for Chapter 2

1. (a) It is the fact that the relation of logical equivalence, $\sim$, is compatible with the propositional connectives that allows us to define internal operations on $\mathcal{F}/\sim$ by the relations given in the statement of the exercise (these relations are independent of the choice of representatives from the equivalence classes of formulas). More precisely, and in conformity with Theorem 1.24, for all formulas F, G, F', and G', if $\mathrm{cl}(F) = \mathrm{cl}(F')$ and $\mathrm{cl}(G) = \mathrm{cl}(G')$, then $\mathrm{cl}(\neg F) = \mathrm{cl}(\neg F')$, $\mathrm{cl}(F \wedge G) = \mathrm{cl}(F' \wedge G')$, $\mathrm{cl}(F \vee G) = \mathrm{cl}(F' \vee G')$, $\mathrm{cl}(F \Rightarrow G) = \mathrm{cl}(F' \Rightarrow G')$ and $\mathrm{cl}(F \Leftrightarrow G) = \mathrm{cl}(F' \Leftrightarrow G')$. (Also see our comments in Section 1.2).

We have the following logical equivalences ($\top$ is a tautology, $\bot$ an anti-logy, the numbers in brackets refer to the list established in Section 1.2):

- $(A \not\Leftrightarrow B) \sim (B \not\Leftrightarrow A)$;   [no. 41]
- $((A \not\Leftrightarrow B) \not\Leftrightarrow C) \sim (A \not\Leftrightarrow (B \not\Leftrightarrow C))$;   [the first of these formulas is $\neg(\neg(A \Leftrightarrow B) \Leftrightarrow C)$ which is logically equivalent to $((A \Leftrightarrow B) \Leftrightarrow C)$ [no. 43], to $(A \Leftrightarrow (B \Leftrightarrow C))$ [no. 58], and hence to the second formula.]
- $(A \not\Leftrightarrow \bot) \sim A$ [no. 48 and no. 41]
- $(A \not\Leftrightarrow A) \sim \bot$ [no. 49 and no. 43]
- $(A \wedge B) \sim (B \wedge A)$ [no. 3]
- $((A \wedge B) \wedge C) \sim (A \wedge (B \wedge C))$ [no. 5]
- $(A \wedge (B \not\Leftrightarrow C)) \sim ((A \wedge B) \not\Leftrightarrow (A \wedge C))$ [we justify this using the method of Exercise 14 from Chapter 1: the polynomial associated with the first formula is $x(1 + 1 + y + z)$; the one associated with the second is $1 + 1 + xy + xz$, which is the same.]
- $(A \wedge \top) \sim A$;   [no. 46]
- $(A \wedge A) \sim A$;   [no. 1].

These equivalences prove that the operation $\not\Leftrightarrow$ on $\mathcal{F}/\sim$ is commutative and associative, that it admits the class $\mathbf{0}$ of antilogies as an identity element, and that every element has an inverse element (itself) relative to this identity element. The structure $\langle \mathcal{F}/\sim, \not\Leftrightarrow \rangle$ is thus a commutative group. Moreover, the operation $\wedge$ is commutative, associative, distributive over $\not\Leftrightarrow$, is idempotent and has an identity element: the class $\mathbf{1}$ of tautologies. Consequently, the structure $\langle \mathcal{F}/\sim, \not\Leftrightarrow, \wedge \rangle$ is a commutative ring with unit that is a Boolean ring.

(b) It is true by definition that for all formulas F and G, $\mathsf{cl}(F) \leq \mathsf{cl}(G)$ if and only if $\mathsf{cl}(F) \wedge \mathsf{cl}(G) = \mathsf{cl}(F)$, which is equivalent to $\mathsf{cl}(F \wedge G) = \mathsf{cl}(F)$, which is to say $(F \wedge G) \sim F$, or also to $\vdash^* ((F \wedge G) \Leftrightarrow F)$. Now the formula $((F \wedge G) \Leftrightarrow F)$ is logically equivalent to $(F \Rightarrow G)$ (see [no. 38]). It follows that

$$\mathsf{cl}(F) \leq \mathsf{cl}(G) \text{ if and only if } \vdash^* (F \Rightarrow G).$$

(Let us observe in passing that the property '$(F \Rightarrow G)$ is a tautology' defines a binary relation on $\mathcal{F}$ that is compatible with the relation of logical equivalence, $\sim$, and that the relation induced by it on the quotient set $\mathcal{F}/\sim$ is the order relation of the Boolean algebra.)

Let us also observe that this is in no case a total ordering: if A is a propositional variable, we have neither $\mathsf{cl}(A) \leq \mathsf{cl}(\neg A)$ nor $\mathsf{cl}(\neg A) \leq \mathsf{cl}(A)$ since neither of the two formulas $(A \Rightarrow \neg A)$ nor $(\neg A \Rightarrow A)$ is a tautology (they are equivalent, respectively, to $\neg A$ and to A).

For the ordering $\leq$, the smallest element is the class $\mathbf{0}$ of antilogies and the greatest element is the class $\mathbf{1}$ of tautologies.

Let $x = \mathsf{cl}(F)$ and $y = \mathsf{cl}(G)$ be two elements of $\mathcal{F}/\sim$. We have (Theorem 1.18)

$$x \cap y = xy = x \wedge y = \mathsf{cl}(F \wedge G)$$

and

$$x^{\complement} = 1 + x = 1 \nLeftrightarrow x = \mathsf{cl}(\neg F).$$

It follows, using de Morgan's laws, that

$$x \vee y = \left(x^{\complement} \cap y^{\complement}\right)^{\complement} = \mathsf{cl}(\neg(\neg F \wedge \neg G)) = \mathsf{cl}(F \vee G) = x \vee y.$$

Hence, the operations of least upper bound, greatest lower bound, and complementation are, respectively, disjunction, conjunction, and negation.

(c) Suppose that P is the finite set $\{A_1, A_2, \ldots, A_n\}$ ($n \in \mathbb{N}^*$). Then the quotient set $\mathcal{F}/\sim$ is also finite and has 2^{2^n} elements (see Section 1.3). This shows that the Boolean algebra $\langle \mathcal{F}/\sim, \nLeftrightarrow, \wedge, \mathbf{0}, \mathbf{1}\rangle$ is atomic (Theorem 2.39) and also that the number of its atoms is 2^n (Theorem 2.50 and Corollary 2.51). We will show that these atoms are the the equivalence classes of the formulas

$$\bigwedge_{1 \le k \le n} \varepsilon_k A_k,$$

obtained as the n-tuple $(\varepsilon_1, \varepsilon_2, \ldots, \varepsilon_n)$ varies over $\{\mathbf{0}, \mathbf{1}\}^n$.

Let us observe from the outset that there are 2^n such classes, which amounts to saying that these formulas under consideration are pairwise logically inequivalent (which is guaranteed by Lemma 1.25). Thus, if we succeed in showing that these classes are atoms, we will be assured that we have thereby found all the atoms of this Boolean algebra.

Consider an n-tuple $(\varepsilon_1, \varepsilon_2, \ldots, \varepsilon_n) \in \{\mathbf{0}, \mathbf{1}\}^n$ and let H denote the associated formula:

$$H = \bigwedge_{1 \le k \le n} \varepsilon_k A_k.$$

We will reuse the notation from Section 1.3. We have $\Delta(H) = \{\delta_{\varepsilon_1 \varepsilon_2 \ldots \varepsilon_n}\}$ (Lemma 1.25); hence, $\mathsf{cl}(H) \ne \mathbf{0}$. It also follows from this that for any formula F such that $\mathsf{cl}(F) \le \mathsf{cl}(H)$ (which means that $\vdash^* (F \Rightarrow H)$ and is obviously equivalent to $\Delta(F) \subseteq \Delta(H)$),

• either $\Delta(F) = \{\delta_{\varepsilon_1 \varepsilon_2 \ldots \varepsilon_n}\} = \Delta(H)$, and so $\mathsf{cl}(F) = \mathsf{cl}(H)$;
• or $\Delta(F) = \emptyset$, and in this case, $\mathsf{cl}(F) = \mathbf{0}$.

We have thus proved that $\mathsf{cl}(H)$ is a non-zero element of the Boolean algebra $\mathcal{F}/\sim$ below which the only elements are itself and $\mathbf{0}$; this means that it is an atom.

2. We will make use of the results from Exercise 1.

Let X and Y be two subsets of E. For every element $x \in X$, we have $x \in X \triangle Y$ if and only if one and only one of the statements $x \in X$ and $x \in Y$ is true; this means precisely that $x \in X \triangle Y$ if and only if the statement $x \in X \nLeftrightarrow x \in Y$ is true. The commutativity of $\triangle$ follows from that of $\nLeftrightarrow$; for all subsets X and Y of E,

$$X \triangle Y = \{x \in E : x \in X \nLeftrightarrow x \in Y\}$$
$$= \{x \in E : x \in Y \nLeftrightarrow x \in X\} = Y \triangle X.$$

The associativity of $\triangle$ is obtained from that of $\nLeftrightarrow$ in an analogous fashion. Moreover,

$$X \triangle \emptyset = \{x \in E : x \in X \nLeftrightarrow x \in \emptyset\} = \{x \in E : x \in X\} = X$$

(here, we have used the fact that $x \in \emptyset$ is false and that false is an identity element for $\nLeftrightarrow$). Also,

$$X \triangle X = \{x \in E : x \in X \nLeftrightarrow x \in X\} = \emptyset.$$

Thus, for the operation $\triangle$, $\emptyset$ is an identity element and every element of $\wp(E)$ is its own inverse.

These remarks show that $\langle \wp(E), \triangle \rangle$ is a commutative group.

The intersection operation on $\wp(E)$ is commutative, associative and has E as an identity element. These facts follow from the corresponding properties of the propositional connective $\wedge$. Moreover,

$$X \cap (Y \triangle Z) = \{x \in E : x \in X \wedge (x \in Y \nLeftrightarrow x \in Z)\};$$
$$(X \cap Y) \triangle (X \cap Z) = \{x \in E : (x \in X \wedge x \in Y) \nLeftrightarrow (x \in X \wedge x \in Z)\};$$

these two sets coincide in view of the distributivity of $\wedge$ over $\nLeftrightarrow$. Thus, intersection distributes over symmetric difference. We have thereby shown that the structure $\langle \wp(E), \triangle, \cap \rangle$ is a commutative ring with unit. It is also a Boolean algebra since for every $X \in \wp(E)$, we have $X \cap X = X$.

It is immediate to verify that the order relation in this Boolean algebra is inclusion, that union and intersection correspond to the operations $\smile$ and $\frown$, and that the complement of an element is its usual complement in the set-theoretic sense.

3. The fact that properties of the kind considered in this exercise are preserved under isomorphism is banal and it is never a problem to verify this. As an example, we will treat question (a), leaving the others to the reader.

   (a) Let $a \in A$ be an atom of $\mathcal{A}$ and let $y \in B$ be less than or equal to $f(a)$ in $\mathcal{B}$. As f is an isomorphism, there is a unique element $x \in A$ such that $y = f(x)$ and this element, x, is such that $\mathbf{0} \leq x \leq a$ since $\mathbf{0} \leq f(x) \leq f(a)$ (Theorem 2.48). But a is an atom; hence either $x = \mathbf{0}$ (and so $y = f(x) = \mathbf{0}$) or else $x = a$ (and so $y = f(x) = f(a)$). It follows that the only elements less than or equal to $f(a)$ are $\mathbf{0}$ and $f(a)$. Given that a is non-zero, that

$f(\mathbf{0}) = \mathbf{0}$ and that f is injective, we can only conclude that $f(a)$ is non-zero and hence that it is an atom of $\mathcal{B}$.

Conversely, if we suppose that it is $f(a)$ that is an atom of $\mathcal{B}$, it suffices to apply what we have just done to the isomorphism f^{-1} from $\mathcal{B}$ into $\mathcal{A}$ to convince ourselves that $a = f^{-1}(f(a))$ is an atom of $\mathcal{A}$.

4. (a) Let $\langle A, \leq, \mathbf{0}, \mathbf{1}\rangle$ be a Boolean algebra. Suppose that it is complete and consider a non-empty subset X of A. Set $Y = \{x \in A : x^{\complement} \in X\}$ and let b denote the greatest lower bound of Y (Y is obviously a non-empty subset of A). It is very easy to prove that $a = b^{\complement}$ is the least upper bound of X: on the one hand, for every $x \in X$, we have $b \leq x^{\complement}$, hence $x \leq b^{\complement} = a$ and a is an upper bound for X; on the other hand, if m is an upper bound of X, then $m^{\complement}$ is a lower bound of Y (trivial to verify), hence $m^{\complement} \leq b$ and $a = b^{\complement} \leq m$, which shows that a is the least of the upper bounds of X. When we interchange the words 'upper' and 'lower' and reverse the inequalities in the preceding, we prove that if every non-empty subset of A has a greatest upper bound, then the Boolean algebra is complete.

(b) Let $\mathcal{A} = \langle A, \leq, \mathbf{0}, \mathbf{1}\rangle$ and $\mathcal{B} = \langle B, \leq, \mathbf{0}, \mathbf{1}\rangle$ be two Boolean algebras and let f be an isomorphism of Boolean algebras from $\mathcal{A}$ onto $\mathcal{B}$. We assume that $\mathcal{A}$ is complete and will show that the same is true of $\mathcal{B}$. Let B be a non-empty subset of Y and let X denote its preimage under f; X is a non-empty subset of A since f is a bijection. It is very easy to verify that if a is the greatest lower bound of X (which exists, by hypothesis), then $f(a)$ is the greatest lower bound of Y.

(c) Let E be a set and let X be a non-empty subset of $\wp(E)$. It is clear that the set $G = \bigcap_{Z \in X} Z$, the intersection of the elements of X, is the greatest lower bound of X.

(d) Let E be an infinite set and let H be an infinite subset of E whose complement in E is infinite. (Concerning the existence of such a set, refer to Chapter 7.) Consider the following subset X of $\wp(E)$:

$$X = \{\{x\} : x \in H\}.$$

If the Boolean algebra of finite or cofinite subsets of E were complete, X would have a greatest upper bound, M. Then M would have to be a finite or cofinite subset of E that is greater than or equal to (i.e. includes) all the elements of X: so we would have $H \subseteq M$; this prevents M from being finite so forces it to be cofinite. Since H, by hypothesis, is not cofinite, the preceding inclusion would be strict so we could find an element $a \in E$ such that $a \in M$ and $a \notin H$. The set $N = M - \{a\}$ would then be, as is M, a cofinite subset of E that would still be an upper bound for X. But in this case, M would not be the least upper bound of X, which is a contradiction. The Boolean algebra under consideration is therefore not complete.

(e) To begin with, let us recall that every finite Boolean algebra is complete (property 6 of Theorem 2.29). It follows that when the set P of propositional variables is finite, then the Boolean algebra $\mathcal{F}/\sim$ of equivalence classes of logically equivalent formulas (which is then itself finite) is complete.

Now suppose that the set P is infinite. Consider two sequences $(p_n)_{n\in\mathbb{N}}$ and $(q_n)_{n\in\mathbb{N}}$ of pairwise distinct elements of P. We construct two sequences of formulas, $(F_n)_{n\in\mathbb{N}}$ and $(G_n)_{n\in\mathbb{N}}$, in the following way:

$$F_0 = p_0 \vee q_0; \quad F_1 = p_1 \vee q_0 \vee q_1; \quad \ldots$$
$$\ldots, F_k = p_k \vee q_0 \vee q_1 \vee \cdots \vee q_k; \quad \ldots$$
$$G_0 = q_0; \quad G_1 = q_0 \vee (q_1 \wedge p_0); \quad \ldots$$
$$\ldots, G_k = q_0 \vee (q_1 \wedge p_0) \vee \cdots \vee (q_k \wedge p_0 \wedge p_1 \wedge \cdots \wedge p_{k-1}); \quad \ldots$$

For every integer n, we have $G_{n+1} = G_n \vee (q_{n+1} \wedge p_0 \wedge p_1 \wedge \cdots \wedge p_n)$, hence $\vdash^* G_n \Rightarrow G_{n+1}$; in other words, $\mathsf{cl}(G_n) \leq \mathsf{cl}(G_{n+1})$, relative to the order relation in the Boolean algebra $\mathcal{F}/\sim$. This inequality is strict, for we can find an assignment of truth values δ_n that satisfies G_{n+1} but that does not satisfy G_n; it suffices to set

$$\delta_n(q_0) = \delta_n(q_1) = \cdots = \delta_n(q_n) = \mathbf{0} \text{ and}$$
$$\delta_n(p_0) = \delta_n(p_1) = \cdots = \delta_n(p_n) = \delta_n(q_{n+1}) = \mathbf{1};$$

the values of δ_n on the other propositional variables may be chosen arbitrarily.

Let n be an integer and consider an assignment of truth values λ such that $\overline{\lambda}(F_n) = \mathbf{0}$. We then have

$$\lambda(p_n) = \lambda(q_0) = \lambda(q_1) = \cdots = \lambda(q_n) = \mathbf{0}$$

and, for every integer $k \geq 1$, $\overline{\lambda}(q_k \wedge p_0 \wedge p_1 \wedge \cdots \wedge p_{k-1}) = \mathbf{0}$. Indeed, if $k \leq n$, $\lambda(q_k) = \mathbf{0}$; and if $k > n$, then p_n occurs in the conjunction $p_0 \wedge p_1 \wedge \cdots \wedge p_{k-1}$ which is therefore not satisfied by λ. So we see that for every m, $\overline{\lambda}(G_m) = \mathbf{0}$. We have thus shown, for all integers m and n, that $G_m \Rightarrow F_n$ is a tautology, or, as well, that $\mathsf{cl}(G_m) \leq \mathsf{cl}(F_n)$. The inequality is strict since, with what we have just proved, we have

$$\mathsf{cl}(G_m) < \mathsf{cl}(G_{m+1}) \leq \mathsf{cl}(F_n).$$

Suppose that the set $\{\mathsf{cl}(F_n) : n \in \mathbb{N}\}$ has a greatest lower bound, $\mathsf{cl}(F)$. According to the preceding, we would then have

$$\text{for all integers } m \text{ and } n,$$
$$\mathsf{cl}(G_m) \leq \mathsf{cl}(F) \leq \mathsf{cl}(F_n),$$

which is to say

$$\vdash^* G_m \Rightarrow F \text{ and } \vdash^* F \Rightarrow F_n. \tag{$\dagger$}$$

Choose an integer r such that for every integer $k \geq r$, neither p_k nor q_k occurs in the formula F (this is possible since a formula can involve only finitely many propositional variables). We will now define two assignments of truth values α and β that are obtained, respectively, by modifying the value assumed by the assignment δ_r, defined above, at the points p_r and q_r:

$\alpha(p_r) = \mathbf{0}$ and $\alpha(x) = \delta_r(x)$ for all variables x other than p_r;

$\beta(q_r) = \mathbf{1}$ and $\beta(x) = \delta_r(x)$ for all variables x other than q_r.

Since p_r and q_r do not occur in F, we obviously have

$$\overline{\alpha}(F) = \overline{\beta}(F) = \overline{\delta_r}(F). \tag{$\dagger\dagger$}$$

On the other hand, it is easy to verify that $\overline{\alpha}(F_r) = \mathbf{0}$ and $\overline{\beta}(G_r) = \mathbf{1}$, which, according to ($\dagger$), requires $\overline{\alpha}(F) = \mathbf{0}$ and $\overline{\beta}(F) = \mathbf{1}$; but this manifestly contradicts ($\dagger\dagger$).

It was therefore absurd to assume the existence of a greatest lower bound for the set $\{\mathrm{cl}(F_n) : n \in \mathbb{N}\}$. So the Boolean algebra under consideration is not complete.

(f) We have seen that the Boolean algebra of subsets of a set is atomic and, in question (c) above, that it is complete. We are guaranteed by question (c) of Exercise 3 and by question (b) above that every Boolean algebra that is isomorphic to a Boolean algebra of subsets of some set is necessarily atomic and complete.

Now let us consider the converse. We will be motivated by the proof of Theorem 2.50, which can be viewed as a special case of the result to be proved here. Let $\mathcal{A} = \langle A, \leq, \mathbf{0}, \mathbf{1} \rangle$ be a Boolean algebra that is atomic and complete. Denote the set of its atoms by E and let φ denote the map from A into $\wp(E)$ which, with each element x in A, associates the set of atoms that are less than or equal to it:

$$\varphi(x) = \{a \in E : a \leq x\}.$$

Let us show that φ is surjective. Let X be a subset of E. If X is empty, then $X = \varphi(\mathbf{0})$ (there is no atom below $\mathbf{0}$). If X is not empty, it has a least upper bound that we will denote by M. Every element of X is an atom that is below M, thus $X \subseteq \varphi(M)$. If b is an atom that does not belong to X, then for every element x in X, we have $x \leq b^{\complement}$ (this follows from Theorem 2.40: x is an atom and we do not have $x \leq b$ since b is also an atom and $x \leq b$ would mean that either $x = b$ or $x \in X$ and $b \notin X$). We conclude from this that $M \leq b^{\complement}$ and, consequently, that b is not below M (for, since b is non-zero, we cannot have $b \leq b^{\complement}$). It follows that every atom that is below M is an element of X, i.e. $\varphi(M) \subseteq X$. To conclude, $X = \varphi(M)$ and φ is surjective.

For all elements x and y of A, if $x \leq y$, then $\varphi(x) \subseteq \varphi(y)$ since any atom that is below x is an atom that is below y.

For all elements x and y of A, if $\varphi(x) \subseteq \varphi(y)$, then $x \leq y$: indeed, if x is not below y, then $xy^{\complement} \neq \mathbf{0}$ (Lemma 2.31; since A is atomic, we can then find an atom $a \in E$ such that $a \leq xy^{\complement}$, which is to say $a \leq x$ and $a \leq y^{\complement}$; but the atom a cannot simultaneously lie below y and $y^{\complement}$; this shows that $a \in \varphi(x)$ and $a \notin \varphi(y)$, hence $\varphi(x) \not\subseteq \varphi(y)$.

Theorem 2.48 permits us to conclude that φ is an isomorphism of Boolean algebras from $\mathcal{A}$ onto $\wp(E)$. Thus, every Boolean algebra that is atomic and complete is isomorphic to the Boolean algebra of subsets of some set. Since every finite Boolean algebra is atomic and complete, Theorem 2.50 is a corollary of what we have just proved.

5. Let b be an element of B that is not an atom of the Boolean algebra $\mathcal{B}$. Then either $b = \mathbf{0}$ and so b is not an atom of $\mathcal{A}$ or else we can find an element $c \in B$ that is non-zero, distinct from b, and that satisfies $c \leq b$; but, since $c \in A$, we have found a non-zero element of A that is strictly below b, which shows that b is not an atom of $\mathcal{A}$.

6. We apply Theorem 2.54. We have $\mathbf{0} \in B$ since $\mathbf{0} \leq 1 + a$; if $x \in B$, we have either $x \geq a$ and so $x^{\complement} \leq 1 + a$, or else $x \leq 1 + a$ and so $x^{\complement} \geq a$; so in all cases, $x^{\complement} \in B$; finally, if x and y are elements of B and if at least one of these is below $1 + a$, then so is their greatest lower bound $x \frown y$; and if this is not the case, then we have $x \geq a$ and $y \geq a$, hence $x \frown y \geq a$; so in all cases, $x \frown y \in B$.

Suppose that $\mathcal{A}$ is complete and consider a non-empty subset X of B. In A, X has a greatest lower bound, m. Let us show that $m \in B$; this will prove that the Boolean sub-algebra that is B is complete. To do this, observe that if at least one of the elements of X is below $1 + a$, we have $m \leq 1 + a$; and otherwise, for every element $x \in X$, $x \geq a$ and so a is below X, hence below m; in all cases, $m \in B$.

7. (a) Set $G = \bigcap_{F \in Z} F$. We will verify that G satisfies the three conditions of Theorem 2.68. Let F_0 be a filter in the (non-empty) set Z. We have $\mathbf{0} \notin F_0$ hence $\mathbf{0} \in G$. As well, we have $\mathbf{1} \in F$ for every $F \in Z$, hence $\mathbf{1} \in G$. This verifies condition (f). Let x and y be two elements of G; for every $F \in Z$, we have $x \in F$ and $y \in F$, hence $x \frown y \in F$; it follows that $x \frown y \in G$ and that condition (ff) is satisfied. Finally, if $x \in G$, $y \in A$, and $x \leq y$, then for every $f \in Z$, we have $x \in F$, hence $y \in F$; this shows that $y \in G$ and that condition (fff) is satisfied. So the set G is a filter on $\mathcal{A}$.

Let a be an element of A that is distinct from $\mathbf{0}$ and $\mathbf{1}$ (we assume such an element exists). Consider the principal filters F_a and F_{1+a} generated by a and $1 + a$, respectively and let $Z = \{F_a, F_{1+a}\}$. The set $K = \cup_{F \in Z} F = F_a \cup F_{1+a}$ is certainly not a filter since $a \in K$, $1 + a \in K$ and $a \frown (1+a) = \mathbf{0} \notin K$.

(b) We set $H = \cup_{F \in Z} F$ and once again apply Theorem 2.69. As before, choose a specific filter $F_0 \in Z$. We have $\mathbf{1} \in F_0$ hence $\mathbf{1} \in H$. For every $F \in Z$, we have $\mathbf{0} \notin F$, hence $\mathbf{0} \notin H$. So H satisfies condition (f). Let x and y be elements of A such that $x \in H$ and $y \geq x$; there exists a filter $F \in Z$ such that $x \in F$, and so $y \in F$; hence $y \in H$. This verifies condition (fff) for H. As we see, the additional hypothesis on Z is not involved in the proofs of these two conditions. We will use it to verify condition (ff): given two elements x and y of H, we have to show that $x \frown y \in H$. There are filters F_1 and F_2 in the set Z such that $x \in F_1$ and $y \in F_2$; as Z is totally ordered by inclusion, we have either $F_1 \subseteq F_2$ or $F_2 \subseteq F_1$; $F = F_1 \cup F_2$ is therefore a filter in Z that contains both x and y; so it contains their greatest lower bound $x \frown y$. We thus have $x \frown y \in F$ and $F \in Z$, from which it follows that $x \frown y \in \cup_{F \in Z} F = H$.

8. Let E^* denote the set of intersections of finitely many elements of E. We can construct E^* in the following way: set $E_0 = E$ and for every $n \in \mathbb{N}$,

$$E_{n+1} = \{Z \in \wp(\mathbb{N}) : (\exists X \in E)(\exists Y \in E_n)(Z = X \cap Y)\};$$

we then have

$$E^* = \bigcup_{n \in \mathbb{N}} E_n.$$

Since the set E is countable, we easily show, by induction, that each of the sets E_n is countable; so their union E^* is also countable (a countable union of countable sets). We may refer to Chapter 7 in Volume 2 for details concerning these questions of cardinality.

We can therefore produce an enumeration of E^*:

$$E^* = \{X_n : n \in \mathbb{N}\}.$$

Let us now show that the filter generated by E is the set

$$F = \{X \in \wp(\mathbb{N}) : (\exists n \in \mathbb{N})(X \supseteq X_n)\}.$$

We can convince ourselves that F is a filter that includes E by examining the proof of Lemma 2.79 (note that the elements of E^* are all non-empty). To continue, suppose that G is a filter that includes E; G must contain any element that is a lower bound (i.e. intersection) of finitely many elements of E, as well as any element that is above such an element. The first condition is equivalent to $G \supseteq E$ and the second, to $G \supseteq F$. We have proved that F is a filter that includes E and that it is itself included in any filter that includes E; accordingly, F is the intersection of all filters that include E or, equivalently, the filter generated by E.

Suppose that F is an ultrafilter. We will distinguish two cases. In the first, we will show that F is trivial and in the second, we will arrive at a contradiction.

We will thereby have established the desired property.

- If there exists an integer n such that the set X_n is finite, then since $X_n \in F$, $\mathbb{N} - X_n \notin F$, which shows that F does not extend the filter of cofinite subsets of $\mathbb{N}$ (the Frechet filter); so F is a trivial ultrafilter (Theorem 2.77).

- In the case where X_n is infinite for every $n \in \mathbb{N}$, we will construct a subset $A \subseteq \mathbb{N}$ such that neither A nor $\mathbb{N} - A$ belongs to F, which shows that F is not an ultrafilter, contrary to our hypothesis. Let us define two sequences $(a_n)_{n \in \mathbb{N}}$ and $(b_n)_{n \in \mathbb{N}}$ of natural numbers by induction as follows:

  a_0 is the least element of X_0 and b_0 is the least element of $X_0 - \{a_0\}$;

  and for every $n \in \mathbb{N}$,

  a_{n+1} is the least element of $X_{n+1} - \{a_0, a_1, \ldots, a_n, b_0, b_1, \ldots, b_n\}$ and

  b_{n+1} is the least element of $X_{n+1} - \{a_0, a_1, \ldots, a_n, a_{n+1}, b_0, b_1, \ldots, b_n\}$.

The fact that all of the sets X_n are infinite guarantees that this is a valid definition.

Set $A = \{a_n : n \in \mathbb{N}\}$ and $B = \mathbb{N} - A$.

It is clear that the set $\{b_n : n \in \mathbb{N}\}$ is included in B and that the sets A and B are both infinite.

For every integer n, we have

$$a_n \in X_n; \quad b_n \in X_n; \quad a_n \notin B; \quad b_n \notin A$$

which shows that the set X_n is not included in A nor in B. It follows that neither of the sets A nor $B = \mathbb{N} - A$ belongs to the filter F. So it is not an ultrafilter.

9. (a) This property was established in Exercise 20 from Chapter 1.

   (b) and (c) Consider the map $\varphi : \mathcal{F}/\sim \longrightarrow \{0, 1\}$ defined for all $x \in \mathcal{F}/\sim$ by

$$\varphi(x) = \begin{cases} 1 & \text{if } \delta_1(F) = 1 \text{ for all formulas } F \in x \\ 0 & \text{if } \delta_1(F) = 0 \text{ for all formulas } F \in x \end{cases}$$

   (there are obviously no other cases).

   We see that φ is nothing more than the assignment of truth values δ_1 'modulo' the relation of logical equivalence $\sim$; it is the map that would be denoted by h_{δ_1} using the notation from Example 2.57 and it is a homomorphism of Boolean algebras from $\mathcal{F}/\sim$ into $\{0, 1\}$. We obviously have

$$J = \{x \in \mathcal{F}/\sim : \varphi(x) = 1\},$$

   which proves that J is an ultrafilter and that φ is its associated homomorphism (Theorem 2.72, $(1') \Leftrightarrow (3')$).

10. (a) Let x be an element of A. To say that $H(x)$ is a singleton is to say that there is a unique homomorphism of Boolean algebras from $\mathcal{A}$ into $\{0, 1\}$ that assumes the value 1 at x.

- Suppose that x is an atom: so we have $x \neq \mathbf{0}$ and $H(x)$ is then non-empty. But if h is an element of $H(x)$, then for all $y \in A$, we have either that $xy = y$, which requires $h(y) = \mathbf{1}$, or else $xy = \mathbf{0}$ which requires $h(y) = \mathbf{0}$ (since $h(xy) = h(x)h(y) = h(y)$). The value of $h(y)$ is thus determined for any element $y \in A$. There is therefore a unique element in $H(x)$: it is a singleton.

- Now suppose that $H(x)$ is a singleton: it is then necessarily an atom in the Boolean algebra $\mathcal{B}(S)$ (since it is an atom in $\wp(S)$ (Exercise 5)). As H is an isomorphism of Boolean algebras, x is an atom of $\mathcal{A}$ (Exercise 3a).

Remark: In the first part of the proof, we could have, in similar fashion, observed that $H(x)$ is an atom in $\mathcal{B}(S)$; but this would have in no way implied that $H(x)$ is a singleton, for an atom in $\mathcal{B}(S)$ is not necessarily an atom in $\wp(S)$ (Exercise 5).

(b) If $\mathcal{A}$ contains an atom a, $H(a)$ is an open (and closed) subset of S that reduces to a single point: consequently, S contains at most one isolated point. Conversely, if h is an isolated point in S, then $\{h\}$ is an open subset of S. As the topology of S is Hausdorff, every singleton is closed. It follows that $\{h\}$ is clopen, i.e. an element of $\mathcal{B}(S)$. As a consequence, there is a (unique) element $a \in A$ whose image under the isomorphism H is $\{h\}$. According to question (a), a is necessarily an atom.

(c) As the clopen sets constitute a basis of open sets for the topological space S, and as every non-empty open set must include at least one open set from a given basis, we see that for a set $X \subseteq S$ to meet every non-empty open set (i.e. for it to be dense in S), it is necessary and sufficient that it meet every non-empty clopen set.

Let I denote the set of isolated points of S.

Suppose that $\mathcal{A}$ is atomic. Let Ω be a non-empty clopen subset of S. There then exists one and only one element x in A such that $H(x) = \Omega$. As Ω is not empty, x is not zero; so we can choose an atom $a \in A$ that is below x. Then $H(a)$ is included in $H(x)$ since H is order-preserving. But according to (a), $H(a)$ is a singleton and its unique element is an isolated point of S; so Ω contains a point from I. This proves that I is dense in S.

Conversely, suppose that I is dense in S. For every non-zero element $x \in A$, $H(x)$ is a non-empty clopen subset of S and, in view of this, it contains at least one isolated point, h. But then $\{h\}$ is the image under H of one (and only one) atom a in A (see the proof of (b)). We have $h(a) = \{h\} \subseteq H(x)$, hence $a \leq x$ (Theorem 2.48). We have found an atom that is below x, i.e. $\mathcal{A}$ is atomic.

11. Suppose first that the Boolean algebra $\mathcal{A} = \langle A, \leq, \mathbf{0}, \mathbf{1} \rangle$ is dense. If b is a non-zero element in A, then, by definition, there is at least one element $c \in A$ such that $\mathbf{0} < c < b$, which shows that b is not an atom. Conversely, suppose there

are no atoms in $\mathcal{A}$ and consider two elements a and b of A such that $a < b$; then $a + b$ is not zero and is not an atom. So there is an element $d \in A$ such that $\mathbf{0} < d < a + b$. Set $c = a \smile d$. It is immediate to verify that we then have $a < c < b$; so $\mathcal{A}$ is dense.

12. (a) Suppose that (y, z) is a bipartition of x. We then have

$$x = y \smile z = y + z + yz = y + z + (y \frown z) = y + z.$$

It follows immediately that $x + y = z$. We have $y \neq x$ since $z \neq \mathbf{0}$, and $y \leq x$ since $x = y \smile z$. As $y \neq \mathbf{0}$, we conclude that $\mathbf{0} < y < x$.

Conversely, if $\mathbf{0} < y < x$ and $z = x + y$, then we have $y \neq \mathbf{0}$, $z \neq \mathbf{0}$ (since $y \neq x$),

$$y \smile z = y + z + yz = y + x + y + yx + y^2 = x + y + xy = x \smile y = x$$

and finally, $y \frown z = y(x + y) = yx + y^2 = y + y = \mathbf{0}$.

Let a be a non-zero element of A. Since a is not an atom, there is an element $b \in A$ such that $\mathbf{0} < b < a$. From all that has preceded, $(b, a + b)$ is a bipartition of a.

(b) We proceed by induction. The choice of u_0 and u_1 is explicitly given in the statement of the exercise.

Suppose that the element $u_{\varepsilon_0 \varepsilon_1 \ldots \varepsilon_{n-1}}$ has been defined in conformity with the imposed conditions (by convention, for the case $n = 0$, $u_{\varepsilon_0 \varepsilon_1 \ldots \varepsilon_{n-1}} = u_\emptyset = 1$). Then this element is non-zero and, according to (a), has at least one bipartition. If the condition $u_{\varepsilon_0 \varepsilon_1 \ldots \varepsilon_{n-1}} \frown a_n \neq \mathbf{0}$ and $u_{\varepsilon_0 \varepsilon_1 \ldots \varepsilon_{n-1}} \frown (1 + a) \neq \mathbf{0}$ is not satisfied, we choose an arbitrary bipartition of $u_{\varepsilon_0 \varepsilon_1 \ldots \varepsilon_{n-1}}$ as the pair $(u_{\varepsilon_0 \varepsilon_1 \ldots \varepsilon_{n-1}0}, u_{\varepsilon_0 \varepsilon_1 \ldots \varepsilon_{n-1}1})$. If the condition is satisfied, then it is easy to verify that the pair $(u_{\varepsilon_0 \varepsilon_1 \ldots \varepsilon_{n-1}} \frown a_n, u_{\varepsilon_0 \varepsilon_1 \ldots \varepsilon_{n-1}} \frown (1 + a_n))$ is a bipartition of $u_{\varepsilon_0 \varepsilon_1 \ldots \varepsilon_{n-1}}$.

(c) Let x and ε be as indicated. First, observe that for all natural numbers m and n, if $n \leq m$, then $u_{\varepsilon_0 \varepsilon_1 \ldots \varepsilon_m} \leq u_{\varepsilon_0 \varepsilon_1 \ldots \varepsilon_n}$. It follows that for every integer $k \in \mathbb{N}$,

(1) if $x \frown u_{\varepsilon_0 \varepsilon_1 \ldots \varepsilon_k} = \mathbf{0}$, then for every integer $p \geq k$, $x \frown u_{\varepsilon_0 \varepsilon_1 \ldots \varepsilon_p} = \mathbf{0}$;

(2) if $x \frown u_{\varepsilon_0 \varepsilon_1 \ldots \varepsilon_k} \neq \mathbf{0}$, then for every integer $q \leq k$, $x \frown u_{\varepsilon_0 \varepsilon_1 \ldots \varepsilon_q} \neq \mathbf{0}$.

Let us next prove that at least one of the conditions (i) or (ii) is satisfied: if (i) is not satisfied, we can find an integer k such that $x \frown u_{\varepsilon_0 \varepsilon_1 \ldots \varepsilon_k} = \mathbf{0}$; according to (1), we then have that for every integer $p \geq k$, $u_{\varepsilon_0 \varepsilon_1 \ldots \varepsilon_p} = \mathbf{0}$; but $u_{\varepsilon_0 \varepsilon_1 \ldots \varepsilon_p} \neq \mathbf{0}$, and

$$u_{\varepsilon_0 \varepsilon_1 \ldots \varepsilon_p} = (x \frown u_{\varepsilon_0 \varepsilon_1 \ldots \varepsilon_p}) \smile ((1 + x) \frown u_{\varepsilon_0 \varepsilon_1 \ldots \varepsilon_p});$$

it follows that $(1 + x) \frown u_{\varepsilon_0 \varepsilon_1 \ldots \varepsilon_p} \neq \mathbf{0}$, which means that condition (ii) is satisfied.

Now let us show that (i) and (ii) cannot be satisfied simultaneously (it is here that that the countability of A intervenes). Since x is one of the elements of A, there is an integer k such that $x = a_k$. Three cases are then possible:

- $a_k \frown u_{\varepsilon_0\varepsilon_1...\varepsilon_{k-1}} = \mathbf{0}$; this contradicts condition (i);
- $(1 + a_k) \frown u_{\varepsilon_0\varepsilon_1...\varepsilon_{k-1}} = \mathbf{0}$; this contradicts condition (ii);
- $a_k \frown u_{\varepsilon_0\varepsilon_1...\varepsilon_{k-1}} \neq \mathbf{0}$ and $(1 + a_k) \frown u_{\varepsilon_0\varepsilon_1...\varepsilon_{k-1}} \neq \mathbf{0}$; in this case,

$$u_{\varepsilon_0\varepsilon_1...\varepsilon_{k-1}0} = a_k \frown u_{\varepsilon_0\varepsilon_1...\varepsilon_{k-1}}, \text{ which implies}$$

$$u_{\varepsilon_0\varepsilon_1...\varepsilon_{k-1}0} \frown (1 + a_k) = \mathbf{0};$$

and

$$u_{\varepsilon_0\varepsilon_1...\varepsilon_{n-1}1} = (1 + a_k) \frown u_{\varepsilon_0\varepsilon_1...\varepsilon_{k-1}}, \text{ which implies}$$

$$u_{\varepsilon_0\varepsilon_1...\varepsilon_{k-1}1} \frown a_k = \mathbf{0}.$$

Thus, if $\varepsilon_k = \mathbf{0}$, then $u_{\varepsilon_0\varepsilon_1...\varepsilon_k 0} \frown (1 + a_k) = \mathbf{0}$ and condition (i) fails, or if $\varepsilon_k = \mathbf{1}$, then $u_{\varepsilon_0\varepsilon_1...\varepsilon_k} \frown a_k = \mathbf{0}$ and it is condition (ii) that fails.

(d) The condition is sufficient: indeed, we have already noted that if $n \leq m$, $u_{\varepsilon_0\varepsilon_1...\varepsilon_m} \leq u_{\varepsilon_0\varepsilon_1...\varepsilon_n}$. It follows that

$$u_{\varepsilon_0\varepsilon_1...\varepsilon_m} \frown u_{\varepsilon_0\varepsilon_1...\varepsilon_n} = u_{\varepsilon_0\varepsilon_1...\varepsilon_m} \neq \mathbf{0}.$$

To show that it is necessary, suppose there exists an integer k such that $0 \leq k \leq n$, $\varepsilon_0 = \xi_0, \ldots, \varepsilon_{k-1} = \xi_{k-1}$ and $\varepsilon_k \neq \xi_k$. We then have $u_{\varepsilon_0\varepsilon_1...\varepsilon_k} \frown u_{\xi_0\xi_1...\xi_k} = \mathbf{0}$, because $(u_{\varepsilon_0\varepsilon_1...\varepsilon_m}, u_{\xi_0\xi_1...\xi_k})$ is a bipartition of $u_{\xi_0\xi_1...\xi_{k-1}}$. On the other hand, $u_{\varepsilon_0\varepsilon_1...\varepsilon_n} \leq u_{\varepsilon_0\varepsilon_1...\varepsilon_k}$ and $u_{\xi_0\xi_1...\xi_m} \leq u_{\xi_0\xi_1...\xi_k}$; hence $u_{\varepsilon_0\varepsilon_1...\varepsilon_n} \frown u_{\xi_0\xi_1...\xi_m} = \mathbf{0}$.

(e) Let x be an element of A. Question (c) shows that for any sequence $f \in \{0, 1\}^{\mathbb{N}}$, we have $f \in h(x)$ or $f \in h(1+x)$, but never both simultaneously. In other words, $h(1 + x)$ is the complement of $h(x)$ in $\{0, 1\}^{\mathbb{N}}$.

For every integer $n \in \mathbb{N}$, set

$$\Gamma_n(x) = \{f \in \{0, 1\}^{\mathbb{N}} : (x \frown u_{f(0)f(1)...f(n)} \neq \mathbf{0}\}, \text{ and}$$

$$V_n = \{(\varepsilon_0, \varepsilon_1, \ldots, \varepsilon_n) \in \{0, 1\}^{n+1} : x \frown u_{\varepsilon_0\varepsilon_1...\varepsilon_n} \neq \mathbf{0}\}.$$

We then have

$$\Gamma_n(x) = \bigcup_{(\varepsilon_0,\varepsilon_1,...,\varepsilon_n)\in V_n} \{f \in \{0, 1\}^{\mathbb{N}} : f(0) = \varepsilon_0 \text{ and}$$

$$f(1) = \varepsilon_1 \text{ and } \ldots \text{ and } f(n) = \varepsilon_n\}$$

and $h(x) = \cap_{n\in\mathbb{N}}\Gamma_n(x)$.

So for every n, $\Gamma_n(x)$ is a finite union of clopen sets from the basis of open sets for the topology on $\{0, 1\}^{\mathbb{N}}$. So it is itself clopen.

The set $h(x)$, an intersection of closed sets, is then closed. But we have seen that the complement of $h(x)$ in $\{\mathbf{0}, \mathbf{1}\}^{\mathbb{N}}$ is $h(\mathbf{1}+x)$; so it too is a closed set. We conclude from this that $h(x)$ is clopen, i.e. is an element of the Boolean algebra $\mathcal{B}(\{\mathbf{0}, \mathbf{1}\}^{\mathbb{N}})$.

Let us prove that h preserves the order and least upper bounds: if x and y are elements of A such that $x \leq y$, then for every sequence $f \in h(x)$ and for every integer $n \in \mathbb{N}$, we have

$$y \cap u_{f(0)f(1)...f(n)} \geq x \cap u_{f(0)f(1)...f(n)} > \mathbf{0},$$

which shows that $f \in h(y)$; hence $h(x) \subseteq h(y)$. It follows that for all elements z and t of A, we have $h(z) \subseteq h(z \smile t)$ and $h(t) \subseteq h(z \smile t)$, hence

$$h(z) \cup h(t) \subseteq h(z \smile t).$$

Conversely, suppose that f is an element of $h(z \smile t)$ and that $f \notin h(z)$: this means that there exists an integer k such that $z \cap u_{f(0)f(1)...f(k)} = \mathbf{0}$. According to remark (1) from (c), we also have that for every integer $p \geq k$, $z \cap u_{f(0)f(1)...f(p)} = \mathbf{0}$. Now, for every integer n, we have

$$(z \cap u_{f(0)f(1)...f(n)}) \smile (t \cap u_{f(0)f(1)...f(n)})$$
$$= (z \smile t) \cap u_{f(0)f(1)...f(n)} \neq \mathbf{0}.$$

It follows that $tc \cap u_{f(0)f(1)...f(p)} \neq \mathbf{0}$ for all $p \geq k$; but remark (2) from (c) shows that this must also be true for the integers $p \leq k$. Finally, for every integer n, we have $t \cap u_{f(0)f(1)...f(n)} \neq \mathbf{0}$, i.e. $f \in h(t)$. Thus $h(z \smile t) \subseteq h(z) \cup h(t)$, which completes the proof that h preserves greatest upper bounds.

From the definition of h, it is immediate that $h(\mathbf{0}) = \emptyset$ and $h(\mathbf{1}) = \{\mathbf{0}, \mathbf{1}\}^{\mathbb{N}}$. Theorem 2.45 and Remark 2.46 allow us to conclude from the preceding that h is a homomorphism of Boolean algebras from $\mathcal{A}$ into $\mathcal{B}(\{\mathbf{0}, \mathbf{1}\}^{\mathbb{N}})$.

This homomorphism is injective: to see this, it suffices to verify that for every non-zero element $a \in A$, $h(a)$ is not empty. To accomplish this, we will define by induction a sequence $f \in \{\mathbf{0}, \mathbf{1}\}^{\mathbb{N}}$ as follows:

$$f(0) = \begin{cases} 0 & \text{if } a \cap u_0 \neq \mathbf{0} \\ 1 & \text{if } a \cap u_0 = \mathbf{0} \end{cases};$$

and for every n,

$$f(n+1) = \begin{cases} 0 & \text{if } a \cap u_{f(0)f(1)...f(n)0} \neq \mathbf{0} \\ 1 & \text{if } a \cap u_{f(0)f(1)...f(n)0} = \mathbf{0} \end{cases}.$$

As a is non-zero, we cannot have $a \cap u_0 = \mathbf{0}$ and $a \cap u_1 = \mathbf{0}$ simultaneously (since $u_1 = \mathbf{1} + u_0$); hence $a \cap u_{f(0)} \neq \mathbf{0}$.

Suppose (induction hypothesis) that $a \frown u_{f(0)f(1)...f(n)} \neq \mathbf{0}$. Since the pair

$$(u_{f(0)f(1)...f(n)0}, u_{f(0)f(1)...f(n)1})$$

is a bipartition of $u_{f(0)f(1)...f(n)}$, we can not have $a \frown u_{f(0)f(1)...f(n)0} = \mathbf{0}$ and $a \frown u_{f(0)f(1)...f(n)1} = \mathbf{0}$. We conclude from this that

$$a \frown a \frown u_{f(0)f(1)...f(n)f(n+1)} \neq \mathbf{0}.$$

We have thereby proved that $f \in h(a)$; hence $h(a) \neq \emptyset$.

It remains to prove that h is surjective. For every integer n and every $(n+1)$-tuple $(\alpha_0, \alpha_1, \ldots, \alpha_n) \in \{0, 1\}^{n+1}$, set

$$\Omega_{\alpha_0\alpha_1...\alpha_n} = \{f \in \{0, 1\}^{\mathbb{N}} : f(0) = \alpha_0, f(a) = \alpha_1, \ldots, f(n) = \alpha_n\},$$

and let us show that $h(u_{\alpha_0\alpha_1...\alpha_n}) = \Omega_{\alpha_0\alpha_1...\alpha_n}$. If $f \in \Omega_{\alpha_0\alpha_1...\alpha_n}$ and $k \in \mathbb{N}$, we have

$$u_{\alpha_0\alpha_1...\alpha_n} \frown u_{f(0)f(1)...f(k)} = \begin{cases} u_{\alpha_0\alpha_1...\alpha_n} & \text{if } k \leq n \\ u_{f(0)f(1)...f(k)} & \text{if } k > n \end{cases};$$

so in both cases, we have a non-zero element of A, which proves that f belongs to $h(u_{\alpha_0\alpha_1...\alpha_n})$. Conversely, if $f \in h(u_{\alpha_0\alpha_1...\alpha_n})$, we have, in particular, that

$$u_{\alpha_0\alpha_1...\alpha_n} \frown u_{f(0)f(1)...f(n)} \neq \mathbf{0};$$

hence, according to question (d), $f(0) = \alpha_0$, $f(a) = \alpha_1, \ldots$, and $f(n) = \alpha_n$, which shows that $f \in \Omega_{\alpha_0\alpha_1...\alpha_n}$.

It is easy to be convinced that the family

$$\mathcal{O} = (\Omega_{\alpha_0\alpha_1...\alpha_n})_{n\in\mathbb{N}, (\alpha_0,\alpha_1,...,\alpha_n)\in\{0,1\}^n}$$

constitutes a basis of open sets for the topology of $\{0, 1\}^{\mathbb{N}}$. Indeed, in Section 2.1.2, we considered basic open sets of the following form: the set of maps from $\mathbb{N}$ into $\{0, 1\}$ that assume given values at a given finite number of points; now such a set is manifestly a (finite) union of sets from the family $\mathcal{O}$.

So let us consider an arbitrary clopen set $V \in \mathcal{B}(\{0, 1\}^{\mathbb{N}})$. V is a union of sets from the family $\mathcal{O}$ but, since we are in a compact space, there is a finite number of sets from the family $\mathcal{O}$ whose union is V. Suppose, for example, that $V = G_1 \cup G_2 \cup \cdots \cup G_p$ where each G_i is a set of the form $\Omega_{\alpha_0\alpha_1...\alpha_n}$. Now each set $\Omega_{\alpha_0\alpha_1...\alpha_n}$ is the image under h of the element $u_{\alpha_0\alpha_1...\alpha_n}$. Hence there exist elements $b_1, b_2, \ldots, b_p$ in A such that $G_1 = h(b_1), G_2 = h(b_2), \ldots$, and $G_p = h(b_p)$. Since h preserves greatest upper bounds, we may conclude that

$$V = h(b_1 \smile b_2 \smile \cdots \smile b_p).$$

It follows that the image of the map h is the whole of $\mathcal{B}(\{0, 1\}^{\mathbb{N}})$.

We have thus proved that every countable atomless Boolean algebra is isomorphic to the Boolean algebra of clopen subsets of the space $\{0, 1\}^{\mathbb{N}}$ (with the discrete topology on $\{0, 1\}$ and the product topology on the space). We may also conclude from this that any Boolean topological space with a countable basis of clopen sets is homeomorphic to the space $\{0, 1\}^{\mathbb{N}}$. This space is often called the Cantor space. It is in fact homeomorphic to the **Cantor ternary set** (the set of real numbers in the interval [0,1] that are of the form

$$\sum_{n=1}^{+\infty} (x_n/3^n),$$

where, for every n, x_n is equal either to 0 or to 2), where this set has the topology induced as a subset of $\mathbb{R}$. Details concerning this can be found in the text of Kelly which we cited earlier in connection with Tychonoff's theorem.

13. (a) The map G is injective: indeed, if δ and λ are two distinct elements of $\{0, 1\}^P$, then for at least one propositional variable A, we will have $\delta(A) \neq \lambda(A)$, hence $h_\delta(\mathsf{cl}(A)) \neq h_\lambda(\mathsf{cl}(A))$, which implies that $h_\delta \neq h_\lambda$, i.e. $g(\delta) \neq g(\lambda)$.

The map G is surjective onto $S(\mathcal{F}/\sim)$: indeed, let h be a homomorphism from $\mathcal{F}/\sim$ into $\{0, 1\}$; define $\delta : P \rightarrow \{0, 1\}$ by $\delta(A) = h(\mathsf{cl}(A))$ for any propositional variable A. We will prove that $h_\delta = h$, in other words, that for every formula $F \in \mathcal{F}$,

$$\overline{\delta}(F) = h(\mathsf{cl}(F)).$$

It suffices, in fact, to prove this for formulas that can be written exclusively with the symbols for connectives $\neg$ and $\wedge$ (and the propositional variables!) since we know that every equivalence class contains such a formula. We argue by induction on the height of formulas:

- for propositional variables, this is the definition;
- if $\overline{\delta}(F) = h(\mathsf{cl}(F))$, then $\overline{\delta}(\neg F) = 1 + h(\mathsf{cl}(F)) = h((\mathsf{cl}(F))^{\complement})$ since h is a homomorphism; but $(\mathsf{cl}(F))^{\complement} = \mathsf{cl}(\neg F)$, hence $\overline{\delta}(\neg F) = h(\mathsf{cl}(\neg F))$;
- if $\overline{\delta}(F) = h(\mathsf{cl}(F))$ and $\overline{\delta}(G) = h(\mathsf{cl}(G))$, then

$$\overline{\delta}(F \wedge G) = \overline{\delta}(F) \times \overline{\delta}(G) = h(\mathsf{cl}(F)) \times h(\mathsf{cl}(G))$$
$$= h(\mathsf{cl}(F) \times \mathsf{cl}(G)) = h(\mathsf{cl}(F \wedge G)).$$

(b) Let T be a subset of $\mathcal{F}$.

- Suppose that $T/\sim$ is a filterbase. Then (Lemma 2.81) there is an ultrafilter $\mathcal{U}$ on $\mathcal{F}/\sim$ that includes $T/\sim$. Let h denote the homomorphism of Boolean algebras from $\mathcal{F}/\sim$ into $\{0, 1\}$ that is associated with this ultrafilter. We then have that $h(x) = 1$ for every $x \in T/\sim$. According to (a), we can find a (unique) assignment of truth values δ on P such that $h_\delta = h$. For every formula $F \in T$, we will have $\mathsf{cl}(F) \in T/\sim$, hence $h_\delta(\mathsf{cl}(F)) = \mathbf{1}$, or in other words, $\overline{\delta}(F) = \mathbf{1}$. We conclude that T is satisfiable.

- Conversely, suppose that T is satisfiable and that δ is an assignment that satisfies it. Then for every formula $F \in T$, we have $h_\delta(\mathsf{cl}(F)) = \mathbf{1}$, which can also be expressed as: for every element x in $T/\sim$, $h_\delta(x) = \mathbf{1}$. It follows that $T/\sim$ is included in the set

$$\mathcal{U} = \{y \in F/\sim: h_\delta(y) = \mathbf{1}\},$$

which is none other than the ultrafilter on $F/\sim$ associated with the homomorphism h_δ. So there does exist an ultrafilter that includes $T/\sim$, which amounts to saying (Lemma 2.81) that $T/\sim$ is a filterbase for the Boolean algebra $\mathcal{F}/\sim$.

(c) We will prove the non-trivial implication in the second version of the compactness theorem. So suppose that T is a contradictory set of formulas from $\mathcal{F}$. According to (b), $T/\sim$ is not a filterbase, i.e. $T/\sim$ does not have the finite intersection property or, equivalently, there exist (a finite number of) formulas $F_1, F_2, \ldots, F_k$ such that the greatest lower bound in $T/\sim$ of $\mathsf{cl}(F_1), \mathsf{cl}(F_2), \ldots, \mathsf{cl}(F_k)$ is $\mathbf{0}$. Now this greatest lower bound is the equivalence class of the formula

$$F_1 \wedge F_2 \wedge \cdots \wedge F_k.$$

It follows that this formula is not satisfied by any assignment of truth values, or again that the set $\{F_1, F_2, \ldots F_k\}$ is a finite subset of T that is contradictory.

14. (a) Let us verify that the order relation $\leq_B$ satisfies the properties listed in Theorem 2.32.

- $\mathbf{0} \in B$ and is obviously the least element; a is the greatest element;
- if x and y are elements of B, then $x \leq a$ and $y \leq a$, hence $x \frown y \leq a$ and $x \smile y \leq a$, which shows that any two elements of B have a greatest lower bound (a least upper bound, respectively) which is their greatest lower bound (least upper bound, respectively) in $\mathcal{A}$;
- since the operations $\frown$ and $\smile$ are the same as in A, they are obviously distributive with respect to one another;
- for any element x in B, we have $a \frown x^\complement \in B$ (since $a \frown x^\complement \leq a$); moreover, it is immediate that $(a \frown x^\complement) \smile x = a$ and that $(a \frown x^\complement) \frown x = \mathbf{0}$.

Thus B, with the order relation $\leq$, is a Boolean algebra which has the same least element and the same operations $\frown$ and $\smile$ as the Boolean algebra $\mathcal{A}$, but whose greatest element and whose complementation operation are different: the greatest element is a and the complement of an element $x \in B$ is $a \frown x^\complement$ (where $x^\complement$ is its complement in $\mathcal{A}$).

(b) Consider the map φ from A into B which, with each element x, associates $x \frown a$. For all elements x and y in A, we have

$$\varphi(x \frown y) = (x \frown y) \frown a = (x \frown a) \frown (y \frown a) = \varphi(x) \frown \varphi(y), \quad \text{and}$$
$$\varphi(x^{\complement}) = x^{\complement} \frown a = (x^{\complement} \frown a) \smile (a^{\complement} \frown a) = (x^{\complement} \smile a^{\complement}) \frown a$$
$$= (x \frown a)^{\complement} \frown a = (\varphi(x))^{\complement} \frown a.$$

Since $(\varphi(x))^{\complement} \frown a$ is the complement of $\varphi(x)$ in the Boolean algebra $\langle B, \leq \rangle$ considered in the previous question, Theorem 2.45 allows us to conclude that φ is a homomorphism of Boolean algebras from A into $\langle B, \leq \rangle$.
This homomorphism is surjective since for all $y \in B$, $y = \varphi(y)$.
The kernel of φ is

$$\{x \in A : x \frown a = \mathbf{0}\} = \{x \in A : x \leq a^{\complement}\},$$

which is precisely the ideal I. The Boolean algebra $\langle B, \leq \rangle$, which is the image of the homomorphism φ, is thus isomorphic to the quotient Boolean algebra A/I (the isomorphism being the map from B into A/I that, with each element x in B, associates the set : $\{y \in A : y \frown a = x\}$).

15. (a) Let X be a subset of E distinct from E. We have

$$\wp(X) = \{Y \in \wp(E) : Y \subseteq X\}.$$

Here, we recognize the principal ideal of A generated by the element X (Example 2.62).
Conversely, let I be a subset of $\wp(E)$ that is an ideal of A. Let X denote the union of all the subsets of E that belong to I (I is non-empty since $\emptyset \in I$):

$$X = \bigcup_{Y \in I} Y.$$

It is obvious that I is included in $\wp(X)$. Since E is finite, so is I and hence X is the union, i.e. the least upper bound, of a non-zero finite number of elements of the ideal I, which proves (Corollary 2.61) that X belongs to I. As a consequence, every subset of X, i.e. everything below X, also belongs to I (part (iii) of Theorem 2.59). Thus $\wp(X)$ is included in I. As we also have the reverse inclusion, we conclude that $I = \wp(X)$. Note that X can not possibly equal E, for this would mean that $I = \wp(E)$ and I would no longer be an ideal.

(b) The kernel I of the homomorphism h is an ideal in the Boolean algebra A (Theorem 2.64). According to question (a), there must then exist a subset $K \subseteq E$ such that $I = \wp(K)$. Uniqueness is automatic: if $I = \wp(K) = \wp(L)$, then $K \subseteq L$ and $L \subseteq K$, so $K = L$. It is the case that for every element Y in $\wp(E)$, $h(Y) = \mathbf{0}$ if and only if $Y \subseteq K$.

Given that h is a homomorphism of Boolean algebras, that $h(K) = \mathbf{0}$ and that $Z = E - K$ is the complement of K in the Boolean algebra $\wp(E)$, we necessarily have $h(Z) = \mathbf{1}$. For all subsets V and W of Z, we have $h(V \cap W) = h(v) \cap h(W)$ (this is true for arbitrary subsets of A) and, on the other hand,

$$h(Z - V) = h(Z \cap (E - V)) = h(Z) \cap h(E - V)$$
$$= \mathbf{1} \cap (h(V))^{\complement} = (h(V))^{\complement}.$$

So we may assert (Theorem 2.46) that the restriction of h to $\wp(Z)$ is a homomorphism of Boolean algebras from $\wp(Z)$ into $\mathcal{C}$. The image $h(\wp(Z))$ of this homomorphism is a Boolean sub-algebra of $\mathcal{C}$ (Theorem 2.53) and we obviously have $h(\wp(Z)) \subseteq h(\wp(E))$. Let us now prove the reverse inclusion: for every element $y \in C$, if $y \in h(\wp(E))$, there exists a subset $V \subseteq E$ such that $y = h(V)$; but $V = (V \cap K) \cup (V \cap Z)$, hence

$$y = h(V \cap K) \smile h(V \cap Z) = \mathbf{0} \smile h(V \cap Z) = h(V \cap Z)$$

(since $V \cap K \subseteq K$); it follows that $y \in h(\wp(Z))$. Moreover, the kernel of h is $I = \wp(K)$ and, since $\wp(K) \cap \wp(Z) = \{\emptyset\}$, we see that the homomorphism restricted to $\wp(Z)$ is injective. Finally, the restriction of h to $\wp(Z)$ is an isomorphism from $\wp(Z)$ onto $h(\wp(E))$.

16. (a) Let $\mathcal{A} = \langle A, \leq, \mathbf{0}, \mathbf{1} \rangle$ be a finite Boolean algebra and let I be an ideal of $\mathcal{A}$. As I is finite and non-empty ($\mathbf{0} \in I$), we can consider the least upper bound a of all the elements of I. By virtue of Corollary 2.61, a belongs to I. Every element of I is obviously below a and every element of A that is below a belongs to I (part (iii) of Theorem 2.59). Thus we have

$$I = \{x \in A : x \leq a\},$$

which means that I is the principal ideal generated by a.

In 15(a) we showed that in the Boolean algebra of subsets of a finite set, every ideal is principal. But we also know that every finite Boolean algebra is isomorphic to the algebra of subsets of some set (Theorem 2.50) and, as it is easy to verify that the property of being a principal ideal is preserved by isomorphism, we realize that in fact, nothing new has been proved here that is not already included in 15(a).

(b) Naturally, we assume that the atoms a_i are pairwise distinct. Let φ denote the map from A into $\wp(\{1, 2, \ldots, k\})$ which, with each element $x \in A$, associates the set

$$\varphi(x) = \{i \in \{1, 2, \ldots, k\} : x \cap a_i = a_i\}.$$

For every $x \in A$ we have

$$x = x \cap \mathbf{1} = (x \cap a_1) \smile (x \cap a_2) \smile \cdots \smile (x \cap a_k).$$

But since the a_i are atoms, each of the elements $x \frown a_i$ is either equal to a_i or equal to $\mathbf{0}$.

It follows that x is the least upper bound of the atoms a_i for which $i \in \varphi(x)$. This proves that the map φ is a bijection: for every subset $J \subseteq \{1, 2, \ldots, k\}$, there exists a unique element $x \in A$ (the least upper bound of the atoms a_j such that $j \in J$) that satisfies $\varphi(x) = J$. We may then conclude that the Boolean algebra is finite: we can even show with ease that φ is an isomorphism of Boolean algebras from $\mathcal{A}$ onto $\wp(\{1, 2, \ldots, k\})$.

(c) If G were not a filterbase, there would exist a finite number of atoms $a_1, a_2, \ldots, a_k$ in $\mathcal{A}$ such that

$$(\mathbf{1} + a_1) \frown (\mathbf{1} + a_2) \frown \cdots \frown (\mathbf{1} + a_k) = \mathbf{0},$$

which translates as: $a_1 \smile a_2 \smile \cdots \smile a_k = \mathbf{1}$ (de Morgan). According to question (b), $\mathcal{A}$ would then be finite, which is contrary to our hypothesis.

(d) If $\mathcal{U}$ is trivial, then it contains at least one atom $a \in A$ (Lemma 2.76); we then have $\mathbf{1} + a \in G$ and $\mathbf{1} + a \notin \mathcal{U}$, hence G is not included in $\mathcal{U}$. If $\mathcal{U}$ is non-trivial, then it does not contain any atom (Lemma 2.76); in this case, for every $x \in G$, $\mathbf{1} + x$ is an atom; hence $\mathbf{1} + x \notin \mathcal{U}$, so $x \in \mathcal{U}$ (Theorem 2.72), which shows that G is included in $\mathcal{U}$.

In the particular case where $\mathcal{A}$ is the Boolean algebra of subsets of an infinite set E, G is the collection of cofinite subsets of E, i.e. the Frechet filter on E. So the result that we have just proved is precisely Theorem 2.77.

(e) It follows immediately from question (a) that if $\mathcal{A}$ is finite, every filter on $\mathcal{A}$ is principal (dual of a principal ideal); in particular, every ultrafilter on $\mathcal{A}$ is principal, i.e. is trivial. Now suppose that $\mathcal{A}$ is infinite. According to question (c) and the ultrafilter theorem (Theorem 2.80), we can find an ultrafilter that includes G. According to (d), such an ultrafilter must be non-trivial.

17. (a) If we take $J = \emptyset$, then $\bigcup_{j \in J} E_j = \emptyset$, hence $\emptyset \in B$. If $X \in B$ and $Y \in B$, there exist subsets J and K of I such that $X = \bigcup_{j \in J} E_j$ and $Y = \bigcup_{k \in K} E_k$. Given that the E_i form a partition of E, then by setting $L = J \cap K$, we have $X \cap Y = \bigcup_{i \in L} E_i$ (indeed, $E_j \cap E_k = \emptyset$ if $j \neq k$ and $E_j \cap E_k = E_j = E_k$ if $j = k$). We conclude from this that $X \cap Y \in B$. On the other hand, it is clear that the complement of X in E is the set $E - X = \bigcup_{j \in I - J} E_j$, which shows that $E - X \in B$. We may thus conclude, using Theorem 2.54, that B is a Boolean sub-algebra of $\mathcal{A}$.

Fix an index $i \in I$ and consider an element X in B such that $X \subseteq E_i$. We have $X = \bigcup_{j \in J} E_j$ for a certain subset J of I. Thus, for every $j \in J$, we have $E_j \subseteq E_i$, which requires (since we are in the presence of a partition) that $E_j = E_i$ or $E_j = \emptyset$. We conclude that $X = E_i$ or $X = \emptyset$, which proves that E_i is an atom of B.

(b) Let U and V be two distinct atoms of $\mathcal{C}$. We have $U \cap V \in \mathcal{C}$ and $U \cap V \subseteq U$, hence either $U \cap V = \emptyset$ or $U \cap V = U$. The second possibility is excluded since, U and V being atoms, we can only have $U \subseteq V$ if $U = V$. It follows that distinct atoms of $\mathcal{C}$ are disjoint subsets of E. On the other hand, the least upper bound of the atoms of $\mathcal{C}$ (i.e. their union) is the set E. Indeed, in a finite Boolean algebra, every non-zero element is the least upper bound of the set of atoms below it (this follows from the proof of Theorem 2.50 and, in the notation used there, we have that $x = M_{h(x)}$ for every non-zero element x of A). We have thereby established that the atoms of $\mathcal{C}$ constitute a partition of the set E.

(c) To each Boolean sub-algebra of $\wp(E)$, there corresponds the partition of E, studied in part (b), formed by the atoms of this sub-algebra. This correspondence is injective: indeed, as we have just recalled, every non-zero element in a finite Boolean algebra is the least upper bound of the atoms below it; thus, if the same set of atoms is associated with two Boolean sub-algebras, these sub-algebras must coincided. This correspondence is also surjective: given a partition of E, we may, as in (a), associate with it a Boolean sub-algebra $\mathcal{B}$ of A and we have seen that the elements of the given partition are the atoms of $\mathcal{B}$. These are, by the way, all the atoms of $\mathcal{B}$ for if there were others, the atoms of $\mathcal{B}$ would no longer constitute a partition of E and (b) would be violated.

18. We have seen (Exercises 3 and 4) that the properties 'is atomic' and 'is complete', as well as their negations, are preserved by isomorphisms of Boolean algebras. As well, we know that every Boolean algebra is isomorphic to a sub-algebra of a Boolean algebra of subsets of some set (Stone's theorem) and that the Boolean algebra of subsets of a set is atomic and complete (Exercise 4). Finally, we know that there exist Boolean algebras that are not complete and others that are not atomic. These remarks permit us to answer questions (a) and (e): consider a Boolean algebra that is non-atomic (respectively, not complete); it is isomorphic to a Boolean sub-algebra of some Boolean algebra A of subsets of a set; A is a Boolean algebra that is atomic and complete and has at least one sub-algebra, $\mathcal{B}$, that is not atomic (respectively, not complete).

The answers to questions (b) and (f) are affirmative: it suffices to consider a finite Boolean algebra. For (b), there are also examples that are infinite (Example 2.55); but for (f), it can be shown that the finite Boolean algebras are the only ones.

The answers to questions (c) and (d) are negative: every Boolean algebra has at least one sub-algebra that contains atoms: the sub-algebra consisting of **0** and **1**, in which **1** is obviously an atom.

19. *Preliminary remarks*: Given a map f from a set E into a set F, let us agree to write $\overline{f^{-1}}$ to denote the 'inverse image map' from $\wp(F)$ into $\wp(E)$ (which, with each subset Y of F, associates the set $\{x \in E : f(x) \in Y\}$; we reserve the notation f^{-1} to denote the inverse (from F into E) of the map f in the case where f is a bijection).

Let us recall some well-known properties of the inverse image map.

As opposed to the 'direct image' map, the inverse image map preserves all the Boolean operations on $\wp(F)$ and $\wp(E)$, which means that for all subsets X and Y of F, we have

$$\overline{f^{-1}}(X \cap Y) = \overline{f^{-1}}(X) \cap \overline{f^{-1}}(Y) \text{ and } \overline{f^{-1}}(F - X) = E - \overline{f^{-1}}(X),$$

or in other words, $\overline{f^{-1}}$ is a homomorphism of Boolean algebras from $\wp(F)$ into $\wp(E)$.

Moreover, we have the following two equivalences:

$\overline{f^{-1}}$ is injective if and only if f is surjective;

$\overline{f^{-1}}$ is surjective if and only if f is injective.

Proof: If f is surjective and if Y and Z are subsets of F such that $\overline{f^{-1}}(Y) = \overline{f^{-1}}(Z)$, we see easily that $Y \cap \text{Im}(f) = Z \cap \text{Im}(f)$, i.e. $Y = Z$, hence $\overline{f^{-1}}$ is injective; if f is not surjective, and if $y \in F - \text{Im}(f)$, then we have $\overline{f^{-1}}(\{y\}) = \overline{f^{-1}}(\emptyset)$, hence $\overline{f^{-1}}$ is not injective; if f is injective and if X is a subset of E, then we have $X = \overline{f^{-1}}(f(X))$ (we have abused notation and written $f(X)$ for the direct image of X under f), hence $\overline{f^{-1}}$ is surjective; finally, if f is not injective and if x and y are distinct elements of E such that $f(x) = f(y)$, then it is clear that $\{x\} \notin \text{Im}(\overline{f^{-1}})$ and $y \notin \text{Im}(\overline{f^{-1}})$, hence $\overline{f^{-1}}$ is not surjective).

Suppose now that $\mathcal{A} = \langle A, +, \times, \mathbf{0}, \mathbf{1}\rangle$ and $\mathcal{A}' = \langle A', +, \times, \mathbf{0}, \mathbf{1}\rangle$.

As well, let us write H_A (respectively, $H_{A'}$) to denote the isomorphism of Boolean algebras from $\mathcal{A}$ onto $\mathcal{B}(S(\mathcal{A}))$ (respectively, from $\mathcal{A}'$ onto $\mathcal{B}(S(\mathcal{A}')))$ constructed using Stone's theorem.

(a) For every homomorphism $\varphi \in \text{Hom}(\mathcal{A}, \mathcal{A}')$, let us define $\Phi(\varphi)$ to be the map from $S(\mathcal{A}')$ into $S(\mathcal{A})$ which, with each homomorphism h from $\mathcal{A}'$ into $\{\mathbf{0}, \mathbf{1}\}$, associates the homomorphism $h \circ \varphi$ from $\mathcal{A}$ into $\{\mathbf{0}, \mathbf{1}\}$ obtained by composition.

Let us prove that $\Phi(\varphi)$ is a continuous map:

We will invoke Lemma 2.5, naturally taking the clopen subsets of $S(\mathcal{A}')$ and $S(\mathcal{A})$ as bases for the open subsets. Let Ω be a basic open subset of $S(\mathcal{A})$. There then exists an element $a \in A$ such that $\Omega = H_A(a) = \{h \in S(\mathcal{A}) : h(a) = \mathbf{1}\}$. Thus we have

$$\Phi(\varphi)^{-1}(\Omega) = \{h' \in S(\mathcal{A}') : ((\Phi(\varphi))(h'))(a) = \mathbf{1}\}$$
$$= \{h' \in S(\mathcal{A}') : (h' \circ \varphi)(a) = \mathbf{1}\}$$
$$= \{h' \in S(\mathcal{A}') : h'(\varphi(a)) = \mathbf{1}\} = H_{A'}(\varphi(a)).$$

So this is a basic open subset of $S(\mathcal{A}')$. Hence $\Omega(\varphi) \in C^0(S(\mathcal{A}'), S(\mathcal{A}))$.

We are now going to define a map Ψ from $C^0(S(\mathcal{A}'), S(\mathcal{A}))$ into $\mathsf{Hom}(\mathcal{A}, \mathcal{A}')$ and we will then show that it is the inverse of the map Φ. For every map $\alpha \in C^0(S(\mathcal{A}'), S(\mathcal{A}))$, set

$$\psi(\alpha) = H_{A'}^{-1} \circ \overline{\alpha^{-1}} \circ H_A.$$

Observe that because α is continuous, the inverse image of any clopen subset of $S(\mathcal{A})$ is a clopen subset of $S(\mathcal{A}')$; in other words, the restriction of the map α^{-1} to $\mathcal{B}(S(\mathcal{A}))$ assumes its values in $\mathcal{B}(S(\mathcal{A}'))$; this validates our definition of $\Psi(\alpha)$. Moreover, as we recalled earlier, $\overline{\alpha^{-1}}$ is a homomorphism. Consequently, the map $\Psi(\alpha)$, which is the composition of three homomorphisms of Boolean algebras (see the diagram below) is indeed an element of $\mathsf{Hom}(\mathcal{A}, \mathcal{A}')$.

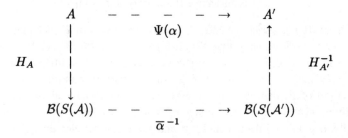

Let us show that $\Psi \circ \Phi$ is the identity on $\mathsf{Hom}(\mathcal{A}, \mathcal{A}')$. For every $\varphi \in \mathsf{Hom}(\mathcal{A}, \mathcal{A}')$, $(\Psi \circ \Phi)(\varphi)$ is the map from A into A' which, with each element $a \in A$, associates

$$
\begin{aligned}
((\Psi \circ \Phi)(\varphi))(a) &= H_{A'}^{-1}\big(\overline{\Phi(\varphi)^{-1}}(\{h \in S(\mathcal{A}) : h(a) = 1\})\big) \\
&= H_{A'}^{-1}(\{g \in S(\mathcal{A}') : ((\Phi(\varphi))(g))(a) = 1\}) \\
&= H_{A'}^{-1}(\{g \in S(\mathcal{A}') : (g \circ \varphi)(a) = 1\}) \\
&= H_{A'}^{-1}(H_{A'}(\varphi(a))) = \varphi(a).
\end{aligned}
$$

Hence,

$$(\Psi \circ \Phi)(\varphi) = \varphi.$$

Next, we show that $\Phi \circ \Psi$ is the identity on $C^0(S(\mathcal{A}'), S(\mathcal{A}))$. For every continuous map α from $S(\mathcal{A}')$ into $S(\mathcal{A})$, $(\Phi \circ \Psi)(\alpha)$ is the map from $S(\mathcal{A}')$ into $S(\mathcal{A})$ which, with each element $g \in S(\mathcal{A}')$, associates

$$
\begin{aligned}
((\Phi \circ \Psi)(\alpha))(g) &= \big(\Phi\big(H_{A'}^{-1} \circ \overline{\alpha^{-1}} \circ H_A\big)\big)(g) \\
&= g \circ H_{A'}^{-1} \circ \overline{\alpha^{-1}} \circ H_A.
\end{aligned}
$$

Now, by definition of $H_{A'}^{-1}$, we have (for example, see the proof of Theorem 2.92) that for all $a \in A$,

$$g\left(H_{A'}^{-1}\left(\overline{\alpha^{-1}}(H_A(a))\right)\right) = 1 \text{ if and only if } g \in \overline{\alpha^{-1}}(H_A(a)),$$

which is equivalent in turn to

$$\alpha(g) \in H_A(a)$$

and to

$$(\alpha(g))(a) = 1.$$

Consequently, for every $a \in A$,

$$g\left(H_{A'}^{-1}\left(\overline{\alpha^{-1}}(H_A(a))\right)\right) = (\alpha(g))(a)$$

(let us not forget that the only possible values are 0 and 1).

We conclude that

$$g \circ H_{A'}^{-1} \circ \overline{\alpha^{-1}} \circ H_A = \alpha(g),$$

which shows that $(\Phi \circ \Psi)(\alpha) = \alpha$.

We have thus established that Φ and Ψ are two bijections that are inverses of one another.

(b) It amounts to the same thing to prove that for every map

$$\alpha \in C^0(S(\mathcal{A}'), S(\mathcal{A})),$$

α is injective (respectively, surjective) if and only if $\Psi(\alpha)$ is surjective (respectively, injective). Now this is a nearly immediate consequence of the definition of Ψ and the properties of the inverse image map that were discussed in the preliminary remarks. Indeed, we have

$$\Psi(a) = H_{A'}^{-1} \circ \overline{\alpha^{-1}} \circ H_A.$$

Since $H_{A'}^{-1}$ and H_A are both bijections, we also have

$$\overline{\alpha^{-1}} = H_A \circ \Psi(a) \circ H_{A'}^{-1}.$$

We conclude that $\Psi(\alpha)$ is injective (respectively, surjective) if and only if $\overline{\alpha^{-1}}$ is; but we have seen that this happens if and only if α is surjective (respectively, injective).

Solutions to the exercises for Chapter 3

1. First we observe that F_1 is a consequence of each of the F_i, that each of the F_i is a consequence of F_2, that F_4 is a consequence of F_3 and that F_6 is a consequence of F_5 (see Exercise 5(b)).

Moreover, it is easy to verify that in every structure whose base set is $\mathbb{N}^*$, in which the symbol g is interpreted by the 'addition' map, and the symbol f by a sequence $u = (u_n)_{n \in \mathbb{N}^*}$ (a map from $\mathbb{N}^*$ into $\mathbb{N}^*$), the following properties will hold:

- F_1 is satisfied if and only if u is a sequence that assumes some value at least twice (i.e. it is not injective);
- F_2 is satisfied if and only if u is a constant sequence;
- F_3 is satisfied if and only if the sequence u is periodic;
- F_4 is satisfied if and only if each value assumed by the sequence u is assumed infinitely many times;
- F_5 is satisfied if and only if the sequence u is stationary (i.e. is constant from some point on);
- F_6 is satisfied if and only if u is a sequence such that for every integer $p \in \mathbb{N}^*$, there exists an index $n \in \mathbb{N}^*$ for which $u_n = u_{n+p}$.

These remarks allow us immediately to answer the questions asked:

(a) The six formulas are satisfied since F_2 is.

(b) Are satisfied: F_1, F_3 (the period is 4) and F_4; are not satisfied: F_2, F_5, and F_6 (there is no integer n for which $u_n = u_{n+1}$).

(c) Are satisfied: F_1, F_5, and F_6; are not satisfied: F_2, F_3, and F_4 (the values 3, 6, 11, and 18 are assumed only once; these are, respectively, u_1, u_2, u_3, and u_4).

(d) F_1 is satisfied (we have $u_2 = u_4 = 2$); the other five formulas are not satisfied: the sequence is neither constant nor periodic nor stationary and it assumes the value 1 only once; finally, if $n > 1$, n, and $n + 1$ can not have the same smallest prime divisor (1 is not a prime number!).

2. Using the distributive rules for quantifiers (Theorem 3.55) as well as some common tautologies, in particular,

$$(A \Rightarrow (B \Rightarrow C)) \Leftrightarrow ((A \wedge B) \Rightarrow C),$$

we obtain the following formulas H_i that are equivalent to the corresponding G_i:

$H_1 : \exists x \forall y (Px \Rightarrow Rxy) \wedge \forall y Py \wedge \forall y \exists z \neg Ryz;$

$H_2 : \forall x \forall z (Rzx \Rightarrow Rxz) \Rightarrow \exists x \forall y Ryz;$

$H_3 : \exists z \forall t Rtz \wedge \forall y \forall x (Rxy \Rightarrow \neg Rxy);$

$H_4 : \exists x (\forall y (Py \Rightarrow Ryx) \wedge \forall y (\forall u (Pu \Rightarrow Ruy) \Rightarrow Rxy));$

$H_5 : \forall x \forall y ((Px \wedge Py \wedge Rxy \wedge \neg Ryx) \Rightarrow \exists z (\neg Rzx \wedge \neg Ryz));$

$H_6 : \exists u \forall x \exists y (Rxy \wedge Pu \wedge Py) \Rightarrow \forall z \exists x Rzx.$

We then see that H_1 is false in any structure in which $\forall y Py$ is false, which is obviously the case for the three structures under consideration. We also see that H_3, which is equivalent to

$$\exists z \forall t\, Rtz \wedge \forall y \forall x \neg Rxy,$$

is contradictory. Finally, we observe that H_6, which is further equivalent to

$$(\exists u\, Pu \wedge \forall x \exists y (Rxy \wedge Py)) \Rightarrow \forall z \exists x\, Rzx,$$

is clearly a universally valid formula.

It follows from these remarks that in the three structures under consideration, G_1 and G_3 are false and G_6 is satisfied. As for $G_2, G_4,$ and G_5, we must examine each structure separately.

(a) The formula $\exists x \forall y\, Rxy$ is satisfied ($\mathbf{0}$ is the least element for $\leq$), hence H_2 is satisfied. If H_4 were satisfied, then the formula $\exists x \forall y (Py \Rightarrow Ryx)$ would also be satisfied and there would exist an integer that is an upper bound for all the even integers, which is absurd; thus H_4 is false. As for H_5, it is satisfied: if m and n are even integers such that $m < n$, then we can find an integer p (for example, one half the sum of m and n) such that $m < p$ and $p < n$.

Conclusion: $G_2, G_5,$ and G_6 are satisfied while $G_1, G_3,$ and G_4 are not.

(b) The formula H_2 is satisfied for reasons that are analogous to those for case (a): the empty set is the least element for the relation $\subseteq$. The formula H_4 is satisfied: it suffices to take x to be the set $\mathbb{N}$; indeed, $\mathbb{N}$ contains all its finite subsets and any subset of $\mathbb{N}$ that contains all the finite subsets of $\mathbb{N}$ is the whole of $\mathbb{N}$. The formula H_5 is also satisfied: if X and Y are finite subsets of $\mathbb{N}$ such that $X \subsetneq Y$ (strict inclusion), then by choosing an integer n that is not in Y (which is always possible), the subset $\{n\}$ of $\mathbb{N}$ is not included in X and does not include Y (note that Y can not be empty).

Conclusion: $G_2, G_4, G_5,$ and G_6 are satisfied while G_1 and G_3 are not.

(c) The formula H_2 is satisfied: indeed, the interpretation of the symbol R is not symmetric, so the formula $\forall x \forall z (Rzx \Rightarrow Rxz)$ is false. If H_4 were satisfied, then the formula $\exists x \forall y (Py \Rightarrow Ryx)$ would also be satisfied, so there would exist a real number that would equal the square of all rational numbers, which is absurd; so H_4 is false. The formula H_5 is satisfied: if x and y are two rational numbers such that $y = x^2$ and $x \neq y^2$, then it suffices to take $z = x$ and we will have $x \neq z^2$ and $z \neq x^2$.

Conclusion: $G_2, G_5,$ and G_6 are satisfied while $G_1, G_3,$ and G_4 are not.

3. (a) We may take

$$F = \forall x\, fx \simeq gx \wedge \forall x \forall y\, fx \simeq fy;$$
$$G = \forall x (\exists y\, x \simeq fy \Rightarrow \exists z\, x \simeq gz);$$
$$H = \exists x \exists y (fx \simeq gy \wedge \forall z \forall t (fz \simeq gt \Rightarrow fz \simeq fx)).$$

We could just as well have taken the formulas F_2 and F_3 from (b) in place of F and G respectively.

(b) For every L-structure $\mathcal{M} = \langle M, \overline{f}, \overline{g} \rangle$, we have:

   o $\mathcal{M} \vDash F_1$ if and only if $\overline{f} = \overline{g}$;

   o $\mathcal{M} \vDash F_2$ if and only if $\overline{f} = \overline{g}$ and $\overline{f}$ is a constant map;

   o $\mathcal{M} \vDash F_3$ if and only if $\mathrm{Im}(\overline{f}) \subseteq \mathrm{Im}(\overline{g})$;

   o $\mathcal{M} \vDash F_4$ if and only if $\mathrm{Im}(\overline{g}) \subseteq \mathrm{Im}(\overline{f})$ and $\overline{g}$ is constant;

   o $\mathcal{M} \vDash F_5$ if and only if $\mathrm{Im}(\overline{f}) \cap \mathrm{Im}(\overline{g})$ is a non-empty set.

Based on these remarks, it is easy to produce the desired models. In each case, the underlying set is $\mathbb{N}$ and we will simply give the interpretations $\overline{f}$ and $\overline{g}$ of f and g:

for a model of $F_1 \wedge \neg F_2$: $\overline{f} = \overline{g} = n \mapsto n + 1$;

for a model of F_2 : $\overline{f} = \overline{g} = n \mapsto 0$;

for a model of $\neg F_1 \wedge F_3$: $\overline{f} = n \mapsto 1$ and $\overline{g} = n \mapsto n + 1$;

for a model of $\neg F_1 \wedge F_4$: $\overline{f} = n \mapsto n + 1$ and $\overline{g} = n \mapsto 1$;

for a model of $\neg F_3 \wedge \neg F_4 \wedge F_5$: $\overline{f} = n \mapsto 2n$ and $\overline{g} = n \mapsto n^2$;

for a model of $\neg F_5$: $\overline{f} = n \mapsto 2n$ and $\overline{g} = n \mapsto 2n + 1$.

4. For G, we may take the formula

$$\exists v_0 \exists v_1 (F \wedge \forall v_2 \forall v_3 (F_{v_2/v_0, v_3/v_1} \Rightarrow (v_2 \simeq v_0 \wedge v_3 \simeq v_1))).$$

Set $H = \exists! v_0 \exists! v_1 F$ and $K = \exists! v_1 \exists! v_0 F$. If the language L has a single binary relation symbol R and if F is the formula

$$R v_0 v_1,$$

then the L-structure $\langle \mathbb{N}, \leq \rangle$ is a model of K but it is not a model of H nor of G. Indeed, the following properties hold:

$$\{a \in \mathbb{N} : \langle \mathbb{N} : v_0 \to a \rangle \vDash \exists! v_1 R v_0 v_1\} = \emptyset$$

(no integer has a unique upper bound);

$$\{b \in \mathbb{N} : \langle \mathbb{N} : v_1 \to b \rangle \vDash \exists! v_0 R v_0 v_1\} = \{0\};$$

(0 is the only natural number that has a unique lower bound (which is, of course, 0)).

So it is true that there is a unique natural number that has a unique lower bound, but it is false that there exists a unique natural number that has a unique upper bound. Since it is obviously false that there exists a unique pair $(a, b) \in \mathbb{N}^2$ such that $a \leq b$, we have here a model of K, of $\neg H$ and of $\neg G$. We obtain a model of H and $\neg K$ by considering the structure $\langle \mathbb{N}, \geq \rangle$. This shows that the formulas G, H, and K are pairwise inequivalent.

5. (a) The answer is negative: in the language whose only symbol is the one for equality, consider the formula $A[x, y] = \neg x \simeq y$; it is clear that in any structure that has at least two elements, the formula $\forall x \exists y A[x, y]$ is satisfied but the formula $\exists y \forall x A[x, y]$ is not.

(b) This time the answer is positive: let $\mathcal{M} = \langle M, \dots \rangle$ be an L-structure that satisfies the formula $\exists y \forall x A[x, y]$ and consider an element a belonging to M such that $\langle \mathcal{M} : y \mapsto a \rangle \vDash \forall x A[x, y]$; then for every b in M, we have

$$\langle \mathcal{M} : y \mapsto a, x \mapsto b \rangle \vDash A[x, y],$$

hence $\langle \mathcal{M} : x \mapsto b \rangle \vDash \exists y A[x, y]$, which shows that $\mathcal{M} \vDash \forall x \exists y A[x, y]$. This result is already included, by the way, in part (9) of Theorem 3.55.

(c) It suffices to apply part (4) of Theorem 3.55 twice.

(d) Subject to a change of bound variables, question (c) shows that the formula

$$\forall u \forall v (A[u, v] \Rightarrow B[u, v]) \Rightarrow \exists x \exists y (A[x, y] \Rightarrow C[x, y])$$

is equivalent to

$$\exists x \exists y ((A[x, y] \Rightarrow B[x, y]) \Rightarrow (A[x, y] \Rightarrow C[x, y])).$$

Applying (c) a second time, we see that if we set

$$G = (A[x, y] \Rightarrow A[y, x]) \Rightarrow$$
$$((A[x, y] \Rightarrow B[x, y]) \Rightarrow (A[x, y] \Rightarrow C[x, y])),$$

then F is equivalent to $\exists x \exists y G$.

6. Let $F[x_1, x_2, \dots, x_m, y_1, y_2, \dots, y_n]$ and $G[x_1, x_2, \dots, x_m, z_1, z_2, \dots, z_p]$ be two formulas in a language L that are universally equivalent. This means that for every L-structure $\mathcal{M} = \langle M, \dots \rangle$ and for all elements $a_1, a_2, \dots, a_m, b_1, b_2, \dots, b_n, c_1, c_2, \dots, c_p$ belonging to M, we have

$$\mathcal{M} \vDash F[a_1, a_2, \dots, a_m, b_1, b_2, \dots, b_n]$$

if and only if   (*)

$$\mathcal{M} \vDash G[a_1, a_2, \dots, a_m, c_1, c_2, \dots, c_p].$$

We need to prove that the formulas

$$F_0 = \forall x_1 \forall x_2 \ldots \forall x_m \forall y_1 \forall y_2 \ldots \forall y_n F \text{ and}$$
$$G_0 = \forall x_1 \forall x_2 \ldots \forall x_m \forall z_1 \forall z_2 \ldots \forall z_p G$$

are universally equivalent. To do this, consider a model $\mathcal{M} = \langle M, \ldots \rangle$ of F_0 and arbitrary elements $a_1, a_2, \ldots, a_m, c_1, c_2, \ldots, c_p$ of M. By choosing arbitrary elements $b_1, b_2, \ldots, b_n$ in $\mathcal{M}$, we have $\mathcal{M} \vDash F[a_1, a_2, \ldots, a_m, b_1, b_2, \ldots, b_n]$ since $\mathcal{M}$ is a model of F_0; thus, by (*), $\mathcal{M} \vDash G[a_1, a_2, \ldots, a_m, c_1, c_2, \ldots, c_p]$, which shows that $\mathcal{M}$ is a model of G_0. In the same way, we show that any model of G_0 is a model of F_0, which proves the desired result.

7. (a) For every i between 1 and 6, we provide a formula G_i that answers the question; we leave the verifications to the reader.

$G_1 = \exists x \forall y Rxy;$

$G_2 = \forall x \forall y ((Rxy \wedge \neg x \simeq y) \Rightarrow \exists z (Rxz \wedge Rzy \wedge \neg z \simeq x \wedge \neg \simeq y));$

$G_3 = \exists x \neg x * x \simeq x;$

$G_4 = \forall x (x * x \simeq c \Rightarrow x \simeq c);$

$G_5 = \exists x \, x * x \simeq d \oplus d;$

$G_6 = \forall x \forall y \forall z (Rxy \vee Ryz \vee Rxz).$

(b) The formula F_1 (where the language is $\{c, \oplus, *\}$) is satisfied in the structure $\langle \mathbb{R}, \mathbf{0}, +, \times \rangle$ (in $\mathbb{R}$, every polynomial of degree 1 has a root); $\neg F_1$ is satisfied in the structure $\langle \mathbb{Z}, \mathbf{0}, +, \times \rangle$ (the polynomial $2X + 1$, for example, does not have a root in $\mathbb{Z}$).

The formula F_2 (in the same language) is satisfied in the structure $\langle \mathbb{C}, \mathbf{0}, +, \times \rangle$ (in $\mathbb{C}$, every polynomial of degree 2 has a root); $\neg F_2$ is satisfied in $\langle \mathbb{R}, \mathbf{0}, +, \times \rangle$ (the polynomial $X^2 + 1$, for example, does not have a root in $\mathbb{R}$).

The formula F_3, which expresses (in the language $\{R\}$) that the interpretation of R is an equivalence relation, is satisfied in the structure $\langle \mathbb{Z}, \equiv_2 \rangle$ while $\neg F_3$ is satisfied in the structure $\langle \mathbb{Z}, \leq \rangle$ (the relation $\leq$ on $\mathbb{Z}$ is not symmetric).

The formula F_4 (in the language $\{*, R\}$) is satisfied in the structure $\langle \mathbb{N}, \times, \leq \rangle$, while $\neg F_4$ is satisfied in $\langle \mathbb{Z}, \times, \leq \rangle$ (the usual order relation is compatible with multiplication in $\mathbb{N}$ but not in $\mathbb{Z}$).

The formula F_5, which is equivalent to $\forall x \forall y \neg (Rxy \wedge Ryx)$, is satisfied in the structure $\langle \mathbb{Z}, < \rangle$ (strict order) while its negation is satisfied in the structure $\langle \mathbb{Z}, \leq \rangle$.

8. (a) F_1 is satisfied by an integer $n \in M$ if and only if n has no other divisor than itself, i.e. if and only if n is prime.

F_2 is satisfied by an integer n if and only if any two divisors of n are comparable for the relation 'divides'. Now any number that is a power of a prime has this property: indeed, if $n = p^k$, and if r and s divide n, then there are integers i and j less than or equal to k such that $r = p^i$ and $s = p^j$ and it is clear that either r divides s or s divides r. On the other hand, an element of M that is not a power of a prime has at least two distinct prime divisors (note that 1 is not in M), neither of which divides the other. So such an element does not satisfy F_2. Thus, the integers in M that satisfy F_2 are precisely the powers of primes.

If an integer $n \in M$ satisfies F_3, it also satisfies the following consequence of F_3 (obtained by letting y equal z in F_3 and invoking a well-known tautology):

$$\forall y((Ryx \wedge Ryy) \Rightarrow Rxy).$$

So every divisor of n must equal a multiple of n, which amounts to saying that every divisor of n must equal n, as for F_1. It follows that a number that is not prime does not satisfy F_3. On the other hand, if n is a prime number, if r divides n and if s divides r and if n, r, and s are in M, then $r = s = n$, hence n divides s; so we see that n satisfies F_3. Thus, the elements of M that satisfy F_3 are precisely the prime numbers.

By 'distributing' the quantifiers according to the usual rules, we easily see that F_4 is universally equivalent to the following formula:

$$\forall t(Rtx \Rightarrow \exists y \exists z(Ryt \wedge Rzy \wedge \neg Rtz)).$$

But it is easy to verify that the formula $\exists y \exists z(Ryt \wedge Rzy \wedge \neg Rtz)$ is also equivalent to

$$\neg \forall y \forall z(Ryt \Rightarrow (Rzy \Rightarrow Rtz)).$$

Now this last formula is none other than $\neg F_{3_{t/x}}$. So the formula F_4 is equivalent to

$$\forall t(Rtx \Rightarrow \neg F_3[t]).$$

In this way, we see that for an element $n \in M$ to satisfy F_4, it is necessary and sufficient that no divisor of n satisfy F_3, which amounts to saying that the elements of M that satisfy F_4 are those that have no prime divisors. But in M, every integer has at least one prime divisor. Hence the set in question is the empty set.

(b) For G, we may take the following formula:

$$Rtx \wedge Rty \wedge Rtz \wedge ((Rux \wedge Ruy \wedge Ruz) \Rightarrow Rut),$$

which is satisfied by a quadruple (a, b, c, d) from M if and only if d is a common divisor of a, b, and c and every common divisor of a, b, and c is a divisor of d.

(c) (1) We change the names of the bound variables in H and then 'bring them to the front' by applying the usual rules. In this way, we obtain, in succession, the following formulas that are equivalent to H, with the last formula in prenex form:

$$\forall x \forall y \forall z ((\exists v (Rvx \wedge Rvy) \wedge \exists w (Rwy \wedge Rwz))$$
$$\Rightarrow \exists t \forall u (Rut \Rightarrow (Rux \wedge Ruz)));$$
$$\forall x \forall y \forall z (\exists v \exists w ((Rvx \wedge Rvy) \wedge (Rwy \wedge Rwz))$$
$$\Rightarrow \exists t \forall u (Rut \Rightarrow (Rux \wedge Ruz)));$$
$$\forall x \forall y \forall z \forall v \forall w \exists t \forall u (((Rvx \wedge Rvy) \wedge (Rwy \wedge Rwz))$$
$$\Rightarrow (Rut \Rightarrow (Rux \wedge Ruz))).$$

(2) Consider the following sub-formula of H: $\exists t \forall u (Rut \Rightarrow (Rux \wedge Ruz))$. It is satisfied in $\mathcal{M}$ when the variables x and z are interpreted respectively by a and c if and only if there exists an integer $d \in M$ all of whose divisors (in M) divide a and c; this amounts to saying that a and c have at least one common divisor in M, or again that a and c are not relatively prime. The interpretation of H in $\mathcal{M}$ is then clear: H is satisfied in $\mathcal{M}$ if and only if for all integers a, b, and c greater than or equal to 2, if a and b are not relatively prime and if b and c are not relatively prime, then a and c are not relatively prime. Now this is manifestly false: $(a, b, c) = (2, 6, 3)$ is a counterexample.

(3) We obtain a model of H by taking $\mathbb{N}$ as the underlying set and interpreting R by the equality relation in $\mathbb{N}$; it is immediate to verify this.

9. (a) Let Ω_1, I_1, and R_1 denote the respective interpretations of Ω, I, and R in the structure $\mathcal{M}$. First, observe that the infinite subsets of $\mathbb{N}$ all have the same cardinality (they are all in one-to-one correspondence with $\mathbb{N}$).

The formula F_1 is not satisfied in $\mathcal{M}$ since we have $\emptyset \subseteq \emptyset$ and $\mathsf{card}(\emptyset) = \mathsf{card}(\emptyset - \emptyset)$; so $\mathcal{M} \vDash \exists x\, Rxx$. Note that for every non-empty subset X of $\mathbb{N}$, we have $(X, X) \notin R_1$ since $\mathsf{card}(X) \geq 0$ while $\mathsf{card}(X - X) = 0$. As the elements of Ω_1 are infinite, hence non-empty, we see that $\mathcal{M} \vDash F_2$. Suppose that X and Y are two elements of Ω_1 and that Z is a subset of $\mathbb{N}$ such that $X \subseteq Z \subseteq Y$; then Z is infinite (since it includes the infinite set X) and $\mathbb{N} - Z$ is infinite (since it includes the infinite set $\mathbb{N} - Y$); hence $Z \in \Omega_1$, which proves that $\mathcal{M} \vDash F_3$. Let X, Y, and Z be three elements of Ω_1 such that $(X, Y) \in R_1$ and $(Y, Z) \in R_1$; then we have $X \subseteq Y \subseteq Z$ and the sets $Y - X$ and $Z - Y$ are infinite; thus $Z - X$ is infinite since $Z - X = (Z - Y) \cup (Y - X)$; it follows that $(X, Z) \in R_1$ and that $\mathcal{M} \vDash F_4$. For all subsets X and Y of $\mathbb{N}$, we can not have both $(X, Y) \in R_1$ and $(Y, X) \in R_1$ unless $X = Y = \emptyset$; as $\emptyset \notin \Omega_1$, we conclude that $\mathcal{M} \vDash F_5$. Let $2\mathbb{N}$ denote the set of even natural numbers; we have $2\mathbb{N} \in \Omega_1$ and $(2\mathbb{N}, \mathbb{N}) \in R_1$ but $\mathbb{N} \notin \Omega_1$, hence $\mathcal{M} \nvDash F_6$. For all subsets X and Y of $\mathbb{N}$ such that $X \in \Omega_1$ and $(Y, X) \in R_1$, Y is a subset of X that must be infinite (otherwise $X - Y$ would be finite and $(X - Y) \cup Y = X$ would be as well); moreover, $\mathbb{N} - Y$, which includes $\mathbb{N} - X$, must also be infinite; hence $Y \in \Omega_1$ and $\mathcal{M} \vDash F_7$. If X is a finite subset of $\mathbb{N}$ whose cardinality is odd, then there is no partition of X into two sets whose cardinalities are equal; then no subset Y of $\mathbb{N}$ can satisfy $(Y, X) \in R_1$; so $\mathcal{M} \nvDash F_8$. In contrast, if X is an element of Ω_1, then we can find, on the one hand, a partition of X into two infinite sets Y and Y' and, on the other hand, a partition of $\mathbb{N} - X$ into two infinite subsets Z_1 and Z_2; by setting $Z = X \cup Z_1$, we see that Y and Z belong to Ω_1 and that $(Y, X) \in R_1$ and that $(X, Z) \in R_1$; it follows that $\mathcal{M} \vDash F_9$. Let X and Y be two elements of Ω_1 such that $(X, Y) \in R_1$; then the set $Y - X$ is infinite (for it is equipotent to X); so we can then partition it into two infinite subsets X_1 and X_2; set $Z = X \cup X_1$; we then have $X \subseteq Z \subseteq Y$ and the sets X, $Z - X = X_1$, $Y - Z = X_2$, and $\mathbb{N} - Z$ are all infinite, which proves that $(X, Z) \notin R_1$, $(Z, Y) \in R_1$, and $Z \in \Omega_1$; it follows that $\mathcal{M} \vDash F_{10}$.

(b) Let D_1 be a subset of $\wp(\mathbb{N})$ and let $\mathcal{M}'$ be the enrichment of the structure $\mathcal{M}$ obtained by interpreting the symbol D by the set D_1. For the four formulas under consideration to be satisfied in $\mathcal{M}'$, it is necessary and sufficient that D_1 be non-empty (formula G_4) and that the inclusion relation restricted to D_1 be a total ordering (formula G_1) that is dense (i.e. for all subsets X and Y in D_1, if $X \subsetneq Y$, then there exists a subset $Z \in D_1$ such that $X \subsetneq Z \subsetneq Y$) (formula G_2) and that has no least or greatest element (formula G_3). We will now construct a subset D_1 of $\wp(\mathbb{N})$ that has these properties. First, we define a subset E_1 of $\wp(\mathbb{Q})$ that has these same properties: this is not difficult; it suffices, for example, to set $J_r = [-r, r] \cap \mathbb{Q}$ for each rational $r > 0$ and to then set $E_1 = \{J_r : r \in \mathbb{Q}_+^*\}$; we have $E_1 \neq \emptyset$ and the inclusion relation on E_1 is a dense total ordering (if $0 < r < s$, and if $t = \frac{r+s}{2}$, then $J_r \subsetneq J_t \subsetneq J_s$) with no least nor greatest element.

We then convert from $\wp(\mathbb{Q})$ to $\wp(\mathbb{N})$: choose a bijection φ from $\mathbb{Q}$ onto $\mathbb{N}$ (one does exist; see Chapter 7); φ induces a bijection Φ from $\wp(\mathbb{Q})$ onto $\wp(\mathbb{N})$ (for every $X \in \wp(\mathbb{Q})$, $\Phi(X) = \{\varphi(x) : x \in X\}$) that is an isomorphism between the ordered structures $\langle \wp(\mathbb{Q}), \subseteq \rangle$ and $\langle \wp(\mathbb{N}), \subseteq \rangle$ (the verification is immediate). It follows that the subset $D_1 = \Phi(E_1)$ of $\mathbb{N}$ answers the question.

10. (a) In the examples that we are about to give, $\underline{0}$ and $\underline{1}$ are constant symbols, f is a unary function symbol, g, $\underline{+}$, and $\underline{\times}$ are binary function symbols, U and V are unary relation symbols, and R is a binary relation symbol. To show that the set $X \subseteq \mathbb{N}$ is the spectrum of a formula F, we must show, on the one hand, that the cardinality of every finite model of F is an element of X and, on the other hand, that for every element n in X, there exists a model of F whose cardinality is n. There is a tendency to omit this second condition, which is, however, indispensable.

(1) $L = \{f\}$; $F = \forall x \forall y (fx \simeq fy \Rightarrow x \simeq y) \wedge \exists x \forall y \neg x \simeq fy$.

(The existence of an injective map from the base set into itself that is not surjective is possible only if this set if infinite; so F does not have any finite models).

(2) Impossible. The base set of any structure is non-empty, hence 0 can not belong to the spectrum of any first order formula.

(3) L reduces to the equality symbol; $F = \forall x \; x \simeq x$.

(4) $L = \{f\}$; $F = \forall x (ffx \simeq x \wedge \neg fx \simeq x)$.

(Let $\langle M, \varphi \rangle$ be a finite model of F. Then φ is an involution (i.e. $\varphi \circ \varphi$ is the identity map) from M onto M that has no fixed points. The binary relation $\approx$ defined on M by

$$a \approx b \text{ if and only if } \varphi(a) = b \text{ or } \varphi(b) = a$$

is an equivalence relation whose equivalence classes each contain exactly two elements. As the classes constitute a partition of M, we see that the cardinality of M is an even number. Conversely, for a non-zero even number $n = 2p$, we obtain a model of F of cardinality n by taking $\{1, 2, \ldots, n\}$ as the base set and by taking the interpretation of f to be the map

$$k \mapsto k + p \quad \text{if } 1 \leq k \leq p;$$
$$k \mapsto k - p \quad \text{if } p + 1 \leq k \leq n.$$

(5) $L = \{U, G\}$; F is the conjunction of the formulas

$$G = \forall x \exists y \exists z (Uy \wedge Uz \wedge x \simeq gyz)$$

and

$$H = \forall x \forall y \forall z \forall t \, ((Ux \wedge Uy \wedge Uz \wedge Ut \wedge gxy \simeq gzt)$$
$$\Rightarrow (x \simeq z \wedge y \simeq t)).$$

(Let $\langle M, U_0, g_0 \rangle$ be a finite model of F. Then the restriction of g_0 to $U_0 \times U_0$ is a bijection from $U_0 \times U_0$ onto M; thus the cardinality of M is that of $U_0 \times U_0$, which is a perfect square. Conversely, for every integer $n \geq 1$, if n is a perfect square, for example $n = p^2$, it is easy to verify that the formula F is satisfied in the L-structure $\langle M, U_0, g_0 \rangle$, where $M = \{0, 1, 2, \ldots, n - 1\}$, $U_0 = \{0, 1, 2, \ldots, p - 1\}$ and g_0 is the map from $M \times M$ into M which, with each pair (a, b), associates $ap + b$ if a and b both belong to U_0 and 0 otherwise.)

(6) $L = \{\simeq\}$; F is the formula

$$\exists x \exists y \exists z (\neg x \simeq y \wedge \neg y \simeq z \wedge \neg x \simeq z \wedge \forall t (t \simeq x \vee t \simeq y \vee t \simeq z)).$$

(7) $L = \{\simeq\}$; $F = \exists x \exists y \exists z \exists t \forall u (u \simeq x \vee u \simeq y \vee u \simeq z \vee u \simeq t)$.

(8) $L = \{\simeq\}$; $F = \exists v_0 \exists v_1 \ldots \exists v_k \bigwedge_{0 \leq i < j \leq k} \neg v_i \simeq v_j$.

(9) $L = \{U, V, g\}$; we set

$$A = \forall x \forall y \; x \simeq y;$$
$$B = \forall x \exists y \exists z (Uy \wedge Vz \wedge x \simeq gyz);$$
$$C = \forall x \forall y \forall z \forall t ((Ux \wedge Vy \wedge Uz \wedge Vt \wedge gxy \simeq gzt)$$
$$\Rightarrow (x \simeq z \wedge y \simeq t));$$
$$D = \exists x \exists y (Ux \wedge Uy \wedge \neg x \simeq y) \wedge \exists z \exists t (Vz \wedge Vt \wedge \neg z \simeq t);$$

and we take

$$F = (A \vee (B \wedge C \wedge D)).$$

(The inspiration comes from (5)). Given a finite model $\langle M, U_0, V_0, g_0 \rangle$ of F, either M is a singleton or else U_0 and V_0 each have at least two elements and the restriction of g_0 to $U_0 \times V_0$ is a bijection from $U_0 \times V_0$ onto M. Thus the cardinality of M is either equal to 1 or else is the product of two integers that are both greater than or equal to 2, i.e. it is, in all cases, not a prime number. Conversely, for every non-zero integer n that is not prime, it is easy to construct a model of F that has n elements).

(10) $L = \{\underline{0}, \underline{1}, \underline{+}, \underline{\times}, R\}$ (the language of fields, together with a binary relation symbol). Let I denote the conjunction of the axioms for a commutative field together with formulas expressing that the interpretation of R is a total order relation; let J be the following formula:

$$\forall x \, R\underline{0}x \wedge \forall x (\exists y (Rxy \wedge \neg x \simeq y) \Rightarrow (Rxx\underline{+1} \wedge \forall t ((Rxt \wedge \neg x \simeq t)$$
$$\Rightarrow Rx\underline{+1}t))).$$

We then set

$$F = I \wedge J.$$

(Let $\langle K, \mathbf{0}, \mathbf{1}, +, \times, \leq \rangle$ be a finite model of F; K is then a finite field with an order relation $\leq$ for which $\mathbf{0}$ is the least element and in which every element a that has a strict upper bound has a least strict upper bound (a successor), namely $a + \mathbf{1}$. Let us agree to denote the strict order associated with $\leq$ by $<$. Recall that the **characteristic** of K is the least integer $m > 0$ such that $m\mathbf{1} = \mathbf{0}$ ($m\mathbf{1}$ denotes the element $\mathbf{1} + \mathbf{1} + \cdots + \mathbf{1}$, with m occurrences of $\mathbf{1}$) and it is easy to show that this characteristic is always a prime number, which we will here denote by p. The satisfaction of the formula J shows that we have (if, for every integer j, we denote the element $j\mathbf{1}$ by j)

$$0 < 1 < 2 < \cdots < p{-}1,$$

(so the cardinality of K is at least equal to p) and that for every integer k with $0 \leq k \leq p - 2$, there is no element of K that is strictly between

K and $k + 1$. Since the order is total and since $\mathbf{0}$ is the least element, any element of K other than $\mathbf{0}, \mathbf{1}, \ldots, p - 1$ must be strictly greater than $p - 1$.

It follows that if the cardinality of K is strictly greater than p, we can find an element b in K for which $p - 1 < b$; $p - 1$ then has a strict upper bound hence, according to formula J, it is strictly less than $p - 1 + 1 = p = \mathbf{0}$ since p is the characteristic of K. We would then have $\mathbf{0} < p - 1$ and $p - 1 < \mathbf{0}$, and hence $\mathbf{0} < \mathbf{0}$ by transitivity, which is absurd. So the cardinality of K is equal to p (and K is isomorphic to the field $\mathbb{Z}/p\mathbb{Z}$). Thus the cardinality of any finite model of F is a prime number.

Conversely, for every prime number p, if we let $\leq$ denote the total ordering on $\mathbb{Z}/p\mathbb{Z}$ defined by: $\overline{0} \leq \overline{1} \leq \cdots \leq \overline{p-1}$ (here, $\overline{k}$ is the equivalence class of k modulo p), then the structure $\langle \mathbb{Z}/p\mathbb{Z}, 0, 1, +, \times, \leq \rangle$ is a model of F of cardinality p. The verification is immediate.)

(b) Let G be a closed formula, in an arbitrary language, whose spectrum is infinite. For every integer $k \geq 1$, let F_k denote the formula

$$\exists v_0 \exists v_1 \ldots \exists v_k \bigwedge_{0 \leq i < j \leq k} \neg v_i \simeq v_j$$

that was previously used in part (8) of (a). Let T be the theory

$$\{G\} \cup \{F_k : k \in \mathbb{N}^*\}.$$

Given an arbitrary integer n, G has at least one model whose cardinality is greater than or equal to $n + 1$ since the spectrum of G is infinite. Such a model is also a model of $\{G\} \cup \{F_1, F_2, \ldots, F_n\}$. This shows that every finite subset of T has a model. By the compactness theorem, T also has a model. But a model of T is nothing other than an infinite model of G.

11. Let T be a non-contradictory theory in a language L such that all its models are isomorphic. All the models of T are then elementarily equivalent (Proposition 3.74); this means that T is complete (Definition 3.81).

12. (a) Let M denote the base set of $\mathcal{M}$. Recall the three conditions that are necessary and sufficient for a subset C to be the base set of a sub-structure of $\mathcal{M}$:

   (1) C is non-empty;
   (2) C contains the interpretations in $\mathcal{M}$ of the constant symbols of L;
   (3) C is closed under the functions that are the interpretations in $\mathcal{M}$ of the function symbols of L.

   It is clear that the intersection of all the subsets of M that satisfy these conditions and that include A will be a set that again includes A (so is non-empty since we assumed that A is non-empty) and satisfies conditions

(2) and (3) above. It is therefore the base set of some sub-structure of $\mathcal{M}$ that includes A. It is the base set of the desired structure $\mathcal{A}$.

The uniqueness of $\mathcal{A}$ is clear: if a sub-structure $\mathcal{B}$ has the indicated properties, then each of the structures $\mathcal{A}$ and $\mathcal{B}$ is a sub-structure of the other, hence $\mathcal{A} = \mathcal{B}$.

(b) If the language consists of the single binary relation symbol R, there is no sub-structure generated by $\emptyset$ in the structure $\langle \mathbb{R}, \leq \rangle$: indeed, the two substructures $\langle \{0\}, \leq \rangle$ and $\langle \{1\}, \leq \rangle$ can not have a common sub-structure, so property (2) must fail. In fact, whenever the language does not contain any symbols for functions or for constants, then there can not be a sub-structure generated by the empty set in any structure whose underlying set contains at least two distinct elements. However, in $\langle \{0\}, \leq \rangle$ for example, there is a sub-structure generated by the empty set: it is the structure itself!

If the language has at least one constant symbol, then in any structure, the empty set generates the same sub-structure as that generated by the set of interpretations of the constant symbols (which in this case is non-empty). For example, if the language is $\{c, f\}$ (one constant symbol and one unary function symbol), then in the structure $\langle \mathbb{N}, \mathbf{0}, n \mapsto n + 1 \rangle$, $\emptyset$ generates the entire structure whereas in the structure $\langle \mathbb{N}, \mathbf{0}, n \mapsto n \rangle$, $\emptyset$ generates the sub-structure $\langle \{0\}, \mathbf{0}, n \mapsto n + 1 \rangle$. Let us give another example, without constant symbols, but with a unary function symbol f. Consider a structure $\mathcal{M}$ in which f is interpreted by the constant function equal to a. In $\mathcal{M}$, there is a sub-structure generated by the empty set: it is $\langle \{a\}, \text{identity} \rangle$.

(c) Let C be the set of interpretations in $\mathcal{M}$ of the constant symbols of the language. If $A \cup C$ is not empty, the substructure $\mathcal{A}$ generated by A has $A \cup C$ as its base set; the interpretations of the constant symbols are the same as in $\mathcal{M}$ and the interpretation in $\mathcal{A}$ of each relation symbol is the restriction to $A \cup C$. In the case where $A = C = \emptyset$, the examples given in (b) show that nothing can be said in general.

(d) If F is satisfied in $\mathcal{M}$, then F is satisfied in every sub-structure of $\mathcal{M}$ (Theorem 3.70), in particular, in every sub-structure of finite type.

Conversely, suppose that F is not satisfied in $\mathcal{M}$.

- If F has no quantifiers, then F is not satisfied in any sub-structure of $\mathcal{M}$ (Theorem 3.69; note that F is closed). Let a be an arbitrary element of $\mathcal{M}$. The sub-structure of $\mathcal{M}$ generated by $\{a\}$ is then a sub-structure of $\mathcal{M}$ of finite type in which F is not satisfied.

- If $F = \forall x_1 \forall x_2 \ldots \forall x_n G[x_1, x_2, \ldots, x_n]$ (where G has no quantifiers and $n \geq 1$), then we can find elements $a_1, a_2, \ldots, a_n$ in $\mathcal{M}$ such that $\mathcal{M} \nvDash G[a_1, a_2, \ldots, a_n]$. Let $\mathcal{A}$ denote the sub-structure (of finite type) of $\mathcal{M}$ generated by $\{a_1, a_2, \ldots, a_n\}$. We have $\mathcal{A} \nvDash G[a_1, a_2, \ldots, a_n]$

(Theorem 3.69). It follows that

$$\mathcal{A} \nvDash \forall x_1 \forall x_2 \ldots \forall x_n G[x_1, x_2, \ldots, x_n],$$

so we have a sub-structure of $\mathcal{M}$ of finite type in which F is not satisfied.

(e) Let L be the language consisting of the equality symbol and let F be the formula $\exists x \exists y \, \neg x \simeq y$; F is satisfied in the structure $\langle \mathbb{N} \rangle$ but is not satisfied in the sub-structure $\langle \{0\} \rangle$ generated by $\{0\}$.

13. (a) We argue by induction on t.

- If t is of height 0, it is either a constant symbol c or a variable x. So one of the the formulas $t \simeq c$ or $t \simeq x$ is universally valid and, *a fortiori*, a consequence of T.

- If $t = fu$, the induction hypothesis presents us with four possibilities concerning the term u:

  * $T \vDash^* u \simeq c$; then $T \vDash^* t \simeq fc$; or $T \vDash^* fc \simeq ffgc(H_3)$, $T \vDash^* ffgc \simeq fgc$ (formula H_1) and $T \vDash^* fgc \simeq c$; it follows that $T \vDash^* t \simeq c$;
  * there exists a variable x such that $T \vDash^* u \simeq x$; then $T \vDash^* t \simeq fx$;
  * there is a variable x such that $T \vDash^* u \simeq fx$; then $T \vDash^* t \simeq ffx$; we may then conclude (formula H_1) that: $T \vDash^* t \simeq fx$;
  * there is a variable x such that $T \vDash^* u \simeq gx$; then $T \vDash^* t \simeq fgx$; it then follows (formula H_3) that: $T \vDash^* t \simeq c$.

- For the case where $t = gu$, the argument is analogous.

(b) Let $\overline{f}$ and $\overline{g}$ denote, respectively, the maps $(a, b) \mapsto (a, b_0)$ and $(a, b) \mapsto (a_0, b)$ from $A \times B$ into $A \times B$. For all pairs $(a, b) \in A \times B$, we have

$$\overline{f}(\overline{f}(a, b)) = \overline{f}(a, b_0) = (a, b_0) = \overline{f}(a, b);$$
$$\overline{g}(\overline{g}(a, b)) = \overline{g}(a_0, b) = (a_0, b) = \overline{g}(a, b);$$
$$\overline{f}(\overline{g}(a, b)) = \overline{f}(a_0, b) = (a_0, b_0) = \overline{g}(a, b_0) = \overline{g}(\overline{f}(a, b));$$

this shows that the formulas H_1, H_2, and H_3 are satisfied in $\mathcal{M}(A, B, a_0, b_0)$. Let (a, b) and (a', b') be two elements of $A \times B$. If $\overline{f}(a, b) = \overline{f}(a', b')$, then $a = a'$; and if $\overline{g}(a, b) = \overline{g}(a', b')$, then $b = b'$, which proves the satisfaction of H_4. Moreover, if $\overline{f}(a, b) = (a, b)$ and if $\overline{g}(a', b') = (ab')$, then $b = b_0$ and $a' = a_0$; in these circumstances, we have $\overline{f}(a, b') = (a, b)$ and $\overline{g}(a, b') = (a', b')$; we have thus found an element whose image under $\overline{f}$ is (a, b) and whose image under $\overline{g}$ is (a', b'); thus H_5 is satisfied in $\mathcal{M}(A, B, a_0, b_0)$.

(c) Let $\mathcal{M} = \langle M, \alpha, \varphi, \psi \rangle$ be an arbitrary model of T. In the first case under (a), we have already seen that $\mathcal{M} \vDash H_6$ and we prove in an analogous fashion that $\mathcal{M} \vDash H_7$.

For every element $x \in M$, if $\varphi(x) = x$, then $\psi(x) = \psi(\varphi(x)) = \alpha$ (according to H_3); conversely, if $\psi(x) = \alpha$, then $\psi(\varphi(x)) = \alpha$ (according to H_3); since we also have $\varphi(\varphi(x)) = \varphi(x)$ (according to H_1), the elements $\varphi(x)$ and x have the same image under φ and the same image under ψ; it follows (according to H_4) that $\varphi(x) = x$. So we see that $\mathcal{M} \models H_8$. The proof that $\mathcal{M} \models H_9$ is analogous.

Let us show that $\mathcal{M} \models H_{10}$. The implication from left to right is evident (take v_1 equal to v_0). Let x be any element of M. If there exists an element y in M such that $x = \varphi(y)$, then $\varphi(x) = \varphi(\varphi(y)) = \varphi(y)$ (according to H_1), so $\varphi(x) = x$. The proof for H_{11} is analogous.

To show that $\mathcal{M} \models H_{12}$, we first note that the implication from right to left follows immediately from H_6 and H_7. In the other direction, let x be an element of M such that $\varphi(x) = x$ and $\psi(x) = x$; we then have $\psi(x) = \alpha$ and $\varphi(x) = \alpha$ according to H_8 and H_9; hence $x = \alpha$.

It is immediate to verify that H_{13} is satisfied in $\mathcal{M}$: if x and y are elements of M such that $\varphi(x) = \psi(y)$, then $\varphi(\varphi(x)) = \varphi(\psi(y))$, hence (according to H_1 and H_3) $\varphi(x) = \alpha$.

(d) Let $\overline{f}$ and $\overline{g}$ be the respective interpretations of f and g in $\mathcal{M}(A, B, a_0, b_0)$ and let $\widetilde{f}$ and $\widetilde{g}$ be their interpretations in $\mathcal{M}(C, D, c_0, d_0)$. We may choose a bijection λ (respectively, μ) from A onto C (respectively, from B onto D) such that $\lambda(a_0) = c_0$ (respectively, $\mu(b_0) = d_0$). The map γ from $A \times B$ into $C \times D$ which, with each pair (x, y), associates the pair $(\lambda(x), \mu(y))$ is an isomorphism between the structures $\mathcal{M}(A, B, a_0, b_0)$ and $\mathcal{M}(C, D, c_0, d_0)$: indeed, γ is first of all a bijection from $A \times B$ onto $C \times D$; also, we have $\gamma(a_0, b_0) = (c_0, d_0)$ and, for all pairs $(a, b) \in A \times B$,

$$\gamma(\overline{f}(a, b)) = \gamma(a, b_0) = (\lambda(a_0), d_0) = \widetilde{f}(\lambda(a), \mu(b)) = \widetilde{f}(\gamma(a, b)),$$

and, analogously, $\gamma(\overline{g}(a, b)) = \widetilde{g}(\gamma(a, b))$.

(e) We begin by observing that $\alpha \in A$ and $\alpha \in B$ (H_6 and H_7); so the structure $\mathcal{M}(A, B, a_0, b_0)$ is well defined (we will continue to denote the interpretations of f and g in this structure by $\overline{f}$ and $\overline{g}$). Let h denote the map from M into $M \times M$ which, with each element x in M, associates the pair $(\varphi(x), \psi(x))$. The satisfaction in $\mathcal{M}$ of the formulas H_{10} and H_{11} shows that $A = \text{Im}(\varphi)$ and $B = \text{Im}(\psi)$; so the values assumed by h are in $A \times B$. The satisfaction of H_4 shows that h is injective (if $(\varphi(x), \psi(x)) = (\varphi(y), \psi(y))$, then $x = y$), and the satisfaction of H_5 shows that h is surjective onto $A \times B$ (if $x = \varphi(x)$ and $y = \psi(y)$, then we can find an element $z \in M$ such that $(x, y) = (\varphi(z), \psi(z)) = h(z)$). So the map h is a bijection from M onto $A \times B$. We will now show that it is a monomorphism from $\mathcal{M}$ into $\mathcal{M}(A, B, a_0, b_0)$. This is a consequence of the following properties:

    ○ $h(\alpha) = (\varphi(\alpha), \psi(\alpha)) = (\alpha, \alpha) = (a_0, b_0)$;

    ○   for every $x \in M$, $h(\varphi(x)) = \overline{f}(h(x))$;

since

$$h(\varphi(x)) = (\varphi(\varphi(x)), \psi(\varphi(x))) = (\varphi(x), \alpha) \quad (H_1 \text{ and } H_3)$$
$$= (\varphi(x), b_0) = \overline{f}(\varphi(x), \psi(x)) = \overline{f}(h(x)).$$

○ for every $x \in M$, $h(\psi(x)) = \overline{g}(h(x))$ (using a similar argument).

The formulas

$$F_n = \exists v_0 \exists v_1 \ldots \exists v_{n-1} \left(\bigwedge_{0 \le i < j < n} \neg v_i \simeq v_j \wedge \bigwedge_{0 \le i < n} f v_i \simeq v_i \right)$$

and

$$G_n = \exists v_0 \exists v_1 \ldots \exists v_{n-1} \left(\bigwedge_{0 \le i < j < n} \neg v_i \simeq v_j \wedge \bigwedge_{0 \le i < n} g v_i \simeq v_i \right)$$

(for $n \ge 1$) then clearly answer the question.

Fix two strictly positive integers n and p. Let T_{np} denote the models of T in which the interpretations of f and g have n and p elements, respectively. Consider two models $\mathcal{M} = \langle M, \alpha, \varphi, \psi \rangle$ and $\mathcal{M}' = \langle M', \alpha', \varphi', \psi' \rangle$ of T_{np}. Let $A = \mathrm{Im}(\varphi)$, $B = \mathrm{Im}(\psi)$, $A' = \mathrm{Im}(\varphi')$, and $B' = \mathrm{Im}(\psi')$; A and A' are equipotent (they both have n elements) and so are B and B' (which each have p elements). According to question (d), the structures $\mathcal{M}(A, B, \alpha, \alpha)$ and $\mathcal{M}(A', B', \alpha' \alpha')$ are isomorphic. Now we have seen that $\mathcal{M}$ is isomorphic to the first of these and $\mathcal{M}'$ to the second. It follows that any two models of the theory T_{np} are isomorphic, which proves (Exercise 11) that this theory is complete (it is non-contradictory by question (b)).

(f) The infinite models of T are those models $\mathcal{M} = \langle M, \alpha, \varphi, \psi \rangle$ of T for which at least one of the sets $\mathrm{Im}(\varphi)$ or $\mathrm{Im}(\psi)$ is infinite. It follows that these are exactly the models of the theory

$$T' = T \cup \{F_k \vee G_k : k \in \mathbb{N}^*\}.$$

The closed formulas that are satisfied in every infinite model of T are thus the closed formulas that are consequences of T'. If F is such a formula, there exists, by the compactness theorem, a finite subset X of T' such that $X \vdash^* F$. There then exists an integer $n \ge 1$ such that $X \subseteq T_n = T \cup \{F_k \vee G_k : 1 \le k \le n\}$ and we have, naturally, $T_n \vdash^* F$. But it is obvious that for every integer k such that $1 \le k \le n$, F_k is a consequence of F_n and G_k is a consequence of G_n; so $F_k \vee G_k$ is a consequence of $F_n \vee G_n$. As a result, the theories T_n and $T \cup \{F_n \vee G_n\}$ are equivalent and we have

$$T \cup \{F_n \vee G_n\} \vdash^* F.$$

The theory T' is not complete: as we have seen, its models are the infinite models of T; now in such models, one of the sets $\mathrm{Im}(\varphi)$ or $\mathrm{Im}(\psi)$

can have an arbitrary finite number of elements; for example, the struc-
tures $\mathcal{M}(\mathbb{N}, \mathbb{N}, 0, 0)$ and $\mathcal{M}(\mathbb{N}, \{0\}, 0, 0)$ are models of T' (they satisfy the
formulas F_k for all $k \geq 1$ but the first satisfies the formula G_2 while the
second does not satisfy this formula; so these structures are not elementarily
equivalent.

(g) Questions (d) and (e) show that the models of T are determined up to
isomorphism by the cardinalities of the sets that are the interpretations of
f and g. Precisely, given a countably infinite model $\mathcal{M}$ of T, there exist
two non-empty sets A and B, each of cardinality less than or equal to $\aleph_0$,
at least one of which is infinite, and two elements $a_0 \in A$ and $b_0 \in B$ such
that $\mathcal{M}$ is isomorphic to the structure $\mathcal{M}(A, B, a_0, b_0)$. For each integer
$n \in \mathbb{N}^*$, set

$$\mathcal{M}_{\infty n} = \mathcal{M}(\mathbb{N}, \{0, 1, \dots, n-1\}, 0, 0);$$
$$\mathcal{M}_{n\infty} = \mathcal{M}(\{0, 1, \dots, n-1\}, \mathbb{N}, 0, 0);$$
$$\mathcal{M}_{\infty\infty} = \mathcal{M}(\mathbb{N}, \mathbb{N}, 0, 0).$$

We see that every countably infinite model of T is isomorphic either to
$\mathcal{M}_{\infty\infty}$ or to one of the $\mathcal{M}_{\infty n}$ or to one of the $\mathcal{M}_{n\infty}$. Naturally, these models
are pairwise non-isomorphic. There are therefore, up to isomorphism, $\aleph_0$
countably infinite models of T.

(h) The models of T'' are the models of T in which the sets of fixed points
of the interpretations of f and g are both infinite. The structure $\mathcal{M}_{\infty\infty}$ is
clearly one such model. According to (g), it is immediate that every count-
ably infinite model of T'' is isomorphic to $\mathcal{M}_{\infty\infty}$. The theory T'' is thus
$\aleph_0$-categorical (Chapter 8); moreover, it is consistent and only has infinite
models; it is therefore complete (Vaught's theorem, Part 2, Chapter 8).

14. (a) Let $\mathcal{M} = \langle M, \overline{f} \rangle$ be a model of A. If an element $a \in M$ satisfies $\overline{f}(\overline{f}(a)) =$
a, then by applying $\overline{f}$ again, we obtain $\overline{f}(\overline{f}(\overline{f}(a))) = \overline{f}(a)$; but since
A is satisfied in $\mathcal{M}$, we have $\overline{f}(\overline{f}(\overline{f}(a))) = a$ and hence $a = \overline{f}(a)$,
which is excluded (also by A). In an analogous manner, if we suppose
$\overline{f}(\overline{f}(a)) = \overline{f}(a)$, we obtain $\overline{f}(\overline{f}(\overline{f}(a))) = \overline{f}(\overline{f}(a))$, i.e. $a = \overline{f}(\overline{f}(a))$
which, as we have just seen, is impossible. So the structure $\mathcal{M}$ satisfies the
following formula G:

$$\forall x (\neg ffx \simeq x \wedge \neg ffx \simeq fx).$$

Furthermore, if we let b denote $\overline{f}(\overline{f}(a))$, we have $\overline{f}(b) = a$ and every
element $c \in M$ that satisfies $\overline{f}(c) = a$ will also satisfy $\overline{f}(\overline{f}(\overline{f}(c))) =$
$\overline{f}(\overline{f}(a))$, so $c = b$. In other words, every element of M has a unique preimage under $\overline{f}$ (which is therefore bijective) and $\mathcal{M}$ satisfies the formula H:

$$\forall x \exists y \forall z (fy \simeq x \wedge (fz \simeq x \Rightarrow z \simeq y)).$$

The rules concerning 'distribution' of quantifiers show that the formula $(G \wedge H)$ is equivalent to the formula proposed in the statement of the exercise. So this formula is satisfied in every model of A, i.e. it is a consequence of A.

(b) Let n be an element of $\mathbb{N}^*$ and let $\mathcal{N} = \langle N, \widetilde{f} \rangle$ be a model of $A \wedge F_n$. Define a binary relation ρ on N as follows:

$$\text{for all } a \text{ and } b \text{ in } N,$$
$$(a, b) \in \rho \text{ if and only if } a = b \text{ or } a = \widetilde{f}(b) \text{ or } a = \widetilde{f}(\widetilde{f}(b)).$$

It is not difficult to verify that this is an equivalence relation. The equivalence class of the element a is its **orbit** under the function $\widetilde{f}$: it contains the three elements a, $\widetilde{f}(a)$, and $\widetilde{f}(\widetilde{f}(a))$ (which are pairwise distinct) and only these. So every equivalence class contains three elements. Since these constitute a partition of N which has n elements, we conclude that n is a multiple of 3.

Conversely, for every integer $p > 0$, we may construct a model $\mathcal{M}_p$ of $A \wedge F_{3p}$ that has $3p$ elements in the following way:

For the underlying set, we take $M_p = \{0, 1, 2\} \times \{0, 1, 2, \ldots, p-1\}$ and as the interpretation of f we take the function f_p defined as follows:

$$\text{for every pair } (i, j) \in M_p,$$
$$f_p(i, j) = (i + 1[\text{mod } 3], j).$$

(c) It is sufficient (Exercise 11) to prove that for every $p \in \mathbb{N}^*$, the models of $A \wedge F_{3p}$ are all isomorphic. So consider an integer $p > 0$ and a model $\mathcal{M} = \langle M, g \rangle$ of $A \wedge F_{3p}$. There are p classes for the equivalence relation ρ defined as in part (b). Denote them by $B_0, B_1, \ldots, B_{p-1}$ and choose an arbitrary element b_i in each B_i.

Now, define a map $\varphi : M \mapsto \{0, 1, 2\} \times \{0, 1, \ldots, p-1\}$ by

$$\varphi(b_i) = (0, i);$$
$$\varphi(g(b_i)) = (1, i) \qquad \text{for all } i \in \{0, 1, \ldots, p-1\}$$
$$\varphi(g(g(b_i))) = (2, i).$$

φ is an isomorphism from $\mathcal{M}$ onto the model $\mathcal{M}_p$ defined in (b) because the classes are in correspondence with one another and for every $a \in M$, we have

$$\varphi(g(a)) = f_p(\varphi(a)).$$

(d) It suffices to generalize the construction of the models $\mathcal{M}_p$ from question (b).

This time, we take $\{0, 1, 2\} \times \mathbb{N}$ as the underlying set. The interpretation of F is the unique function defined on this set that simultaneously extends all the f_p (where the definition of f_p is the same except that the index j can assume any integer value).

(e) We generalize (c). Consider a countably infinite model of A. This time there is a countably infinite number of equivalence classes for ρ; denote them by $\{B_n : n \in \mathbb{N}\}$. Again, select a sequence $\{b_n : n \in \mathbb{N}\}$ such that $b_n \in B_n$ for all n. We define the isomorphism from this model onto the one defined in part (d) exactly as φ is defined in (c), the only difference being that n ranges over $\mathbb{N}$. The verification is simple. All the countably infinite models of A are isomorphic (since they are all isomorphic to the model that we have constructed).

15. PRELIMINARIES: First, observe that F is equivalent to the conjunction of the following formulas:

$$\forall x \forall y (dx \simeq dy \Rightarrow x \simeq y) \quad \forall x \forall y (gx \simeq gy \Rightarrow x \simeq y) \quad \forall x \exists u \, x \simeq du$$

$$\forall x \exists v \, x \simeq gv \qquad \qquad \forall x \, \neg dx \simeq gx \qquad \qquad \forall x \, dgx \simeq gdx.$$

It follows that for F to be satisfied in a structure $\mathcal{M} = \langle M, \overline{d}, \overline{g} \rangle$, it is necessary and sufficient that $\overline{d}$ and $\overline{g}$ be bijections that commute and that, at any given point, never assume the same value. As well, if $\mathcal{M} \models F$, then for every natural number m, we have

$$\text{for every element } a \in M, \; \overline{d^m}(\overline{g}(a)) = \overline{g}(\overline{d^m}(a)).$$

This is obvious if $m = 0$; suppose it is true for $m = k$; then

$$\overline{d^{k+1}}(\overline{g}(a)) = \overline{d^k}(\overline{d}(\overline{g}(a))) = \overline{d^k}(\overline{g}(\overline{d}(a))) \quad \text{[since } \overline{g} \text{ and } \overline{d} \text{ commute]}$$

$$= \overline{g}(\overline{d^k}(\overline{d}(a))) \quad \text{[by the induction hypothesis]}$$

$$= \overline{g}(\overline{d^{k+1}}(a)).$$

(a) We argue by induction on the height of the term t. Since there are no symbols for constants in L, a term t of height 0 is a variable, y for example, and since y can also be written $d^0 g^0 y$ (this is the same term), the formula $\forall y \, t \simeq d^0 g^0 y$ is universally valid; in particular, it is a consequence of T.

If $t = du$ and if $T \vdash^* \forall x \, u \simeq d^m g^n x$, then $T \vdash^* \forall x \, t \simeq d^{m+1} g^n x$.

If $t = gu$ and if $T \vdash^* \forall x \, u \simeq d^m g^n x$, then in every model $\mathcal{M} = \langle M, \overline{d}, \overline{g} \rangle$ of T, we have that for all $a \in M$, $\overline{g}(\overline{d^m}(\overline{g^n}(a))) = \overline{d^m}(\overline{g^{n+1}}(a))$ (preliminaries), which shows that $T \vdash^* \forall x \, t \simeq d^m g^{n+1} x$.

(We will have observed, although this was not explicitly used in the proof, that every term of L can be written in the form

$$d^{m_1} g^{n_1} d^{m_2} g^{n_2} \ldots d^{m_k} g^{n_k} x,$$

where x is a variable and where the m_i and n_i are natural numbers that are all non-zero, except possibly m_1 and n_k).

(b) $\mathcal{M}_0$ is a model of F since the maps s_d and s_g are bijections that commute $[s_d(s_g(i, j)) = s_g(s_d(i, j)) = (i + 1, j + 1)]$ and do not assume the same value at any point $[(i, j + 1) \neq (i + 1, j)]$. Also, for all i and j in $\mathbb{Z}$ and for all natural numbers m and n, we have

$$s_{d^m}(s_{g^n}(i, j)) = (i + n, j + m).$$

It follows that for $(m, n) \neq (0, 0)$, we have

$$s_{d^m}(s_{g^n}(i, j)) \neq (i, j) \text{ and } s_{d^m}(i, j) = (i, j+m) \neq (i+n, j) = s_{g^n}(i, j).$$

Thus, $\mathcal{M}_0$ is a model of each of the formulas $F_{mn}((m, n) \neq (0, 0))$ and, consequently, is a model of T.

(c) The map h_{ab} is a bijection whose inverse bijection is h_{-a-b}. Moreover, for all pairs $(i, j) \in \mathbb{Z} \times \mathbb{Z}$, we have

$$h_{ab}(s_d(i, j)) = (i + a, j + 1 + b) = s_d(h_{ab}(i, j)); \text{ and}$$
$$h_{ab}(s_g(i, j)) = (i + 1 + a, j + b) = s_g(h_{ab}(i, j)).$$

So h_{ab} is an automorphism of $\mathcal{M}_0$.

In fact, it can be shown that there are no automorphisms of $\mathcal{M}_0$ other than the ones in the family of h_{ab}. (*Hint*: given an automorphism h of $\mathcal{M}_0$, let h_1 and h_2 denote the maps from $\mathbb{Z} \times \mathbb{Z}$ into $\mathbb{Z}$ obtained by composing h with the two projections [which means that for i and j in $\mathbb{Z}$, $h(i, j) = (h_1(i, j), h_2(i, j))$]; then show that h_1 does not depend on the second coordinate, that h_2 does not depend on the first coordinate, and that the functions of one variable that are naturally associated with h_1 and h_2 are bijections on $\mathbb{Z}$ that commute with the successor function.)

(d) We already know that $\mathbb{Z} \times \mathbb{Z}$ and $\emptyset$ are definable.

Let A be a subset of $\mathbb{Z} \times \mathbb{Z}$ that is distinct from $\mathbb{Z} \times \mathbb{Z}$ and $\emptyset$. Let i, j, k, and l be elements of $\mathbb{Z}$ such that $(i, j) \in A$ and $(k, l) \notin A$. Set $a = k = i$ and $b = l - j$. We have $h_{ab}(i, j) = (k, l)$, which shows that A is not invariant under the automorphism h_{ab} and hence (Theorem 3.95) is not definable. So $\mathbb{Z} \times \mathbb{Z}$ and $\emptyset$ are the only subsets of $\mathbb{Z} \times \mathbb{Z}$ that are definable in $\mathcal{M}_0$.

16. PRELIMINARIES: We describe a method that we will then use several times for determining all the definable subsets of the base set (or of some cartesian power of this set) in a model $\mathcal{M} = \langle M, \ldots \rangle$ of a language L.

Suppose that we have determined a finite number of subsets $A_1, A_2, \ldots, A_n$ of M^k that are all definable in $\mathcal{M}$ and that constitute a partition of the set M^k. Suppose as well that for every index i between 1 and n and for all elements $\alpha = (a_1, a_2, \ldots, a_k)$ and $\beta = (b_1, b_2, \ldots, b_k)$ belonging to A_i, there exists an automorphism of the structure $\mathcal{M}$ that sends α to β, i.e. that sends a_1 to b_1, a_2 to b_2, $\ldots$, a_k to b_k (this property is automatically satisfied for those A_i that contain only a single element; the automorphism in question reduces to the identity).

Under these circumstances, according to Theorem 3.95, for every subset X of M^k that is definable in $\mathcal{M}$, each of the A_i is either included in X or else is disjoint from X. It is then easy to conclude that every subset of M^k that is definable in $\mathcal{M}$ must be a union of sets chosen from among $A_1, A_2, \ldots, A_n$. In other words, the Boolean algebra of subsets of M^k that are definable in $\mathcal{M}$ is the sub-algebra of $\wp(M^k)$ generated by $A_1, A_2, \ldots, A_n$. This sub-algebra has 2^n elements (refer to Exercise 17 and to Corollary 2.51).

(a) Let A be a non-empty subset of $\mathbb{Z}/n\mathbb{Z}$ that is definable in $\mathcal{M}_1$ and let k be an element of A. For every element $h \in \mathbb{Z}/n\mathbb{Z}$, the map $\varphi : x \mapsto x + h - k$ is a bijection of $\mathbb{Z}/n\mathbb{Z}$ onto itself that commutes with the map $x \mapsto x + 1$ (which means that for all $x \in \mathbb{Z}/n\mathbb{Z}$, $\varphi(x + 1) = \varphi(x) + 1$); so φ is an automorphism of the structure $\mathcal{M}_1$. Since A is definable, it follows that h, which is equal to $\varphi(k)$, belongs to A (Theorem 3.95). Hence $A = \mathbb{Z}/n\mathbb{Z}$ since the only definable subsets of $\mathbb{Z}/n\mathbb{Z}$ that are definable in $\mathcal{M}_1$ are $\emptyset$ and $\mathbb{Z}/n\mathbb{Z}$.

In exactly the same way, (by replacing $\mathcal{M}_1$ by $\mathcal{M}_2$ and $x \mapsto x + 1$ by $x \mapsto x + 2$) we prove that the only subsets of $\mathbb{Z}/n\mathbb{Z}$ that are definable in $\mathcal{M}_2$ are $\emptyset$ and $\mathbb{Z}/n\mathbb{Z}$.

(b) In the structure $\mathcal{N}_1$, the set $\{0\}$ is defined by the formula $gv_0v_0 \simeq v_0$, and the set $\{1, 2\}$ by the negation of this formula. Also, the map $x \mapsto x + x$ is clearly an automorphism of the structure that interchanges the elements 1 and 2. We conclude, from the preliminaries, that there are four subsets of $\mathbb{Z}/3\mathbb{Z}$ definable in $\mathcal{N}_1$: $\emptyset$, $\mathbb{Z}/3\mathbb{Z}$, $\{0\}$, and $\{1, 2\}$.

In the structure $\mathcal{N}_2$, the sets $\{0\}$, $\{3\}$, $\{2, 4\}$, and $\{1, 5\}$ are defined respectively by

$$H_0 : \quad gv_0v_0 \simeq v_0;$$
$$H_3 : \quad ggv_0v_0gv_0v_0 \simeq gv_0v_0 \wedge \neg H_0;$$
$$H_{24} : \quad \exists v_1 gv_1v_1 \simeq v_0 \wedge \neg H_0;$$
$$H_{15} : \quad \neg H_0 \wedge \neg H_3 \wedge \neg H_{24}.$$

Moreover, the map $x \mapsto -x$ is an automorphism of $\mathcal{N}_2$ that interchanges 2 and 4, on the one hand, and 1 and 5 on the other hand. So the circumstances described in the preliminaries are fulfilled and we may conclude that the subsets of $\mathbb{Z}/6\mathbb{Z}$ that are definable in $\mathcal{N}_2$ are the 16 elements of the algebra generated by $\{0\}$, $\{3\}$, $\{2, 4\}$, and $\{1, 5\}$, namely:

$\emptyset$, $\{0\}$, $\{3\}$, $\{2, 4\}$, $\{1, 5\}$, $\{0, 3\}$, $\{0, 2, 4\}$, $\{0, 1, 5\}$, $\{2, 3, 4\}$, $\{1, 3, 5\}$,
$\{1, 2, 4, 5\}$, $\{0, 2, 3, 4\}$, $\{0, 1, 3, 5\}$, $\{0, 1, 2, 4, 5\}$, $\{1, 2, 3, 4, 5\}$, $\mathbb{Z}/6\mathbb{Z}$.

Finally, we consider the structure $\mathcal{N}_3$. The sets $\{0\}$, $\{1\}$, $\{-1\}$, and $\mathbb{R}_+$ are
defined, respectively, by the formulas:

$$
\begin{aligned}
F_0 : & \quad \forall v_1 \, g v_1 v_0 \simeq v_0; \\
F_1 : & \quad \forall v_1 \, g v_1 v_0 \simeq v_1; \\
F_{-1} : & \quad \forall v_1 \, g v_1 g v_0 v_0 \simeq v_1 \wedge \neg F_1; \\
F_{\mathbb{R}_+} : & \quad \exists v_1 \, g v_1 v_1 \simeq v_0.
\end{aligned}
$$

All Boolean combinations of these four sets are also definable
(Theorem 3.94), in particular, $\mathbb{R}_+^* - \{1\}$ and $\mathbb{R}_-^* - \{-1\}$.

If α is a non-zero real number, the map $\psi[\alpha]$ from $\mathbb{R}$ into $\mathbb{R}$ which, with
each real x, associates

$$
\begin{aligned}
0 & \quad \text{if } x = 0; \\
x^\alpha & \quad \text{if } x > 0; \\
-(-x)^\alpha & \quad \text{if } x < 0;
\end{aligned}
$$

is a bijection which commutes with the map $(x, y) \mapsto xy$ (for all reals x
and y, $\psi[\alpha](x, y) = \psi[\alpha](x) \, \psi[\alpha](y)$); so this is an automorphism of the
structure $\mathcal{N}_3$. Let a and b be two elements of $\mathbb{R}_+^* - \{1\}$; $\ln b / \ln a$ is then
a non-zero real number and the automorphism $\psi[\ln b / \ln a]$ sends a to b.
We obtain the same conclusion when we replace $\mathbb{R}_+^* - \{1\}$ by $\mathbb{R}_-^* - \{-1\}$,
assume that a and b belong to $\mathbb{R}_-^* - \{-1\}$, and consider the automorphism
$\psi[\ln(-b) / \ln(-a)]$.

In this way, we see that the five sets $\{0\}$, $\{1\}$, $\{-1\}$, $\mathbb{R}_+^* - \{1\}$, and
$\mathbb{R}_-^* - \{-1\}$, which are definable in $\mathcal{N}_3$ and which constitute a partition
of $\mathbb{R}$, satisfy the conditions described in the preliminaries. It follows that
the Boolean algebra of subsets of $\mathbb{R}$ that are definable in $\mathcal{N}_3$ is the subalgebra of $\wp(\mathbb{R})$ generated by $\{0\}$, $\{1\}$, $\{-1\}$, $\mathbb{R}_+^* - \{1\}$, and $\mathbb{R}_-^* - \{-1\}$. It
contains 32 elements.

(c) Question (1) was answered just after Theorem 3.95 where we saw that the
only subsets of $\mathbb{R}$ definable in $\langle \mathbb{R}, \leq \rangle$ are $\emptyset$ and $\mathbb{R}$. We will again invoke the
preliminaries to answer question (2). Consider the following three subsets
of $\mathbb{R}^2$:

$$
A_1 = \left\{ (x, y) \in \mathbb{R}^2 : x = y \right\}, \text{ which is defined by the formula}
$$
$$
v_0 \simeq v_1;
$$

$A_2 = \{(x, y) \in \mathbb{R}^2 : x < y\}$, which is defined by the formula
$$Rv_0v_1 \wedge \neg v_0 \simeq v_1;$$

$A_3 = \{(x, y) \in \mathbb{R}^2 : x > y\}$, which is defined by the formula
$$Rv_1v_0 \wedge \neg v_0 \simeq v_1.$$

Let $\alpha = (a, b)$ and $\beta = (c, d)$ be two elements of $\mathbb{R}^2$. If they both belong to A_1, then $a = b$ and $c = d$ and the map $x \mapsto x+c-a$ is an automorphism of the structure $\langle \mathbb{R}, \leq \rangle$ that sends α to β. If α and β belong to A_2 (respectively, to A_3), then $a < b$ and $c < d$ (respectively, $a > b$ and $c > d$); the real $\frac{d-c}{b-a}$ is strictly positive and the map

$$x \mapsto \frac{d-c}{b-a}x + \frac{bc - ad}{b-a}$$

is an automorphism of $\langle \mathbb{R}, \leq \rangle$ that sends α to β. So the conditions described in the preliminaries are satisfied and we see that there are eight subsets of $\mathbb{R}^2$ that are definable: $\emptyset$, $\mathbb{R}^2$, A_1, A_2, A_3 and the following three sets:

$$A_2 \cup A_3 = \{(x, y) \in \mathbb{R}^2 : x \neq y\};$$

$$A_1 \cup A_3 = \{(x, y) \in \mathbb{R}^2 : x \geq y\};$$

$$A_1 \cup A_2 = \{(x, y) \in \mathbb{R}^2 : x \leq y\}.$$

17. (a) We take $\mathbb{Z}/(n+1)\mathbb{Z} = \{0, 1, \ldots, n\}$ as the base set and for the interpretation of R, we take the binary relation $\overline{R}$ defined by

for all elements a and b in $\mathbb{Z}/(n + 1)\mathbb{Z}$,
$(a, b) \in \overline{R}$ if and only if $b = a + 1$;

in other words, $\overline{R}$ is addition in $\mathbb{Z}/(n+1)\mathbb{Z}$. The $(n+1)$-tuple $(0, 1, \ldots, n)$ is an $(n + 1)$-cycle for $\overline{R}$ hence the structure we have just defined does not satisfy F_{n+1}. Moreover, it is easy to verify that if $2 \leq k \leq n$, then there are no k-cycles in this structure; this shows that it satisfies the formulas $F_2, F_3, \ldots, F_n$.

(b) If $T \vdash^* G$, then, by the compactness theorem, there is a finite subset T' of T such that $T' \vdash^* G$. We can find an integer $p \geq 2$ such that $T' \subseteq \{F_2, F_3, \ldots, F_p\}$; naturally, this implies that $\{F_2, F_3, \ldots, F_p\} \vdash^* G$, which means precisely that G is satisfied in any L-structure that does not have any cycles of length less than or equal to p.

(c) Let G be a closed formula that is a consequence of T and let p be an integer greater than or equal to 2 such that G is satisfied in any L-structure that does not have any cycles of length less than or equal to p (see question (b)). Consider a model of the formula

$$F_2 \wedge F_3 \wedge \cdots \wedge F_p \wedge \neg F_{p+1};$$

(part (a) guarantees that such a model exists); G is satisfied in this model which contains at least one cycle of length $p + 1$.

(d) We argue by contradiction. Let T_0 be a finite theory that is equivalent to T and let H be the conjunction of the formulas of T_0. The theory T is then equivalent to $\{H\}$. In particular, the formula H is a consequence of T; hence (question (c)), it has at least one model that contains a cycle; but such a model is not a model of T (whose models must not contain any cycles). This contradicts the equivalence of T and $\{H\}$.

18. We will suppose the existence of a theory T of L that has the indicated property and arrive at a contradiction. Consider the theory (in the language L')

$$T' = T \cup \{F_n : n \in \mathbb{N}\}.$$

So a model of T' must be an L'-structure in which the interpretation of R is a well-ordering and in which the set of interpretations of the c_n constitutes an 'infinite descending chain', i.e. a non-empty subset of the base set that does not have a least element modulo R. This situation is obviously contradictory and we conclude that T' is contradictory. From the compactness theorem for predicate calculus, we can find a finite subset T'' of T' that is contradictory. So there is a natural number N such that $T'' \subseteq T \cup \{F_n : n \leq N\}$. The theory $T \cup \{F_n : n \leq N\}$ is then itself contradictory. However, consider the L_0-structure $\mathcal{M}_0 = \langle \mathbb{N}, \leq \rangle$ (in which R is interpreted by the usual order relation on $\mathbb{N}$). According to our hypothesis, $\mathcal{M}_0$ can be enlarged to an L-structure $\mathcal{M}$ that is a model of T. $\mathcal{M}$ can be further enlarged to an L'-structure $\mathcal{M}'$ by interpreting each symbol c_n by the integer $N + 1 - n$ if $n \leq N + 1$ and by 0 if $n > N + 1$. It is clear that $\mathcal{M}'$ is a model of the formulas $F_0, F_1, \ldots, F_N$ and also a model of T. Thus we have a model of the theory T_N, which is absurd.

So the property 'is a well-ordering' is not finitely axiomatizable.

19. Suppose that $F[c_1, c_2, \ldots, c_k]$ is a consequence of T. Let $\mathcal{M} = \langle M, \ldots \rangle$ be an L-structure that is a model of T. For all elements $a_1, a_2, \ldots, a_k$ of M, we can enrich $\mathcal{M}$ to an L'-structure $\mathcal{M}'$ by interpreting each of the constant symbols c_i (for $1 \leq i \leq k$) by the element a_i; $\mathcal{M}'$ is also a model of T (Lemma 3.58) and hence, by our hypothesis, a model of $F[c_1, c_2, \ldots, c_k]$. It follows (by applying Proposition 3.45 k times) that

$$\langle \mathcal{M}' : x_1 \to a_1, x_2 \to a_2, \ldots, x_k \to a_k \rangle \vDash F;$$

but this is clearly equivalent to

$$\mathcal{M} \vDash F[a_1, a_2, \ldots, a_k],$$

which allows us to conclude that the formula $\forall x_1 \forall x_2 \ldots \forall x_k F[x_1, x_2, \ldots, x_k]$ is satisfied in $\mathcal{M}$. This formula, which is true in every L-structure that is a model of T, is therefore a consequence of the theory T.

20. (a) According to Theorem 3.91, the stated hypothesis means that the theory $T \cup \Delta(\mathcal{M})$ (in the language L_M) does not have a model. By the compactness theorem, we conclude that some finite subset E of this theory does not have a model. Let K denote the conjunction of the formulas from $\Delta(\mathcal{M})$ that belong to E (there are only finitely many). The theory $T \cup \{K\}$ is then contradictory. Note that any conjunction of formulas from $\Delta(M)$ is again a closed, quantifier-free formula of L_M that is satisfied in $\mathcal{M}^*$ (the natural extension of $\mathcal{M}$ to the language L_M). Thus $K \in \Delta(M)$. So there exists a quantifier-free formula $H = H[x_1, x_2, \ldots, x_n]$ of the language L and parameters $a_1, a_2, \ldots, a_n$ in M such that $K = H[a_1, a_2, \ldots, a_n]$ (we have omitted the underlining and avoided the distinction between the parameters and the corresponding constant symbols of the language L_M). To say that $T \cup \{K\}$ is a contradictory theory in the language L_M is equivalent to saying that the formula $\neg K$ is a consequence of the theory T:

$$T \vdash^* \neg H[a_1, a_2, \ldots, a_n].$$

Using the result from Exercise 19, we conclude that the formula

$$\forall x_1 \forall x_2 \ldots \forall x_n \neg H[a_1, a_2, \ldots, a_n]$$

is a consequence of the theory T. Since we also have that

$$\mathcal{M} \models H[a_1, a_2, \ldots, a_n]$$

(because $K \in \Delta(M)$), the formula

$$\forall x_1 \forall x_2 \ldots \forall x_n \neg H[x_1, x_2, \ldots, x_n]$$

is not satisfied in $\mathcal{M}$. So the formula $G = \neg H$ answers the question.

(b) Suppose there exists an extension $\mathcal{N}$ of $\mathcal{M}$ that is a model of T. Then $\mathcal{N}$ also satisfies all the closed formulas that are consequences of T, in particular, those that are universal; so $\mathcal{N}$ is a model of $U(T)$. But any universal closed formula that is satisfied in $\mathcal{N}$ must also be satisfied in the substructure $\mathcal{M}$ of $\mathcal{N}$ (Theorem 3.70); it follows that $\mathcal{M}$ is a model of $U(T)$.

Conversely, suppose that $\mathcal{M}$ is a model of $U(T)$. Then there cannot exist a quantifier-free formula $G[x_1, x_2, \ldots, x_n]$ of L such that

$$T \vdash^* \forall x_1 \forall x_2 \ldots \forall x_n G[x_1, x_2, \ldots, x_n] \text{ and}$$
$$\mathcal{M} \nvDash \forall x_1 \forall x_2 \ldots \forall x_n G[x_1, x_2, \ldots, x_n].$$

So we cannot be in the situation described in the previous question. The conclusion is that there exists at least one extension of $\mathcal{M}$ that is a model of T.

(c) If $\mathcal{M}$ has an extension that is a model of T, then every substructure of $\mathcal{M}$ has this same property since an extension of $\mathcal{M}$ is also an extension of any

SOLUTIONS

substructure of $\mathcal{M}$. In particular, every substructure of $\mathcal{M}$ of finite type has an extension that is a model of T.

To prove the converse, we will use the equivalence established in (b): it suffices to prove that $\mathcal{M}$ is a model of $U(T)$, knowing that all its substructures of finite type are. But this result follows immediately from part (d) of Exercise 12: every formula of $U(T)$ is universal; if it is true in every substructure of $\mathcal{M}$ of finite type, then it is true in $\mathcal{M}$; hence $\mathcal{M}$ is a model of $U(T)$.

21. (a) We argue by contradiction. Suppose that all the elements of A have the same type; suppose also that B is a subset of M that is definable in $\mathcal{M}$ by a formula $F[v]$ of L and is such that $B \cap A \neq \emptyset$ and $A \not\subseteq B$. If we choose elements a and b in A such that $a \in B$ and $b \notin B$, then we have that $\mathcal{M} \vDash F[a]$ and $\mathcal{M} \nvDash F[b]$; It follows that $F \in \theta(a)$ and $F \notin \theta(b)$, hence $\theta(a) \neq \theta(b)$; this contradicts that fact that a and b have the same type.

(b) Again, we argue by contradiction. If $\theta(a) \neq \theta(h(a))$, then there exists a formula $f[v_1] \in \mathcal{F}_1$ such that $\mathcal{M} \vDash F[a]$ and $\mathcal{M} \nvDash F[h(a)]$; but this situation contradicts Theorem 3.72.

(c) In $\mathcal{M}_1$, all elements have the same type: if a and b are elements of $\mathbb{Z}$, the map $x \mapsto x + b - a$ is an automorphism of $\mathcal{M}_1$ that sends a to b; so a and b have the same type according to the preceding question (we find ourselves in the situation that was already studied in Section 3.6.1 and again in part (c) of Exercise 16).

In $\mathcal{M}_2$, 0, and 1 do not have the same type: indeed, the formula $\exists w\, f w \simeq w$ is satisfied by 1 but not by 0.

In $\mathcal{M}_3$, all elements have the same type: if a and b are elements of $\mathbb{Z}$, the map $x \mapsto x + b - a$ is an automorphism of $\mathcal{M}_3$ that sends a to b; so a and b have the same type according to question (b).

In $\mathcal{M}_4$, 0, and 1 do not have the same type: indeed, the formula $g v v \simeq v$ is satisfied by 0 but not by 1.

(d) Let $\mathcal{M} = \langle M, \ldots \rangle$ be a model of T. We define a map φ from M into $\{0, 1\}^n$ in the following way: for every element $a \in M$ and for all i between 1 and n, we set

$$\varepsilon_i = \begin{cases} 1 & \text{if } F_i \in \theta(a); \\ 0 & \text{if } F_i \notin \theta(a); \end{cases}$$

and define

$$\varphi(a) = (\varepsilon_1, \varepsilon_2, \ldots, \varepsilon_n).$$

Let a and b be two elements of M such that $\varphi(a) = \varphi(b)$; this means that for all i between 1 and n, $F_i \in \theta(a)$ if and only if $F_i \in \theta(b)$, or equivalently that the formula $\bigwedge_{1 \leq k \leq n} (F_i[v_0] \Leftrightarrow F_i[v_1])$ is satisfied in $\mathcal{M}$ by the pair (a, b).

Given that $\mathcal{M}$ is a model of T and that we have assumed that the formula G is a consequence of T, it follows that we must have $a = b$, which proves that the map φ is injective. So the cardinality of M is at most that of $\{0, 1\}^n$ which is 2^n.

(e) As indicated, let us add two new constant symbols c and d to the language L and let us consider the following theory S_1 of the enriched language L_1:

$$S_1 = S \cup \{F[c] \Leftrightarrow F[d] : F \in \mathcal{F}_1(L)\} \cup \{\neg c \simeq d\}.$$

It is clear that in every model of S_1, the interpretations of c and d are two distinct elements that have the same type in the reduct of this model to the language L. So to answer the proposed question, it suffices to show that S_1 is a consistent theory in L_1. Suppose it is not. Then by the compactness theorem, there would be a finite subset of S_1 that is contradictory; so there would exist a finite number of formulas $F_1, F_2, \ldots, F_n$ of L with one free variable such that the theory

$$S_0 = S \cup \{F_i[c] \Leftrightarrow F_i[d] : 1 \leq i \leq n\} \cup \{\neg c \simeq d\}$$

is contradictory. It amounts to the same thing to say that the formula

$$\bigwedge_{1 \leq i \leq n} (F_i[c] \Leftrightarrow F_i[d]) \wedge c \simeq d$$

is a consequence of S. We may then apply the result from Exercise 19 and, relative only to the language L, conclude that

$$S \vdash^* \forall v_0 \forall v_1 \left(\bigwedge_{1 \leq i \leq n} (F_i[v_0] \Leftrightarrow F_i[v_1]) \Rightarrow v_0 \simeq v_1 \right).$$

According to question (d), this requires that any model of S can have at most 2^n elements. But we assumed that S had at least one infinite model; so we have arrived at a contradiction.

We could give a shorter proof of the property that we just obtained by invoking a theorem that will be proved in Chapter 8 of Part 2, the ascending Löwenheim–Skolem theorem, which asserts that when a theory has a countably infinite model, then it has models of every infinite cardinality. So let $\mathcal{M} = \langle M, \ldots \rangle$ be a model of S and suppose that there is no pair of elements of M that have the same type; this means that the map $a \mapsto \theta(a)$ from M into the set of subsets of $\mathcal{F}_1$ is injective (indeed, if it were not, we could find two distinct elements a and b in M for which $\theta(a) = \theta(b)$, i.e. distinct elements having the same type). As a consequence, the cardinality of the model $\mathcal{M}$ is bounded by the cardinality of the set $\wp(\mathcal{F}_1)$. So it would suffice to take a model of S of cardinality strictly greater than that of $\wp(\mathcal{F}_1)$ (such a model would exist by the Löwenheim–Skolem theorem cited above,

since we have assumed that S has at least one infinite model) to guarantee
the existence of at least one pair of distinct elements having the same type.

(f) According to what we have just proved, a theory T can satisfy the required
conditions only if it does not have any infinite model. We could consider,
for example, in the language L that has only two distinct constant symbols
c and d, the theory T consisting of the single formula

$$\forall v_0 (v_0 \simeq c \lor v_0 \simeq d) \land \neg c \simeq d.$$

It is obvious that every model of T has exactly two elements, $\overline{c}$ and $\overline{d}$, and
that these two elements do not have the same type since the formula $v_0 \simeq c$
belongs to $\theta(\overline{c})$ but does not belong to $\theta(\overline{d})$.

(g) The language L consists of a constant symbol c and a unary function symbol
f. Consider the L-structure $\mathcal{N}$ whose base set is $\mathbb{N}$, in which c is interpreted
by 0 and f by the successor function. This structure is infinite and contains
no pair of distinct elements having the same type: indeed, if n and p are
integers such that $0 \leq n < p$, the formula $v_0 \simeq f^p c$ belongs to $\theta(p)$ but
does not belong to $\theta(n)$; so n and p do not have the same type.

Solutions to the exercises for Chapter 4

1. There are cases in which $F \Rightarrow \forall v F$ is not universally valid. For example,
in a language that contains a unary predicate symbol P, take $F = Pv$. The
universal closure of $Pv \Rightarrow \forall v Pv$ is equal to $\forall v(Pv \Rightarrow \forall v Pv)$ which is
itself equivalent to $\exists v Pv \Rightarrow \forall v Pv$. This last formula is false in a structure in
which the interpretation of P is neither the empty set nor the entire underlying
set.

2. (a) In a language that contains a unary predicate symbol P, take the formula
$F = Pw$. Then $\forall v F \Rightarrow \forall w F_{w/v}$ is equal to $\forall v Pw \Rightarrow \forall w Pw$ which
is logically equivalent to $Pw \Rightarrow \forall w Pw$. We saw in Exercise 1 that this
formula is not universally valid.

(b) This time, the language has a binary predicate symbol R and $F = \exists w Rvw$.
Then $\forall v F \Rightarrow \forall w F_{w/v}$ is equal to

$$\forall v \exists w Rvw \Rightarrow \forall w \exists w Rww,$$

which is logically equivalent to $\forall v \exists w Rvw \Rightarrow \exists w Rww$. This formula is
false, for example, in the structure whose base set is $\mathbb{N}$ and in which R is
interpreted by the strict order relation.

3. We begin by writing a derivation of $\exists v_0 F$ in T, followed by a derivation of of
$\forall v_0 (F \Rightarrow G)$, always in T (these two proofs exist by hypothesis). We complete
this with the sequence of formulas in Table S4.1, which constitute a derivation
of $\exists v_0 G$ from $\exists v_0 F$ and $\forall v_0 (F \Rightarrow G)$.

Table S4.1

(1)	$\forall v_0(F \Rightarrow G) \Rightarrow (F \Rightarrow G)$	Example 4.8
(2)	$(F \Rightarrow G)$	by modus ponens, since $\forall v_0(F \Rightarrow G)$ appears earlier
(3)	$\forall v_0 \neg G \Rightarrow \neg G$	Example 4.8
(4)	$(F \Rightarrow G) \Rightarrow ((\forall v_0 \neg G \Rightarrow \neg G) \Rightarrow (\forall v_0 \neg G \Rightarrow \neg F))$	tautology
(5)	$(\forall v_0 \neg G \Rightarrow \neg G) \Rightarrow (\forall v_0 \neg G \Rightarrow \neg F)$	from (2) and (4) by modus ponens
(6)	$\forall v_0 \neg G \Rightarrow \neg F$	from (3) and (5) by modus ponens
(7)	$\forall v_0(\forall v_0 \neg G \Rightarrow \neg F)$	from (6) by generalization
(8)	$\forall v_0(\forall v_0 \neg G \Rightarrow \neg F) \Rightarrow (\forall v_0 \neg G \Rightarrow \forall v_0 \neg F)$	axiom schema (b)
(9)	$\forall v_0 \neg G \Rightarrow \forall v_0 \neg F$	from (7) and (8) by modus ponens
(10)	$\exists v_0 F \Leftrightarrow \neg \forall v_0 \neg F$	axiom schema (a)
(11)	$(\exists v_0 F \Leftrightarrow \neg \forall v_0 \neg F) \Rightarrow (\exists v_0 F \Rightarrow \neg \forall v_0 \neg F)$	tautology
(12)	$\exists v_0 F \Rightarrow \neg \forall v_0 \neg F$	from (10) and (11) by modus ponens
(13)	$\neg \forall v_0 \neg F$	from (12) by modus ponens, since $\exists v_0 F$ appears earlier
(14)	$(\forall v_0 \neg G \Rightarrow \forall v_0 \neg F) \Rightarrow (\neg \forall v_0 \neg F \Rightarrow \neg \forall v_0 \neg G)$	tautology
(15)	$\neg \forall v_0 \neg F \Rightarrow \neg \forall v_0 \neg G$	from (14) and (9) by modus ponens
(16)	$\neg \forall v_0 \neg G$	from (15) and (13) by modus ponens
(17)	$\exists v_0 G \Leftrightarrow \neg \forall v_0 \neg G$	axiom schema (a)
(18)	$(\exists v_0 G \Leftrightarrow \neg \forall v_0 \neg G) \Rightarrow (\neg \forall v_0 \neg G \Rightarrow \exists v_0 G)$	tautology
(19)	$\neg \forall v_0 \neg G \Rightarrow \exists v_0 G$	from (17) and (18) by modus ponens
(20)	$\exists v_0 G$	from (16) and (19) by modus ponens

4. Here is a proof of $F \vee \forall v G$ from $\forall v(F \vee G)$:

(1) $\forall v(F \vee G) \Rightarrow (F \vee G)$	axiom schema (c)
(2) $\forall v(F \vee G)$	formula of the theory
(3) $(F \vee G)$	from (1) and (2) by modus ponens
(4) $(F \vee G) \Rightarrow (\neg F \Rightarrow G)$	tautology
(5) $\neg F \Rightarrow G$	from (3) and (4) by modus ponens
(6) $\forall v(\neg F \Rightarrow G)$	by generalization
(7) $\forall v(\neg F \Rightarrow G) \Rightarrow (\neg F \Rightarrow \forall v G)$	axiom schema (b) (v is not free in F)
(8) $(\neg F \Rightarrow \forall v G)$	from (6) and (7) by modus ponens
(9) $(\neg F \Rightarrow \forall v G) \Rightarrow (F \vee \forall v G)$	tautology
(10) $(F \vee \forall v G)$	from (8) and (9) by modus ponens

5. (a) We define $\varphi(F)$ by induction on F:

   ○ if F is an atomic formula, then $\varphi(F) = \mathbf{1}$ (in fact, here, we could just as well have chosen $\varphi(F) = \mathbf{0}$)

   ○ if not, we use conditions (1), (2), (3), and (4) as the definition, after noting that they are compatible with one another.

   (b) The verification is immediate: if F is a tautology, then $\varphi(F) = \mathbf{1}$ thanks to conditions (3) and (4); if F is of the form $\exists v G \Leftrightarrow \neg \forall v \neg G$, then $\varphi(\exists v G) = \mathbf{1}$

by condition (2), $\varphi(\neg\forall v\neg G) = \mathbf{1}$ by condition (1) and hence $\varphi(\exists v G \Leftrightarrow \neg\forall v\neg G) = \mathbf{1}$ by condition (4). For the formulas of schema (b) or (c), the same type of argument applies.

(c) According to condition (4) when the connective α is $\Rightarrow$, if $\varphi(F \Rightarrow G) = \mathbf{1}$ and $\varphi(F) = \mathbf{1}$, we must have $\varphi(G) = \mathbf{1}$.

(d) Let $(F_1, F_2, \ldots, F_n)$ be a derivation of $F(F = F_n)$ that does not invoke the rule of generalization. This means that for all i between 1 and n, one of the following two situations must hold:

- F_i is an axiom,
- F_i follows by modus ponens from two formulas that precede it, i.e. there exist j and k less than i such that $F_j = F_k \Rightarrow F_i$.

We can show immediately by induction on the integer i, using the results from parts (b) and (c), that $\varphi(F_i) = \mathbf{1}$. Hence $\varphi(F) = \mathbf{1}$.

Now consider the formula $F = \forall v(G \Rightarrow G)$ where G is an arbitrary formula. This formula is clearly provable (it is obtained by generalization from the tautology $G \Rightarrow G$). However, $\varphi(F) = \mathbf{0}$ and hence it cannot be derived without the rule of generalization.

6. (a) We adapt the definition from the previous exercise in an obvious way.

(b) One must first verify that φ assumes the value $\mathbf{1}$ for the tautologies and for the formulas of axiom schemas (a) and (b); this follows from the conditions imposed on φ. Finally, if F is obtained by generalization, then F begins with a universal quantifier so $\varphi(F) = \mathbf{1}$.

This allows us to show, as before, that if F has a proof that does not make use of axiom schema (c), then $\varphi(F) = \mathbf{1}$.

(c) Let F be a formula of axiom schema (c) such that $\varphi(F) = \mathbf{0}$, for example, $F = \forall v_0 \exists v_1 G \Rightarrow \exists v_1 G$, where G is an arbitrary formula. This formula is provable but, since $\varphi(F)$ is not equal to $\mathbf{1}$, it cannot be proved without appealing to schema (c).

7. Let $(F_1, F_2, \ldots, F_n)$ be a derivation of F $(F = F_n)$ that does not appeal to axiom schema (a). We will show that $(F_1^*, F_2^*, \ldots, F_n^*)$ is a derivation of F^*.

- If F_i is a tautology, then so is F_i^*. To see this, note that there exists a propositional tautology $P[A_1, A_2, \ldots, A_k]$ depending on the propositional variables $A_1, A_2, \ldots, A_k$ and formulas $G_1, G_2, \ldots, G_k$ such that

$$F_i = P[G_1, G_2, \ldots, G_k].$$

So $F_i^* = P[G_1^*, G_2^*, \ldots, G_k^*]$ which shows that F_i^* is a tautology.

- If F_i belongs to axiom schema (b), say $F_i = \forall v(H \Rightarrow G) \Rightarrow (H \Rightarrow \forall v G)$ where v is a variable that has no free occurrence in H, then

$$F_i^* = \forall v(H^* \Rightarrow G^*) \Rightarrow (H^* \Rightarrow \forall v G^*)$$

so F_i^* is also an axiom.

- The argument is similar if F_i belongs to axiom schema (c).
- If F_i is obtained by modus ponens from F_j and F_k with j and k less than i, then $F_j = F_k \Rightarrow F_i$, hence $F_j^* = F_k^* \Rightarrow F_i^*$ is obtained by modus ponens from F_j^* and F_k^*.
- If F_i is obtained by generalization from F_j ($j < i$), then F_i^* is obtained by generalization from F_j^*.

This shows that if F can be proved without an appeal to axiom schema (a), then $\vdash F^*$. For example, if P is a unary predicate symbol, $F = \exists v Pv \Leftrightarrow \neg \forall v \neg Pv$ is certainly provable (it is an axiom) but it is not provable without schema (a): otherwise, $F^* = \forall v Pv \Leftrightarrow \neg \forall v \neg Pv$ would also be provable, which is not the case since F^* is not universally valid (it is universally equivalent to $\forall v Pv \Leftrightarrow \exists v Pv$).

8. We are asked to prove, using Herbrand's method, that if T does not have a model, then T is not consistent. As in the section on Herbrand's method, we assume that each formula F_n can be written in the form

$$F_n = \forall v_1 \exists v_2 \forall v_3 \ldots \forall v_{2k-1} \exists v_{2k} B_n[v_1, v_2, \ldots, v_{2k}]$$

where k is an integer and B_n is a formula without quantifiers.

Let $\mathcal{T}$ denote the set of terms of the language and Θ_i denote the set of sequences of length i of elements of $\mathcal{T}$. We then introduce, for all n and i, a map $\alpha_{i,n}$ from Θ_i into $\mathbb{N}$ in such a way that the following properties are satisfied:

(1) if v_m occurs in one of the terms $t_1, t_2, \ldots, t_i$, then $\alpha_{i,n}(t_1, t_2, \ldots, t_i) > m$;

(2) if $j < i$ and $(t_1, t_2, \ldots, t_i)$ is a sequence that extends $(t_1, t_2, \ldots, t_j)$, then $\alpha_{i,n}(t_1, t_2, \ldots, t_j) < \alpha_{i,n}(t_1, t_2, \ldots, t_i)$;

(3) if τ and σ are two distinct sequences whose respective lengths are i and j and if n and m are arbitrary integers, then $\alpha_{i,n}(\tau) \neq \alpha_{j,m}(\sigma)$.

The codings from Chapter 6 show how such functions can be constructed without difficulty.

By definition, an **avatar** of F_n will be a formula of the form

$$B_n[t_1, v_{\alpha_{1,n}(t_1)}, t_2, v_{\alpha_{2,n}(t_1,t_2)}, \ldots, t_k, v_{\alpha_{k,n}(t_1,t_2,\ldots,t_k)}].$$

Denote the set of all avatars of F_n by A_n. To prove the two lemmas that follow, it more or less suffices to recopy the proofs of Theorems 4.38 and 4.39.

Lemma 1: If $\bigcup_{n \in \mathbb{N}} A_n$ is propositionally satisfiable, then $\{F_n : n \in \mathbb{N}\}$ has a model.

Lemma 2: If I is a finite subset of $\mathbb{N}$ and if $\bigcup_{n \in I} A_n$ is not propositionally satisfiable, then $\neg \bigwedge_{n \in I} F_n$ is provable.

These two lemmas permit us to prove the desired theorem.

9. Here are some possible refutations.

(a)

$(A \wedge B) \Rightarrow$	from $(A \wedge B) \Rightarrow C$ and $C \Rightarrow$ by cut on C
$B \Rightarrow$	from $(A \wedge B) \Rightarrow$ and $\Rightarrow A$ by cut on A
$\Box$	from $B \Rightarrow$ and $B \Rightarrow$ by cut on B

(b)

$(A \wedge B) \Rightarrow$	from $(A \wedge B) \Rightarrow C$ and $C \Rightarrow$ by cut on C
$(A \wedge A) \Rightarrow$	from $(A \wedge B) \Rightarrow$ and $A \Rightarrow B$ by cut on B
$A \Rightarrow$	from $(A \wedge A)$ by simplification
$\Box$	from $A \Rightarrow$ and $\Rightarrow A$ by cut on A

(c)

$(B \wedge C) \Rightarrow$	from $(A \wedge B) \Rightarrow$ and $C \Rightarrow A$ by cut on A
$B \Rightarrow$	from $(B \wedge C) \Rightarrow$ and $\Rightarrow C$ by cut on C
$\Rightarrow (B \vee B)$	from $D \Rightarrow B$ and $\Rightarrow (D \vee B)$ by cut on D
$\Rightarrow B$	from $\Rightarrow (B \vee B)$ by simplification
$\Box$	from $B \Rightarrow$ and $\Rightarrow B$ by cut on B

(d)

$(A \wedge A \wedge B) \Rightarrow C$	from $(A \wedge B) \Rightarrow (C \vee D)$ and $(A \wedge D) \Rightarrow$ by cut on D
$(A \wedge B) \Rightarrow C$	from $(A \wedge A \wedge B) \Rightarrow C$ by simplification
$B \Rightarrow (C \vee C)$	from $(A \wedge B) \Rightarrow C$ and $\Rightarrow (A \vee C)$ by cut on A
$B \Rightarrow C$	from $B \Rightarrow (C \vee C)$ by simplification
$\Rightarrow (C \vee C)$	from $B \Rightarrow C$ and $\Rightarrow (B \vee C)$ by cut on B
$\Rightarrow C$	from $(C \vee C)$ by simplification
$\Rightarrow E$	from $\Rightarrow C$ and $C \Rightarrow E$ by cut on C
$\Rightarrow F$	from $\Rightarrow C$ and $C \Rightarrow F$ by cut on C
$(E \vee F) \Rightarrow$	from $\Rightarrow C$ and $(C \wedge E \wedge F) \Rightarrow$ by cut on C
$F \Rightarrow$	from $\Rightarrow E$ and $(E \wedge F) \Rightarrow$ by cut on E
$\Box$	from $F \Rightarrow$ and $\Rightarrow F$ by cut on F

10. Let $\mathcal{A}$ denote the set of propositional variables. We define a map from the set of reduced clauses into the set of maps from $\mathcal{A}$ into $\{0, 1, 2\}$: to each clause $\mathcal{C}$ corresponds the map $\alpha_{\mathcal{C}}$ from $\mathcal{A}$ into $\{0, 1, 2\}$ defined by

$$\alpha_{\mathcal{C}}(A) = 0 \text{ if and only if } A \text{ appears in the premiss of } \mathcal{C};$$
$$\alpha_{\mathcal{C}}(A) = 1 \text{ if and only if } A \text{ appears in the conclusion of } \mathcal{C};$$
$$\alpha_{\mathcal{C}}(A) = 2 \text{ if and only if } A \text{ does not appear in } \mathcal{C}.$$

This map has an inverse: it is the map which, with a given map α from $\mathcal{A}$ into $\{0, 1, 2\}$, associates the clause whose premiss is the conjunction of the propositional variables A for which $\alpha(A) = 0$, taken in the order of increasing index, and whose conclusion is the disjunction of the propositional variables A

for which $\alpha(A) = 1$, also taken in the order of increasing index. Therefore, there are as many reduced clauses as there are maps from $\mathcal{A}$ into $\{0, 1, 2\}$, namely 3^n.

11. Suppose that there are, in all, n propositional variables appearing in S. Let P be a clause in S. We may begin by supposing that any given propositional variable occurs at most once in P: for if some variable occurs in both the premiss and the conclusion of P, then P is a tautology and we can ignore it; if not, we can reduce to this case by simplification. Let M be the number of variables appearing in P; m is greater than or equal to 3 by hypothesis. For P to be false, all the variables in its premiss must be true and all the variables in its conclusion must be false. Among the 2^n assignments of truth values to P, there are at most 2^{n-m} (i.e. at most one-eighth of them, since $2^{n-m} \leq 2^n \cdot \frac{1}{8}$) that could make P false.

   If we perform this calculation for each of the seven clauses in S, we see that there are at most seven-eighths of the assignments of truth values that could make one of the clauses in S false. So there are some left over that must satisfy all the clauses in S.

12. (a) Here is the desired refutation:

$$
\begin{array}{ll}
C_5 = \Rightarrow B & \text{from } C_3 \text{ and } C_4 \text{ by cut on } A \\
C_6 = C \Rightarrow & \text{from } C_2 \text{ and } C_5 \text{ by cut on } B \\
C_7 = (A \wedge B) \Rightarrow & \text{from } C_1 \text{ and } C_6 \text{ by cut on } C \\
C_8 = A \Rightarrow & \text{from } C_5 \text{ and } C_7 \text{ by cut on } B \\
\square & \text{from } C_4 \text{ and } C_8 \text{ by cut on } A
\end{array}
$$

   (b) From C_1 and C_3, we derive $(A \wedge A) \Rightarrow C$ by cut on B, then $A \Rightarrow C = \mathcal{D}$ by simplification. The set $\{C_2, C_4, \mathcal{D}\}$ is satisfied by the assignment of truth values δ defined by $\delta(A) = \delta(C) = \mathbf{1}$ and $\delta(B) = \mathbf{0}$; therefore it is not refutable.

   Our conclusion is that in a derivation of the empty clause from a finite set of clauses Γ, it is not legitimate to replace two arbitrarily chosen clauses by a clause that can be obtained from them by cut. What is possible (and which is what we did to prove Theorem 4.50), is to perform this operation systematically for all pairs of clauses that permit the elimination of some given variable (and only these pairs).

13. We will be content to provide the details for (a), (c), and (d) and to simply state the answer for the others.

   (a) 1.B. Simplification. The system (a) is equivalent to
   $(v_0, g\, gv_5v_1\, hv_4), (g\, hb\, gv_2v_3\, g\, hb\, v_0)$.

   1.C. Reduction. First set $\tau_1(v_0) = g\, gv_5v_1\, hv_4$ then unify:
   $(g\, hb\, gv_2v_3, g\, hb\, ggv_5v_1hv_4)$.

   2. A and B. Simplification and clean-up; we obtain in succession
   $(hb, hb), (gv_2v_3, g\, gv_5v_1\, hv_4)$;
   $(v_2, gv_5v_1), (v_3, hv_4)$.

2.C. Reduction. We can perform two reductions simultaneously by setting $\tau_2(v_2) = g v_5 v_1$ and $\tau_2(v_3) = h v_4$. We then obtain the empty system.

The substitution $\tau = \tau_2 \circ \tau_1$ is a principal unifier. We may calculate $\tau(v_0) = g\, g v_5 v_1\, h v_4$, $\tau(v_2) = g v_5 v_1$, $\tau(v_3) = h v_4$.

(b) There is no unifier.

(c) **1.B. Simplification.** We obtain:

$(v_4, f\, f v_3 v_9\, f v_{11} v_{11})$,

$(g\, g f g v_1 v_6 f v_5 v_{12} g f v_{10} a v_2\, f v_7 v_8, g\, g v_2 g f v_{10} a f v_6 v_0\, v_4)$.

1.C. Reduction. Set $\tau_1(v_4) = f\, f v_3 v_9\, f v_{11} v_{11}$ and the reduced system is

$(g\, g f g v_1 v_6 f v_5 v_{12} g f v_{10} a v_2\, f v_7 v_8, g\, g v_2 g f v_{10} a f v_6 v_0\, f f v_3 v_9 f v_{11} v_{11})$.

2.B. Simplification.

$(g\, f g v_1 v_6 f v_5 v_{12} g f v_{10} a v_2, g\, v_2\, g f v_{10} a f v_6 v_0), (f v_7 v_8, f\, f v_3 v_9\, f v_{11} v_{11})$.
    Then

$(f\, g v_1 v_6\, f v_5 v_{12}, v_2), (g\, f v_{10} a\, v_2, g\, f v_{10} a\, f v_6 v_0)$,

$(v_7, f v_3 v_9), (v_8, f v_{11} v_{11})$.

2.C. Reduction. Set $\tau_2(v_2) = f\, g v_1 v_6\, f v_5 v_{12}$, $\tau_2(v_7) = f v_3 v_9$, $\tau_2(v_8) = f v_{11} v_{11}$. We then obtain

$(g\, f v_{10} a\, f g v_1 v_6 f v_5 v_{12}, g\, f v_{10} a\, f v_6 v_0)$.

3. A and B. Simplification and clean-up. We obtain in succession

$(f v_{10} a, f v_{10} a), (f\, g v_1 v_6\, f v_5 v_{12}, f v_6 v_0)$;

$(g v_1 v_6, v_6), (f v_5 v_{12}, v_0)$.

It is impossible to unify $(g v_1 v_6, v_6)$ so system (c) does not have a unifier.

(d) **1. B. Simplification.**

$(v_2, g\, g v_9 v_3\, g v_9 v_{10}), (f\, f f g v_4 v_1 f v_3 v_5 f g v_7 b v_0\, g v_6 v_8,$
$f\, f v_0 f g v_7 b f v_{11} v_{12}\, v_2)$.

1. C. Reduction. We set $\tau_1(v_2) = g\, g v_9 v_3\, g v_9 v_{10}$ and we obtain

$(f\, f f g v_4 v_1 f v_3 v_5 f g v_7 b v_0\, g v_6 v_8, f\, f v_0 f g v_7 b f v_{11} v_{12}\, g g v_9 v_3 g v_9 v_{10})$.

2. B. Simplification.

$(f\, f g v_4 v_1 f v_3 v_5\, f g v_7 b v_0, f\, v_0\, f g v_7 b f v_{11} v_{12}), (g v_6 v_8, g\, g v_9 v_3\, g v_9 v_{10})$;

$(f\, g v_4 v_1\, f v_3 v_5, v_0), (f\, g v_7 b\, v_0, f\, g v_7 b\, f v_{11} v_{12}), (v_6, g v_9 v_3)$,

$(v_8, g v_9 v_{10})$.

2. C. Reduction. We set $\tau_2(v_0) = f\, g v_4 v_1\, f v_3 v_5$, $\tau_2(v_6) = g v_9 v_3$, $\tau_2(v_8) = g v_9 v_{10}$ and we obtain

$(f\, g v_7 b\, f g v_4 v_1 f v_3 v_5, f\, g v_7 b\, f v_{11} v_{12})$.

3. A and B. Simplification. We obtain in succession

$(g v_7 b, g v_7 b), (f\, g v_4 v_1\, f v_3 v_5, f v_{11} v_{12})$;

$(g v_4 v_1, v_{11}), (f v_3 v_5, v_{12})$.

3. C. Reduction. We set $\tau_3(v_{11}) = g v_4 v_1$ and $\tau_3(v_{12}) = f v_3 v_5$. The system we then obtain is empty and $\tau = \tau_3 \circ \tau_2 \circ \tau_1$ is a principal unifier of system (d). We may calculate

$$\tau(v_0) = f\, g v_4 v_1\, f v_3 v_5,\ \tau(v_2) = g\, g v_9 v_3\, g v_9 v_{10},\ \tau(v_6) = g v_9 v_3,$$
$$\tau(v_8) = g v_9 v_{10},\ \tau(v_{11}) = g v_4 v_1 \text{ and } \tau(v_{12}) = f v_3 v_5.$$

(e) There is no unifier (we notice this from the occurrence test after having completely decomposed the terms).

14. Consider the inverse α^{-1} of α (it is a permutation of V) and let σ' denote the substitution that extends it. It is immediate to verify that $\sigma \circ \sigma'$ and $\sigma' \circ \sigma$ are substitutions that equal the identity on V, and hence are equal to the identity: therefore σ has an inverse so it bijective.

For the converse, we begin by observing that if σ is a substitution and t is a term, then the length of t, $\lg[t]$, is less than or equal to that of $\sigma(t)$ (easily proved by induction on t). Suppose that τ and τ' are two substitutions such that both $\tau \circ \tau'$ and $\tau' \circ \tau$ are equal to the identity. Then, for every variable v,

$$\lg[\tau(v)] \le \lg[\tau'(\tau(v))] = 1;$$

since $\tau'(\tau(v)) = v$, $\tau(v)$ cannot be a constant symbol, so it must be a variable. So the restriction of τ to V is a map from V into V and we see, without difficulty, that the restriction of τ' to V is the inverse map.

15. (a) By induction on t. If t is a variable or a constant symbol, there is nothing to prove. If t is of the form $f t_1 t_2 \ldots t_n$ where n is an integer and f is an n-ary function symbol, then the equality $t = \sigma(t) = f \sigma(t_1) \sigma(t_2) \ldots \sigma(t_n)$ implies $t_1 = \sigma(t_1), t_2 = \sigma(t_2), \ldots, t_n = \sigma(t_n)$ by the unique readability theorem. We then invoke the induction hypothesis.

(b) This follows immediately from (a).

(c) Let $v \in A$; then v has an occurrence in a term of the form $\pi(t)$. But $\pi = \sigma \circ \sigma_1 \circ \pi$ and so, according to (b), $\sigma \circ \sigma_1(v) = v$. This shows that $\sigma_1(v)$ cannot be a constant symbol nor a term whose length is greater than 1 (see Exercise 14). Hence it is a variable and the restriction of σ_1 to A is a surjective map from A onto B. But since for every variable $v \in A$, $\sigma \circ \sigma_1(v) = v$, this map is also injective.

Let A' denote the complement of A in V and B' the complement of B in V. Then A' and B' have the same number of elements and we can find a bijection τ from B' onto A'. Let τ_1 denote the inverse bijection. We define the substitutions σ' and σ_1' by

$$\begin{aligned}
\sigma'(v) &= \sigma(v) && \text{if } v \in B; \\
\sigma'(v) &= \tau(v) && \text{if } v \in B'; \\
\sigma_1'(v) &= \sigma_1(v) && \text{if } v \in A; \\
\sigma_1'(v) &= \tau_1(v) && \text{if } v \in A'.
\end{aligned}$$

Then properties (i), (ii), and (iii) in the statement of the exercise are evident. Let us consider (iv): for example, we need to show that for every variable v, we have $\sigma' \circ \pi_1(v) = \sigma \circ \pi_1(v)$; but this is clear since all the variables that occur in $\pi_1(v)$ (which is equal to $\sigma_1 \circ \pi(v)$) are in B and σ and σ' agree on B.

(d) First of all, it is clear that if σ is a bijective substitution from $\mathcal{T}$ into $\mathcal{T}$, then $\sigma \circ \tau$ is also a principal unifier. Conversely, suppose that π' is another principal unifier of S. Then there are substitutions σ and σ' such that $\pi = \sigma \circ \pi_1$ and $\pi_1 = \sigma_1 \circ \pi$. But, according to (c) there also exists a bijective substitution σ_1' such that $\pi_1 = \sigma_1' \circ \pi$.

16. To apply the rule of resolution, we must begin by separating the two clauses. We obtain

$$Sv_2 \Rightarrow (Pv_2 \lor Rv_2); \quad (Pv_0 \land Pfv_1) \Rightarrow Qv_0v_1.$$

Before applying the cut rule, we can unify Pv_2 and Pv_0 or else Pv_2 and Pfv_1.

In the first case, the principal unifier is $\tau(v_0) = v_2$ and $\tau(v_i) = v_i$ for $i \neq 2$. The result is

$$(Sv_2 \land Pfv_1) \Rightarrow (Rv_2 \lor Qv_0v_1).$$

In the second case, the principal unifier is $\tau(v_2) = fv_1$ and $\tau(v_i) = v_i$ for $i \neq 2$. The result is

$$(Sfv_1 \land Pv_0) \Rightarrow (Rfv_1 \lor Qv_0v_1).$$

17. To obtain the Skolem forms of these formulas, we must add two constant symbols, say a and b, to the language. We obtain the following formulas:

$$\forall v_1(Pa \land (Rv_1 \Rightarrow Qav_1));$$
$$\forall v_0 \forall v_1(\neg Pv_0 \lor \neg Sv_1v \neg Qv_0v_1);$$
$$Rb \land Sb.$$

When we rewrite these in the form of clauses, we obtain the set:

(1) $\Rightarrow Pa$
(2) $Rv_1 \Rightarrow Qav_1$
(3) $(Pv_0 \land Sv_1 \land Qv_0v_1) \Rightarrow$
(4) $\Rightarrow Rb$
(5) $\Rightarrow Sb.$

Using the principal unifier $\tau(v_0) = a$ and $\tau(v_i) = v_i$ for $i \neq 0$, we can unify Pa in (1) with Pv_0 in (3) and obtain, by the cut rule:

(6) $(Sv_1 \land Qav_1) \Rightarrow.$

We can then apply the cut rule to (2) and (6) (to be strictly rigorous, we would first have to separate these two formulas, for example, replacing (2) by $Rv_2 \Rightarrow Qav_2$, and then return to the two original formulas with the help of a unifier!):

(7) $(Rv_1 \land Sv_1) \Rightarrow.$

Now we can unify Rb and Rv_1 (the principal unifier is $\tau(v_1) = b$) and apply the cut rule to (4) and (7): the result is

(8) $Sb \Rightarrow$

which, together with (5), yields the empty clause.

18. We must refute the set $\{F_1, F_2, F_3, \neg G\}$. So we begin by putting F_1 and $\neg G$ in prenex form; this produces, respectively:

$$\forall v_0 \forall v_1 (Rv_0 v_1 \Rightarrow Rv_0 f v_0) \text{ and}$$
$$\forall v_0 \forall v_1 \forall v_2 (\neg Rv_0 v_1 \vee \neg Rv_1 v_2 \vee \neg Rv_2 v_0).$$

Then, we must add Skolem functions to the language in order to rewrite these formulas as clauses; we need a unary function symbol g and a constant symbol a. We obtain

(1) $Rv_0 v_1 \Rightarrow Rv_0 f v_0$

(2) $\Rightarrow Rv_0 g v_0$

(3) $\Rightarrow Rffaa$

(4) $(Rv_0 v_1 \wedge Rv_1 v_2 \wedge Rv_2 v_0) \Rightarrow$.

We may apply the rule of resolution to (1) and (2). To separate these clauses, we replace (2) by $\Rightarrow Rv_3 g v_3$ then we unify $Rv_3 g v_3$ with $Rv_0 v_1$ (the unifier is $\tau(v_0) = v_3, \tau(v_1) = g v_3$) and we obtain

(5) $\Rightarrow Rv_3 f v_3$.

We can then unify $Rv_3 f v_3$ with $Rv_0 v_1$ which occurs in the premiss of (4) (the unifier is $\tau(v_0) = v_3, \tau(v_1) = f v_3$) and, after resolution, we have

(6) $(Rf v_3 v_2 \wedge Rv_2 v_3) \Rightarrow$.

Replace (5) by $\Rightarrow Rv_0 f v_0$ to separate it from (6), then unify $Rv_0 f v_0$ with $Rv_2 v_3$ (the unifier is $\tau(v_2) = v_0$ and $\tau(v_3) = f v_0$). The result is

(7) $Rffv_0 v_0 \Rightarrow$

and after unifying with (3) ($\tau(v_0) = a$), we obtain the empty clause.

Bibliography

First, we will suggest a list (no doubt very incomplete) of works concerning mathematical logic. These are either general treatises on logic or more specialized books related to various topics that we have presented. There is one exception: the book edited by J. Barwise, whose ambition was to produce, at the time it was published, a survey of the state of the art. Some of these titles are now out of print but we have included them none the less because they should be available in many university libraries.

J.P. Azra and **B. Jaulin**, *Récursivité*, Gauthiers-Villars, 1973.

J. Barwise (editor), *Handbook of mathematical logic*, North-Holland, 1977.

J.L. Bell and **A.B. Machover**, *A course in mathematical logic*, North-Holland, 1977.

J.L. Bell and **A.B. Slomson**, *Models and ultraproducts*, North-Holland, 1971.

E.W. Beth, *Formal methods*, D. Reidel, 1962.

C.C. Chang and **J.H. Keisler**, *Model theory*, North-Holland, 1973.

A. Church, *Introduction to mathematical logic*, Princeton University Press, 1996.

P. Cohen, *Set theory and the continuum hypothesis*, W.A. Benjamin, 1966.

H. Curry, *Foundation of mathematical logic*, McGraw-Hill, 1963.

D. van Dalen, *Logic and structures*, Springer-Verlag, 1983.

M. Davis, *Computability and unsolvability*, McGraw-Hill, 1958.

F. Drake, *Set theory*, North-Holland, 1979.

H.D. Ebbinghaus, **J. Flum** and **W. Thomas**, *Mathematical logic*, Springer-Verlag, 1984.

R. Fraïssé, *Cours de logique mathématique*, Gauthier-Villars, 1972.

J.Y. Girard, *Proof theory*, Bibliopolis, 1987.

P. Halmos, *Lectures on Boolean algebras*, D. Van Nostrand, 1963.

P. Halmos, *Naive set theory*, Springer-Verlag, 1987.

D. Hilbert and **W. Ackermann**, *Mathematical logic*, Chelsea, 1950.

K. Hrbacek and **T. Jech**, *Introduction to set theory*, Marcel Dekker, 1984.

T. Jech, *Set theory*, Academic Press, 1978.

S. Kleene, *Introduction to metamathematics*, Elsevier Science, 1971.

G. Kreisel and **J.L. Krivine**, *Eléments de logique mathématique*, Dunod, 1966.

J.L. Krivine, *Théorie des ensembles*, Cassini, 1998.

K. Kunen, *Set theory*, North-Holland, 1985.

R. Lalement, *Logique, réduction, résolution*, Masson, 1990.

R.C. Lyndon, *Notes on logic*, D. Van Nostrand, 1966.

A.I. Mal'cev, *The metamathematics of algebraic systems*, North-Holland, 1971

J. Malitz, *An introduction to mathematical logic*, Springer-Verlag, 1979.

Y. Manin, *A course in mathematical logic* (translated from Russian), Springer-Verlag, 1977.

M. Margenstern, *Langage Pascal et logique du premier ordre*, Masson, 1989 and 1990.

E. Mendelson, *Introduction to mathematical logic*, D. Van Nostrand, 1964.

P.S. Novikov, *Introduction à la logique mathématique* (translated from Russian), Dunod, 1964.

P. Odifreddi, *Classical recursion theory*, North-Holland, 1989.

J.F. Pabion, *Logique mathématique*, Hermann, 1976.

R. Péter, *Recursive functions*, Academic Press, 1967.

B. Poizat, *Cours de théorie des modèles*, Nur al-Mantiq wal-Ma'rifah (distributed by Offilib, Paris), 1985.

D. Ponasse, *Logique mathématique*, OCDL, 1967.

W. Quine, *Mathematical logic*, Harvard University Press, 1981.

W. Quine, *Methods of logic*, Harvard University Press, 1989.

H. Rasiowa and **R. Sikorski**, *The mathematics of metamathematics*, PWN–Polish Scientific Publishers, 1963.

A. Robinson, *Complete theories*, North-Holland, 1956.

A. Robinson, *Introduction to model theory and to the metamathematics of algebra*, North-Holland, 1974.

H. Rogers, *Theory of recursive functions and effective computability*, McGraw-Hill, 1967.

J.B. Rosser, *Logic for mathematicians*, McGraw-Hill, 1953.

J.R. Shoenfield, *Mathematical logic*, Addison-Wesley, 1967.

K. Shütte, *Proof theory*, Springer-Verlag, 1977.

W. Sierpinski, *Cardinal and ordinal numbers*, PWN–Polish Scientific Publishers, 1965.

R. Sikorski, *Boolean algebras*, Springer-Verlag, 1960.

R. Smullyan, *First order logic*, Springer-Verlag, 1968.

R.I. Soare, *Recursively enumerable sets and degrees*, Springer-Verlag, 1987.

J. Stern, *Fondements mathématiques de l'informatique*, McGraw-Hill, 1990.

P. Suppes, *Axiomatic set theory*, D. Van Nostrand, 1960.

P. Suppes, *Introduction to logic*, D. Van Nostrand, 1957.

A. Tarski, *Introduction to logic and to the methodology of deductive sciences*, Oxford University Press, 1965.

A. Tarski, **A. Mostowski** and **R. Robinson**, *Undecidable theories*, North-Holland, 1953.

R.L. Vaught, *Set theory*, Birkhaüser, 1985.

Below, for the benefit of the eclectic reader whose curiosity may have been piqued, we supplement this bibliography with references to books which, while related to our subject overall, are of historical or recreational interest.

L. Carroll, *The game of logic*, Macmillan, 1887.

M. Gardner, *Paradoxes to puzzle and delight*, W.H. Freeman, 1982.

K. Gödel, *Collected works* (published under the supervision of S. Feferman), Oxford University Press, 1986.

J. van Heijenoort, *From Frege to Gödel, a source book in mathematical logic (1879–1931)*, Harvard University Press, 1967.

A. Hodges, *Alan Turing: the enigma*, Simon & Schuster, 1983.

R. Smullyan, *What is the name of this book?*, Prentice Hall, 1978.

J. Venn, *Symbolic logic*, Chelsea, 1971 (first edition: 1881).

Index